Power Electronics and Power Systems

The Power Electronics and Power Systems book series encompasses power electronics, electric power restructuring, and holistic coverage of power systems. The series comprises advanced textbooks, state-of-the-art titles, research monographs, professional books, and reference works related to the areas of electric power transmission and distribution, energy markets and regulation, electronic devices, electric machines and drives, computational techniques, and power converters and inverters. The series features leading international scholars and researchers within authored books and edited compilations. All titles are peer reviewed prior to publication to ensure the highest quality content.

To inquire about contributing to the series, please contact:

Dr. Joe Chow

Administrative Dean of the College of Engineering and Professor of Electrical, Computer and Systems Engineering

Rensselaer Polytechnic Institute

Jonsson Engineering Center, Office 7012

110 8th Street

Troy, NY USA

Tel: 518-276-6374

chowj@rpi.edu

Kai Sun
Editor

Cascading Failures in Power Grids

Risk Assessment, Modeling, and Simulation

Springer

Editor
Kai Sun
Electrical Engineering and Computer Science
University of Tennessee, Knoxville
Knoxville, TN, USA

ISSN 2196-3185 ISSN 2196-3193 (electronic)
Power Electronics and Power Systems
ISBN 978-3-031-47999-1 ISBN 978-3-031-48000-3 (eBook)
https://doi.org/10.1007/978-3-031-48000-3

This Springer imprint is published by the registered company Springer Nature Switzerland AG
The registered company address is: Gewerbestrasse 11, 6330 Cham, Switzerland

Preface

The power grid is one of the most intricate engineering systems created by humans. Keeping the lights on is the paramount task for grid operators and power system engineers. Over half a century has passed since one of the largest historical power blackouts happened in the northeastern region of North America in 1965. Today's power grids are incomparable to the ones from back then, in terms of their systematic complexity, technological advancement, and operational sophistication. Since the 1990s, the power industry has undergone remarkable developments in the widespread use of computer-aided tools and modern sensing and communication technologies for more reliable and economical delivery of electricity from power plants to distributed users. We should rightfully possess a robust and smart grid immunizable against a blackout.

Is this the reality? Every few years, we hear of power blackouts occurring worldwide. While natural disasters like hurricanes and tornadoes are unavoidable forces collapsing power grids from time to time, a substantial number of blackouts were not solely due to foreseeable natural disasters and have been found avoidable based on post-event analysis. In fact, a blackout is usually the end of an intricate "cascading" process like a chain reaction, which can take hours for failures to initiate and propagate in the grid until reaching a point of no return. If cascading is predicted or detected early, timely control actions can prevent or mitigate it, thereby avoiding grid collapse and its blackout. These capabilities are facilitated by advanced auxiliary systems that have been progressively deployed over the past decades to support power system sensing, computing, and control.

It is important to note that all deployed auxiliary systems themselves become part of the grid's intricacy. Today's power grid has become an extremely complex system integrating a tremendous number of power system elements forming the grid and auxiliary elements supporting its normal operation. Any issues arising from these elements can impact the grid. For instance, every power system device is safeguarded against damage by protective relays, and can be automatically tripped from the grid when it is overloaded or operating unsafely. Then, its load is shifted onto other devices and may threaten their safety, potentially causing more trips when the grid has no sufficient safety margin. This is how cascading is developed. Moreover, any element with auxiliary systems, such as sensors or computer programs, can also malfunction, potentially causing misoperations of protective relays and controllers that initiate or contribute to cascading failures.

Prevention and mitigation of cascading failures is the central theme of this book. In fact, the power industry has consistently made efforts against cascading failures. Many experiences and lessons drawn from historical blackouts have now crystallized into industry paradigms and standards. Compared to the last century, today's power grids face many new challenges, such as the penetration of intermittent renewable energy resources and their power electronic interfaces and controllers, and the increase in extreme climate conditions due to the global warming trend. Meanwhile, considering the rapid growth in big data, high-performance computing, and artificial intelligence, this book chooses to focus on three areas with the potential for breakthroughs in a near future: risk assessment, modeling, and simulation of cascading failures.

1. Risk assessment utilizes both real and simulated outage data to assess the risks of cascading failures and power outages, considering the probabilities of occurrence and the resulting losses. These techniques will assist grid planners and operators in prevention of cascading failures.
2. Cascading modeling primarily develops models dedicated to understand the mechanisms of cascading failures. This includes probabilistic models that support risk assessment and help understand interdependency and distribution of failures, as well as physical models for more effective cascading failure simulations.
3. Cascading failure simulation considers the various stages a cascading process may go through and the dynamic behavior of the grid across different time scales, addressing the balance between efficient and accurate simulation methods.

All chapters of the book cover one or multiple of these areas. I hope that readers will find this book useful in understanding cascading failures, and the methodologies, analyses, and practices introduced by its chapters will inspire them to contribute to preventing and mitigating cascading failures against blackouts.

I would like to thank Professor Joe H. Chow, the editor of Springer's Power Electronics and Power Systems book series, for his encouragement in the development of this book. I would like to thank the contributors: Ian Dobson, Milorad Papic, Shuchen Huang, Feng Liu, Qinfei Long, Jinpeng Guo, Yunhe Hou, Eduardo Cotilla-Sanchez, Shengwei Mei, Shaowei Huang, and my former postdoctoral researchers and student, Junjian Qi, Rui Yao, and Wenyun Ju, for writing these excellent chapters.

This book is dedicated to the memory of Professor Da-Zhong Zheng and Professor Qiang Lu at Tsinghua University. Without their guidance and mentorship two decades ago, my fascinating journey into the realms of power system dynamics and control, tackling the challenges and complexities of cascading failures, would never have begun. Last, I am deeply grateful to my parents, Xiuxian and Weixue

Sun, for their unconditional love and support from the beginning of my journey, and to my wife Fang and my sons Yi and Rei for their constant encouragement and inspiration.

Knoxville, TN, USA
October 2023

Kai Sun

Contents

Chapter 1
Introduction of Cascading Failures

Kai Sun

1.1 Introduction

Electric power grids are among the most complex networked engineering systems. They play a crucial role in delivering reliable electric power to consumers. The operation of a power grid needs to deal with any potential disturbances and maintain the stability and dynamic performance of the system. Among the various disturbances, cascading failures are considered the most serious and extreme threats to power grid operations and can lead to major stability issues or even system collapse.

In North America, the North American Electric Reliability Corporation (NERC) is responsible for ensuring the reliability and security of the bulk power systems, by whom "cascading" is referred to as "the uncontrolled successive loss of bulk electric system facilities triggered by an incident (or condition) at any location resulting in the interruption of electric service that cannot be restrained from spreading beyond a pre-determined area by studies" [1]. In general, a cascading failure occurs as a series of dependent failures of individual components, beginning from an initial failure, progressively weakening the grid in a chain reaction-like manner, and ending with system collapse. In this context, "individual components" include not only electrical components but also control, protection and communication devices, as well as human actions, as they all contribute to the interconnected, complex system supporting the reliable operation of the power grid. A cascading failure is not effectively prevented or mitigated; it can result in a widespread blackout, leading to significant economic and social consequences.

The studies on cascading failures in power grids include assessing their risks, developing mathematical models to uncover the underlying mechanisms of cascad-

K. Sun (✉)
University of Tennessee, Knoxville, TN, USA
e-mail: kaisun@utk.edu

K. Sun (ed.), *Cascading Failures in Power Grids*, Power Electronics and Power Systems, https://doi.org/10.1007/978-3-031-48000-3_1

ing, simulating historical cascading events and potential scenarios, and designing preventive or mitigative control schemes.

1.2 Blackouts

The outcome of a cascading failure can lead to the power outage of a local area or even a system-wide blackout. Blackouts can have far-reaching impacts, affecting millions of people and incurring substantial costs. Once a blackout happens, it may take a considerable amount of time, ranging from several hours to days, to black start and restore the system and recover power supplies to affected customers.

1.2.1 Historical Blackouts

While power blackouts are generally considered as uncommon, they have occurred persistently throughout history. Notable large-scale blackouts are outlined in Table 1.1, detailing their impacts, durations, and primary causes reported in the literature.

Figure 1.1 compares the durations and numbers of affected customers for most of the blackout events in Table 1.1. Complete power system restoration following a blackout can vary in duration, ranging from several hours to several days, and sometimes even longer. Typically, larger blackouts require a longer period for restoration.

1.2.2 Learnings

The causes of these historical blackouts share some similarities. For instance, they were often the results of cascading failures, and the initial failures were triggered by factors such as line flashovers due to tree contacts, unfavorable weather conditions, human or computer errors, inadequate system monitoring, and mis-operations of protective relays. Analyses on these historical blackouts have suggested the following improvements for electric utility companies and regional transmission organizations to prevent and mitigate cascading failures:

1. Ensuring timely maintenance and updates of transmission infrastructure
2. Improving measurement and communication systems with cybersecurity
3. Strengthening grid monitoring and situational awareness
4. Enhancing contingency analysis concerning system vulnerability
5. Designing effective emergency control, remedial actions, and protection schemes
6. Providing adequate training to system operators for managing major power outages and system restoration

Table 1.1 Historical blackouts

Known as	Date	Location (customers affected)	Duration	Events and causes
1965 Northeast blackout [2]	November 9, 1965	Ontario in Canada and eight northeastern states in the USA (over 30 million people)	13 hours	High load demands during cold weather caused a transmission line to trip in Ontario due to a mis-programmed protective relay, leading to overloads, line trips, power plant shutdowns, and system islanding
1977 New York City blackout [3]	July 13, 1977	New York City and parts of northeastern USA (nine million)	25 hours	Two lightning strikes hit transmission towers, causing several 345 kV circuit breakers to trip and fail to reclose. The operator was unaware of the situation due to inadequate control room displays
July-1996 Western North America blackout [4]	July 2, 1996	West regions of the USA and parts of Canada and Mexico (two million)	6 hours	A 345 kV line flashover caused by a tree contact, along with a relay ground element mis-operation resulted in two line trips, initiating a remedial action scheme that removed two generating units, and then a Zone 3 relay mis-operation opened a 230 kV line, leading to the removal of more units and ultimately voltage collapse due to the loss the steady-state equilibrium
August 1996 Western North America blackout [4]	August 10, 1996	West regions of the USA, Canada, and Mexico (7.5 million)	6 hours	Five 500 kV lines were tripped within 2 hours due to tree contacts, in addition to an out-of-service circuit breaker, leading to the tripping of dozens of lines and more than ten generating units. This caused growing power oscillations, voltage collapse, and, ultimately, the uncontrolled separation of the Western Interconnection into five islands
1999 Southern Brazil blackout [5]	March 11 to June 22, 1999	Southern Brazil (over 75 million)	Several months	A lightning strike caused most of 440 kV circuits at a substation to trip, triggering a chain reaction of line trips and power plant shutdowns due to the limited grid infrastructure and under- or over-frequency issues, ultimately resulting in system separation into islands
2003 US-Canadian blackout [6]	August 14, 2003	Northeastern and midwestern states in the USA and Ontario in Canada (55 million)	2 hours to 4 days	A power plant shutdown followed by a line trip due to contact with a tree, coupled with inadequate situational awareness due to a state estimator problem, alarm system failure, and lack of coordination among system operators, resulted in more line trips, voltage instability, and eventual system separation

(continued)

Table 1.1 (continued)

Known as	Date	Location (customers affected)	Duration	Events and causes
2003 Italy blackout [7]	September 28, 2003	All of Italy and part of Switzerland (56 million)	12 hours	A line trip caused by falling trees resulted in two other lines becoming heavily loaded and tripped. Then, cascading line trips caused Italy to lose its imported power from France and Switzerland, followed by under-frequency protections and the islanding of Italy from the rest of Europe
2006 European blackout [8]	November 4, 2006	Several European countries including Germany, France, Italy, Span, Portugal, Belgium, and Netherlands (15 million)	2 hours	A 380 kV line in Germany was switched off without timely analysis and communication with neighboring systems, which was followed by the switch-off of a second 380 kV line and mis-operations due to lack of coordination among system operators. As a result, several interconnection lines were tripped, leading to the separation of the European northeast, west, and southwest regions
2011 Southwest blackout [9]	September 8, 2011	Arizona, Southern California, and parts of Mexico (about 2.7 million)	12 hours	A mis-operation at a capacitor bank caused a single 500 kV line to trip. This broke a major transmission corridor from Arizona to the San Diego area during peak-load hours, leading to power-flow redistributions, voltage deviations, and equipment overloads. Then, multiple lines, transformers, and generating units tripped, triggering region-wide automatic load shedding and an intertie separation scheme and resulting in the tripping of major nuclear units and, eventually, a complete blackout of San Diego and Mexico's Baja California Control area
2012 Indian blackouts [10]	July 30–31, 2012	Northern, Eastern and North-eastern regions of India (over 600 million)	2 days	Multiple outages weakened interregional corridors and led to overloading and the tripping of a 400 kV tie line, resulting in power swing, system separation, and blackouts
2015 Ukraine blackout [11]	December 23, 2015	Two oblasts of Ukraine (0.23 million)	6 hours	A synchronized and coordinated cyber-attack compromised the SCADA systems of three Ukrainian regional power distribution companies, leading to isolation and shutdown of multiple substations
2019 Argentina, Paraguay and Uruguay blackout [12]	June 16, 2019	Most of Argentina, all of Uruguay, and parts of Paraguay (48 million)	1 day	The disconnection of three 500 kV lines due to a short-circuit fault or mis-operation overloaded remaining lines in Argentina, leading to system collapse

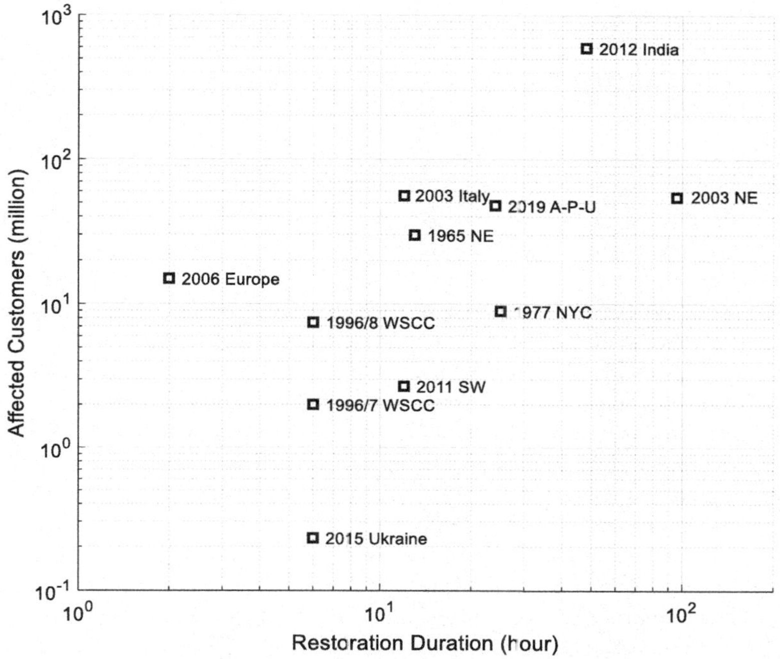

Fig. 1.1 Restoration durations and affected customers in historical blackout events

7. Facilitating timely information sharing and coordination among operators of regional transmission organizations

The initial step before developing preventive or mitigative controls for a power grid is to gain a comprehensive understanding of its potential cascading failures. This involves assessing the risks associated with cascading scenarios, creating models to uncover the underlying mechanisms behind these scenarios, and conducting simulations of cascading scenarios at acceptable accuracy to evaluate their impacts.

In the rest of this chapter, Sects. 1.3, 1.4, and 1.5 will introduce the prevention and mitigation measures against a cascading failure and the system restoration procedure after a cascading failure. Subsequentially, Sects. 1.6, 1.7, and 1.8 will discuss specific challenges related to risk assessment, modeling, and simulation of cascading failures, which will be further addressed by the remaining chapters of the book.

1.3 Prevention

1.3.1 Contingency Analysis

Contingency analysis, as an essential tool for identifying vulnerabilities of a power system, has potential to reveal probable cascading scenarios. According to security criteria, it assesses the system's responses to a predetermined list of critical contingencies, conducted through either power-flow analysis or dynamic simulations.

However, traditional contingency analysis mainly focuses on high-likelihood N-1 and N-2 contingencies, which are single-circuit or two-circuit faults, and often neglects less probable N-k contingencies (where $k > 2$) unless these circuit failures exhibit high interdependence or close geographic proximity, such as the tripping of multiple circuits connecting the same substation. This indicates the inadequacy of traditional contingency analysis in finding cascading scenarios since cascading failures typically involve a chain reaction of multiple events and failures or, in other words, N-k contingencies having large k.

Expanding the contingency list to include more N-k contingencies will inevitably lead to increased computational burdens in simulations. Therefore, knowing which k circuits are more likely to fail in the early stage of a cascading failure can help exclude less significant circuits, thus narrowing down the most relevant N-k contingencies. Thus, this contingency prescreening procedure can be adopted: first, conduct planning studies to identify more vulnerable areas or find patterns of N-k contingencies that more likely to initiate cascades from historical cascading events [13]; second, calculate power flows for N-k contingencies selected from these vulnerable areas, and rank these contingencies according to the criticalities of their post-contingency power flows. The contingencies that result in either divergent power flows or power-flow solutions with smallest security margins are chosen for further verification through, e.g., dynamic simulations.

With the continuous advancement of computing technology to support grid operations and planning studies, the contingency list will be able to incorporate a greater number of N-k contingencies. Thus, more cascading patterns might become identifiable by contingency analysis and addressed in the system planning stage although it is always computationally challenging to address all combinations for large k.

1.3.2 Interdependence of Failures

Notably, the occurrence of interdependent failures among multiple components in a power grid is not infrequent and can be modeled using tools such as interaction graphs [14, 15] or influence graphs [16]. Unlike the grid's physical topology, an interaction graph models potential interdependences among historical or simulated failures of power system components. The nodes of the graph represent a type of power system components such as power lines, which may be ranked by their vulnerabilities and occurrences in cascading failures, while the edges of the graph signify key links, namely, most likely interactions, between the failures of critical components. Consequently, the probabilities of the various possible paths of potential cascading failures can be determined. By intentionally and proactively weakening key links connected to a failed component in the graph, the failure propagation from this component to others can be effectively stopped, thus effectively mitigating the spreading of a cascading failure.

Interaction graphs as data-driven statistical tools can augment the scope of contingency analysis beyond N-1 and N-2 contingencies. Constructed offline from historical and simulated event data, they identify the possible dependencies among failures and the probabilities of their occurrences. Power system planning engineers can then opt for those graph paths with higher risks than others to explore in-depth N-k contingency studies. Such paths indicate the most vulnerable set of components. This can facilitate risk-based probabilistic contingency analysis, extending conventional deterministic contingency analysis.

While comprehensive identification of all potential cascading scenarios remains a challenge in a large-scale power grid due to the vast combinations of individual component failures and the unpredictable nature of resultant power outages, simulating cascading failures, at least in the initial stages involving multiple failures, is feasible and can become a routine task of offline system planning studies.

1.4 Mitigation

1.4.1 Protection and Remedial Actions

Protective relays constitute the primary line of defense against faults to safeguard equipment from damage using local measurements. However, their uncoordinated protective actions might not effectively mitigate an ongoing cascading failure. Instead, when an unsafe power line or generating unit is tripped to exit from operation according to the setting of its protective relay, the tripping itself becomes a new failure impacting the grid and potentially results in more equipment being overloaded or operating under less safe or unsafe conditions and then being tripped. This can more likely happen especially during an abnormal condition such as a cascading failure. For instance, a line tripped by its protection might induce overloading in other lines due to power-flow redistribution, and a generator tripped by its protection could result in a system frequency drop due to the loss of its active power generation and might also lead to over- or under-excitation conditions in other generators due to the loss of its reactive power control. All these safety issues can trigger more protective actions to protect respective equipment. However, such local protective actions, if not coordinated, may weaken the transmission network and, ultimately, lead to the propagation of failures.

Hence, the coordination of protective and control actions at the system level is crucial in mitigating a cascading failure. In this critical situation, the grid's control center has limited time (few minutes or even several seconds) to response. An example of system-level mitigation strategies is the employment of an automated remedial action scheme (RAS), also known as a special protection system (SPS). A remedial action scheme is designed offline to detect predetermined system conditions and execute automatic, corrective measures. The main objective is to maintain the power system's stability and an acceptable operating condition, focusing its protection on the whole system rather than individual components [18]. Corrective measures with a remedial action scheme include, but not limited to, curtailing or tripping generating units or other sources, reconfiguring the system, undervoltage load shedding (UVLS), underfrequency load shedding (UFLS), selective relay tripping and blocking under power swings or out-of-step conditions, and remote controls of reactive compensators or FACTS (flexible AC transmission system) devices. These actions are automatically taken upon detecting an event or observing an anomaly from monitored system responses or variables.

Typically, a remedial action scheme is integrated in the power grid's control center and interfaced with the SCADA and energy management system (EMS). The scheme acquires remote measurements from transmission substations via a wide-area measurement system with secure communication channels, examines its control logic, and sends control signals to remote protection or actuation devices.

1.4.2 Wide-Area Measurements

A lesson learned from the US-Canadian blackout in 2003 emphasizes the crucial role of real-time situational awareness utilizing wide-area measurements. Knowing the system's real-time dynamics is vital for detecting and mitigating cascading failures. Analyses on the 2003 blackout have revealed that timely alarms could have been activated to initiate corrective actions if the angular separation between the Cleveland area and the rest of the system had been monitored by comparing synchronized phasor measurements from both sides. As an example of corrective actions, controlled system separation with proactive load shedding for an amount of load less than the actual loss incurred during the cascading might have avoided the catastrophic consequences of the 2003 blackout.

After two decades, phasor measurement units (PMUs) have become widely installed in transmission systems, in contrast to the situation in 2003. A wide-area measurement system (WAMS) based on PMUs and other synchrophasor devices can significantly enhance the capabilities of today's power grids in both preventing and mitigating cascading failures. PMUs supply the grid's control center with real-time GPS-synchronized phasor measurements of voltages and currents at a high sampling rate (e.g., 30–60 Hz for a 60 Hz AC power system), facilitated by advanced communication infrastructures. With the increasing integration of inverter-based resources into power transmission systems, electromagnetic transient dynamics and other fast dynamics are introduced. Emerging measurement devices that offer synchronized, phasor measurements at higher sampling rates or even continuous point-on-wave measurements have been developed. The integration of these emerging measuring devices into a WAMS will further boost its monitoring capabilities against major power system disturbances including cascading failures.

Real-time wide-area measurements significantly improve a power system's real-time observability, enabling event detection and localization [19], oscillation monitoring [20], dynamic state estimation [21], and early warning of potential angular and voltage instabilities [22, 23]. These functionalities can enhance operators' situational awareness, facilitate the coordination of distributed protective relays and power system controllers, and thus enable the development of a system-wide, adaptive remedial action scheme for more effective mitigation of cascading failures.

Synchronized wide-area measurements with time stamps are also essential for forensic analysis of power grid disturbances and blackouts [9]. After a cascading event, the PMUs or GPS-synchronized digital fault recorders installed at critical locations can provide recorded synchronized data with time stamps, offering a chronological record of voltage, current, phase angle, and frequency measurements across the power grid. This record is invaluable for comprehending the sequence of component failures, protective relay actions, and system responses during the event and for identifying the root causes and propagation patterns of cascading failures.

1.4.3 Controlled System Separation

Once a cascading failure initiates, propagates to trip equipment, and weakens the power transmission system, abnormalities may arise in voltage and frequency. When the system's stability cannot be maintained such as voltage collapse occurring in an area or some generators going out of step, the power grid may lose its integrity due to cascading line trips, and, as a result, it will separate into electrical islands. This phenomenon is referred to as "uncontrolled system separation." These islands are inadvertently created due to uncoordinated protective relays, and as a result, some of them may experience imbalances in local generation and load. As a result, some islands will have to reject excessive generation, while other islands will need to shed a significant amount of load. Moreover, cascading can continue in some of islands.

As the ultimate line of defense against a blackout, "controlled system separation," which is also called "controlled system islanding," serves as an effective strategy to mitigate cascading failures especially in their late stages. A controlled system separation scheme must proactively divide the power transmission system into distinct electrical islands or isolate the cascade within a single island before it spreads to the entire system. The configuration of each island is strategically designed to ensure that its generation matches the local load, thus allowing each island to continue supplying electricity to local customers and preventing a blackout of the entire system [24, 25]. After all equipment failures are resolved, the transmission system can be restored by resynchronizing all islands. This process of island resynchronization demands significantly less effort compared to a black start procedure after a system-wide blackout. Especially, controlled system separation allows a majority of loads to be preserved within electrical islands. Controlled system separation as a remedial action scheme significantly enhances grid resilience against extreme scenarios.

At present, controlled system separation is not widely implemented by the power industry. A power grid equipped with a controlled separation scheme is typically one that has previously experienced uncontrolled separation events due to inherent weaknesses of the transmission network or external threats such as extreme weather conditions like storms or hurricanes. Usually, a controlled separation scheme can be built upon existing out-of-step protection schemes, integrating selective relay tripping and blocking functions to achieve desired formations of electrical islands. Conventionally, the out-of-step tripping function is designed to trip an unstable generator or a tie line connected to an unstable group of generators, while the blocking function prevents unnecessary trips that may occur during unstable power swings. Out-of-step protective relays are typically placed at individual generators or transmission corridors, where conditions are straightforward to detect from local measurements. If a highly reliable WAMS is deployed, out-of-step relays can be situated and coordinated at multiple locations to manage controlled system separation along desired boundaries and create electrical islands with minimized imbalances of generation and load [26].

1.5 Restoration

Once a cascading failure ends, the grid operator must initiate power system restoration to bring service back to the affected customers. Generally, the system restoration process comprises three stages: preparation, generation and network restoration, and load restoration [27, 28].

In the preparation stage, the operator assesses the extent of the power outage, verifies the communication systems in affected areas, identifies and disconnects damaged equipment, and determines the status of generators. Then, in the generation and network restoration stage, if the outage is confined to a small area, the operator can sequentially energize transmission lines that feed the outage area and crank disconnected generators. In the case of a system-wide blackout, a black-start procedure should be performed, which needs to first start one or multiple generators with black-start capabilities, such as hydraulic units, gas turbine generators, or the on-site small diesel units connected to large generating units. These black-start generators can energize connected transmission lines and crank non-black-start generators in a prioritized order [29]. Meanwhile, the customers characterized as the critical type of loads, such as hospitals and airports, need to be picked up as soon as possible during the stage of generation and network restoration. Finally, in the last load restoration stage, the rest of loads can be gradually restored [30].

The entire process of system restoration following a blackout can span from several hours to several days, as illustrated in Fig. 1.1. In situations where the network is restored in parallel as electrical islands respectively from multiple black-start generators, synchronization of these islands is needed as one of the last steps [31]. If controlled system separation as a mitigation strategy is employed to isolate the cascading failure or split the system into strategically designed electrical islands in which power supply to most of load is continued, most tasks of the black-start procedure can be skipped so that power system restoration can significantly be expedited and focus on synchronizing separated electrical islands, restoring the remaining loads, and bringing the system back to its normal operation.

1.6 Risk Assessment

While blackouts are infrequent on a well-designed power grid, they have significant consequences. It becomes apparent that the risk of cascading failures faced by a power grid is not negligible and should be addressed seriously. In fact, the rapidly growing electricity demands, coupled with a relatively slower expansion of new transmission infrastructure, often lead to power grids operating near their operating limits with reduced safety margins. This increases the risk of failures. Furthermore, the increasing penetration of renewable energy resources introduces more complexities and uncertainties to power system operations. All these factors can heighten the risk of a cascading failure and its consequences, including major power outages or complete system blackouts.

1.6.1 Risk Indices

For a power system with n components, all possible cascading scenarios involving successive k failures are k-permutations of n, which are totaled to $n!/(n\text{-}k)! = n(n\text{-}1)\ldots(n\text{-}k + 1)$, or approximately n^k. This can be an incredibly large number for interconnected power systems having thousands or tens of thousands of components. For instance, for $n = 1000$, the total number of cascading scenarios with 10 successive failures is 10^{30}!

Analyzing the causal relationships and connections among all failures during a specific cascading event can be achieved through post-event analysis. However, foreseeing these ahead of time can be challenging. The permutations and combinations of potential failures across numerous power system components pose a significant challenge in assessing the risk of cascading failures in large-scale power grids.

While one might consider the likelihood of a particular cascading scenario occurring among all possible scenarios to be rare, its consequences and costs can be substantial once it indeed happens. It is neither practical nor cost-effective for stakeholders to address all possible cascading scenarios in their prevention and mitigation strategies. Hence, we have to concentrate available resources and personnel on preparing for the scenarios carrying high risks. This leads to an essential question: how can the risk of cascading failures be defined and assessed without detailed information on power system dynamics or failure interactions?

Generally, the risk associated with a failure depends on the product of its probability and cost [32]. Accordingly, we may define a risk index for a specific cascading scenario to be the product of a probability index and a cost index: $Risk = Probability \times Cost$. The former reflects the likelihood or frequency of the scenario occurring, while the latter quantifies the resultant total loss it might incur. The risk index can be estimated for either a particular cascading scenario or one kind of cascading scenarios that lead to similar blackout sizes. Then, the risk index relies on the combined total probability and the average cost for all scenarios falling into that kind.

The outcomes of risk assessment are important for prioritizing efforts aimed at preventing or mitigating cascading failures. Some risk indices estimated through cascading risk assessment studies can be applied in long-term power system planning to help determine the optimal upgrade plan for the transmission network to enhance its reliability and stability. Other risk indices can be utilized in short-term planning or grid operation studies, such as day-ahead analyses, by predicting potential cascading patterns or system vulnerabilities under various contingencies, allowing for preemptive measures. In fact, selected risk indices could potentially be evaluated online utilizing wide-area measurements, enhancing operators' real-time situational awareness and decision support.

1.6.2 Challenges

Challenges in assessing the risk of cascading failures are notable. One challenge arises from having limited computing resources available to find and examine numerous cascading scenarios. The commonly used approach involves employing the Monte Carlo method, by which sample events on *N-k* contingencies can be repeatedly generated based on prior or estimated probabilities of individual component failures. However, this method requires a large number of samples for cascading scenarios. Unlike *N*-1 and *N*-2 contingencies in routine planning studies, cascading scenarios are much less frequent. Consequently, this results in significant computational burdens when attempting to cover all possible cascading scenarios. Additionally, as discussed later, simulating cascading scenarios itself poses considerable computational challenges due to the need for extended simulation periods and complex dynamics of the power system at multiple timescales.

Another challenge is the lack of information regarding the interdependence or conditional probabilities of component failures. A power system is a complex dynamic system with numerous devices. Many equipment or control failures within each cascading scenario inherently rely on each other. Treating them as independent component failures in the Monte Carlo method could generate an estimated blackout-size probability distribution with an exponential tail, which is different from power-law tails observed from outage data as explained later. This will lead to underestimating the probability and risk of large blackouts. Understanding the conditional probabilities of failures is crucial for accurately gauging the level of risk and can also significantly reduce the necessary number of samples for risk assessment. However, this requires detailed knowledge on the dynamics and controls of the power system under cascading failures.

1.6.3 Micro-perspectives on Risk

It is important to highlight that risk assessment for cascading failures should not solely focus on determining an overarching risk index for the entire system. Instead, it should offer a certain level of details and insights that aid in comprehending the intricacies of cascading failures and their prevention and mitigation. Besides a macrolevel risk index, there are occasions where a microlevel perspective on the risk is crucial.

Firstly, in a power system, the probability distribution representing the likelihood of cascading failures across various scales can vary with the system's steady-state condition [32]. Although it is not possible to entirely prevent cascading failures under low-load conditions, they become less frequent due to increased operating margins of components and hence reduced impacts when a few elements are lost. This probability distribution characterizes the occurrence frequencies of resulting blackouts across different sizes in terms of the total loss of load or the number of equipment failures. Consider these three load levels:

1. At an extremely low load level, the probability distribution has a rapid decline as the size increases, showing an exponential tail. This decline emerges because the likelihood of large blackouts diminishes approximately in an exponential manner within a robust power system.
2. As the load level grows toward a critical threshold, the likelihood of large blackouts increases, shifting the probability distribution toward an approximate power-law shape.
3. At a high load level, cascading scenarios are more likely to generate a heavy tail in the probability distribution for large blackouts. This implies that such substantial blackouts can occur more frequently than a power law or exponential law might suggest.

The log-log plots of the blackout-size probability distribution for these three load levels are illustrated in Fig. 1.2.

The shape of the probability distribution is more informative to power system engineers than the average black size and offers greater insights into the occurrence frequencies of both small and large blackouts at a specific load level. When accounting for different loading levels through respective probability distributions,

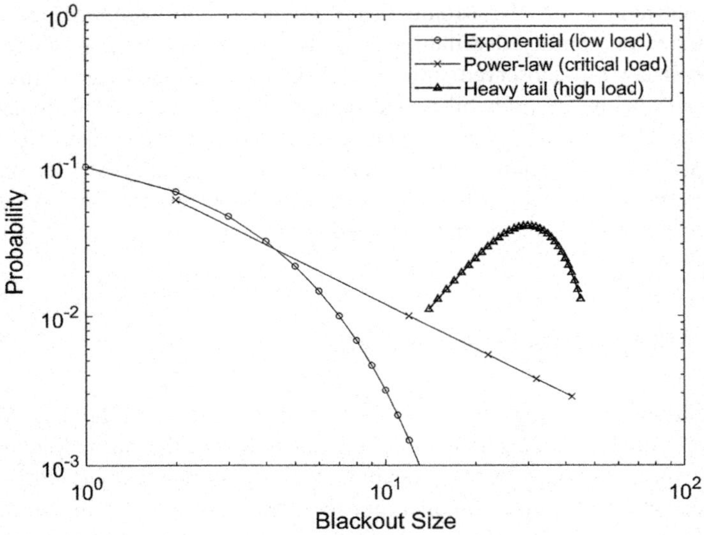

Fig. 1.2 Log-log plots of the blackout-size probability distribution for three load levels

more comprehensive information becomes available. Even if these distributions are combined into a single curve for a range of loading levels, three distinct regions might appear on the curve as the load loss or blackout size increases progressively: an exponential region is followed by a power-law region with significantly heightened probability, concluding with a heavy tail region. Hence, solely presenting an overall risk index derived from the curve's average risk is insufficient to address the criticality of the system's condition in relation to cascading failures.

Second, the risk is influenced by how the grid is dynamically operated and controlled, even with a constant load level. For instance, a power grid's active and reactive powers can be regulated through either a centralized or decentralized control strategy. With the load level fixed, varying the control strategy can alter the probability distribution curve of cascading failures and subsequent blackouts. For instance, a control strategy that reduces power exchange between regions while relying more on local resources could weaken interregional coupling, thereby reducing the risk of extensive cascading failures and the blackout size. On the other hand, a control strategy aimed at minimizing load shedding during cascading events might decrease small-scale power outages but potentially result in a large blackout, as strategic and timely load shedding can effectively mitigate cascading failures. As shown in [32], these two extreme examples often yield different probability distribution curves while potentially sharing the same average blackout size and overall risk value.

In addition, over an extended timeframe, the risk of cascading failures relies on how the power grid is planned and upgraded to accommodate the load growth. On the one hand, power transmission companies often tend to maximize the use of the existing transmission network for economic gains, which increases system stress and consequently raises the likelihood of operating the transmission network near or at its critical loading threshold. On the other hand, system upgrades targeted at potential cascading scenarios can alleviate the system stress. These opposing forces may gradually shape the probability distribution toward a power-law distribution of blackout sizes, while other studies propose the consideration of these opposing forces in blackout mitigation strategies [32]. For stakeholders, understanding the risks associated with cascading failures is not about avoiding critical loading altogether; instead, the risk assessment seeks to propose a practical trade-off between the economic and reliability aspects of a power grid.

1.7 Modeling

1.7.1 Requirements

Cascading failures stand out among the most complex phenomena in a power system. A cascading process may span various timescales of power system behaviors and dynamics, encompassing faults, overloads, protective actions, generator oscillations, transient instability, reactive power controls, voltage collapse, software and communication problems, insufficient situational awareness, human errors, and other factors. Precisely modeling all power system behaviors and dynamics inherent in a cascading process, especially for a large-scale power grid, could be expensive or even impractical. Hence, a dedicated cascading model is essential to risk assessment and simulation of cascading failures.

It is important to note that the term "cascading model" does not necessarily correspond to a power system model that can be used for simulating cascading failures. Rather, we aim to model the phenomenon and process of cascading itself. In other words, the model should specifically describe how a cascading failure initiates, develops, and spreads across a power system. This involves two requirements:

1. A cascading model needs to consider additional factors that are not typically included in a power system model for conventional contingency simulations. These factors are such as interactions between equipment failures, protective and control actions, and even human reactions.
2. A cascading model may simplify irrelevant details and dynamics that do not aid in understanding cascading mechanisms. Essentially, the model aims to provide a credible outline or overall analysis for each cascading scenario at an acceptable level of detail.

1.7.2 Challenges

To authentically capture a cascading process using a cascading model, we may encounter a number of challenges such as:

1. *Large size of a power system:* Interconnected power systems can reach massive scales with numerous components. For instance, the transmission system model of the Eastern Interconnection comprises over 70,000 buses. Handling such big models is the first challenge faced by the developer of a cascading model for this interconnection in order to understand cascading events, such as the 1965 Northeast blackout and 2003 US-Canadian blackout.
2. *Intricate nature of a power system:* Power systems are not only large but also very complex. The components within a power system can exhibit intricate and closely interconnected interactions that might not be fully understood. From a complex system perspective, a large blackout stems less likely from independent events, but rather from the inherent interdependence and tight coupling of components in the system. Consequently, the interactions between components pose challenges in modeling the patterns of failure propagation.
3. *Complex cascading mechanisms*: After a cascading failure starts, its spread is influenced by various power system operational and protection mechanisms, including thermal ratings, voltage stability, frequency protection, transient stability, oscillation, and human responses. Therefore, a cascading model often needs to target specific causes and mechanisms of interest, rather than all of them.
4. *Evolution of a power system*: A power system is generally in a constant state of evolution. As economies and populations grow, the demand for electricity also increases. This leads to heavier loads on transmission lines, reducing their margins and increasing the overall stress on the system. Consequently, the risk of cascading failures also rises. To mitigate this risk, various engineering responses are implemented, such as timely equipment upgrades and maintenance, as well as the adoption of new control strategies. This interplay of opposing processes creates a dynamic equilibrium where the power system can behave like a self-organized critical system [33, 34]. Neglecting the evolving nature of a real-life power system would undermine the credibility of risk assessment or simulation of its cascading failures.
5. *External factors*: Numerous external elements not belonging to the power grid can also contribute to the initiation and propagation of cascading failures. Focusing solely on the power grid might lead to the oversight of crucial risks. For instance, inadequate pruning or clearance of trees beneath certain lines could lead to line tripping due to contact with trees, even when the lines are not overloaded, which can more likely happen on days with high temperatures and low wind. Overhead lines during such days sag more toward the ground to increase the chance of tree contact. Tree-related outages can significantly affect system reliability and can account for a substantial portion of preventable power outages in tree-rich areas. The growth or falling of trees onto overhead lines is widely considered a primary cause of electric power outages [35].
6. *Difficulty in model benchmarking and validation*: Over the past two decades, a number of models have been proposed within academia to investigate cascading failures in power systems. The benchmarking and validation of these cascading models play a crucial role in assessing how well the cascading failures they model correspond to actual events [36]. However, cascading failures are

infrequent occurrences, and developers of such models often struggle to access real outage data from the power industry in a timely manner due to data shortage or confidentiality concerns. Additionally, validating a cascading model using simulated cascading scenarios is far from straightforward. This process encounters challenges in achieving accurate simulations of cascading failures, as further discussed in the subsequent section.

7. *Growing renewable resources*: Last but not least, many power grids across the world are witnessing an increasing penetration of renewables in both transmission and distribution systems. Unlike conventional generators that are typically well-controlled, the majority of renewable resources, such as wind and solar farms, exhibit an intermittent nature. When these resources are integrated into the grid through power electronic inverters, the rapid electromagnetic transient dynamics associated with power electronic controls further amplify the already intricate power system dynamics. Consequently, these inverter-based renewable resources will introduce new dynamics, uncertainties, and risks to grid operations and might become new triggers for cascading failures.

In the practical modeling of cascading failures, a balance between model accuracy and simplicity becomes necessary. A cascading model needs to prioritize specific aspects in alignment with its intended purpose, unless a precise simulation of cascading failures is the primary goal. As suggested by some researchers, potential compromises could involve statistically modeling the overall progression of cascading failures while overlooking intricate interaction details [37]. Another approach could be concentrating the modeling efforts on the most probable or high-risk failures based on existing knowledge or the current system conditions. Simplified power system models might also be adopted to elucidate the overarching characteristics of cascading failures. Furthermore, some models might focus on analyzing only a subset of the cascading mechanisms or examining just the initial phase of the failure sequence, up until a point of no return.

Various cascading models have been developed. Most of them fall into two categories: probabilistic models and physical models. Probabilistic models primarily rely on statistical failure data and often do not necessitate knowledge of the system's topology. In contrast, physical models are designed to replicate and simulate cascading processes or critical cascading mechanisms.

1.7.3 Probabilistic Models

When considering the probability distribution or risk associated with a cascading failure and subsequent blackout, a probabilistic model can be constructed directly through statistical analysis of datasets containing historical or simulated failures. In this context, comprehensive knowledge of the network topology and the complexities of the power system model is not essential. As a result, cascading failure modeling can assume a generic cascading process and simplify power system-related characteristics. In the following, several probabilistic models are briefly introduced.

Branching processes [38, 39] are standard probabilistic models used to analyze the progression of random events that generate new events, forming a branching structure. They can find applications in diverse fields including biology, physics, computer science, and engineering. In a branching process, each event can lead to multiple subsequent events, resulting in a tree-like structure where the number of events grows over time. This concept proves particularly valuable for studying phenomena like the spread of diseases, the growth in populations, information dissemination through networks, and cascading failures in power grids and other engineering systems. By providing insights into the probabilities of different outcomes and the overall behavior of an evolving system, a branching process model facilitates an understanding of complex and dynamic processes without necessitating detailed simulations.

When applied to cascading failures, the theory of branching processes can offer analytic formulas for the probability distribution of the total number of failures [40]. Typically, the model involves defining two random parameters with respective probability distributions. The first parameter is the number of initial failures, and the second parameter is the number of child failures that each parent failure tends to generate during the progress of failure propagation. Starting from a random number of initial failures, subsequent failures manifest in stages or generations. This occurs as the propagation leads to a random number of child failures through branches from each parent failure. Higher values of these two random numbers increase the likelihood of larger numbers failures, specifically large-size blackouts. These two parameters can be estimated from historical or simulated failures so as to enable computing the distribution of the blackout size.

When considering multiple types of failures, such as line outages, bus outages, and load shedding, a multitype branching process model can be employed to estimate a joint distribution of two or more types. This approach prevents underestimating losses and the extent of failure propagation when analyzing each individual type [41]. For instance, a system might exhibit a closer proximity to criticality when the interdependence of various types of outages is taken into account, compared to what a single-type branching process model would indicate.

Branching processes serve as a foundational probabilistic framework for understanding the progression of cascading failures. By capturing the generative nature of cascading events and their subsequent effects, a cascading model based on branching processes can contribute to risk assessment. However, as a probabilistic model, it does not retain information about the network topology or power flows of the power grid and does not attempt to specify how failures propagate in the system in detail, such as which, where, or why lines outage.

The *CASCADE model* [42, 43] is another probabilistic model on cascading failures of a system comprising many randomly loaded identical components. In this model, an initial disturbance can trigger failures of some components as they exceed their loading limits. Each component failure leads to a fixed load increase for other components. As more components fail, the remaining components become progressively more loaded and more likely to fail. The probability distribution associated with the number of failed components follows an extended quasibinomial distribution. Explicit formulas for this distribution are derived through a recursion process.

Initially, the CASCADE model was used to investigate critical loading in power transmission systems and the power-law tails in the probability distributions of the blackout size. However, the model does not consider various time intervals between failures, the intricate structure of the power grid, and the diverse types and interactions of the power system components. It can be well approximated by a Galton-Watson branching process model that deals with failures described by discrete numbers such as line outages [44]. As a high-level probabilistic model, it can still identify general features of loading-dependent cascading failures and provide estimates for probabilistic distributions on the number of failed components, which are important to risk assessment.

Examples of more informative probabilistic models include *interaction graph* [14, 15] and *influence graph* models [16, 17, 45]. In comparison to high-level probabilistic models such as branching process models, an interaction graph offers more detailed insights into the propagation of failures for a particular power system. Its enhanced capability arises from its construction utilizing a dataset from historical or simulated failures on the system. The graph models component failures by nodes and their interactions and influences by edges, and it does not represent the actual grid topology. As probabilistic models, interaction graphs assess statistical consequences and interdependences of component failures based on datasets. These quantified consequences help identify a set of critical components, i.e., the nodes of the graph, whose failures could potentially trigger the most severe outcomes in terms of indices crucial to grid operators. These indices include the count or scale of subsequent failures and the extent of load shedding. Furthermore, the interdependences among failures, particularly those linked to critical components, illuminate the intricate ways in which failures interact and propagate, and they are represented by edges of the graph.

By representing the most critical components and their connections through nodes and edges in an interaction graph or an influence graph, critical propagation patterns and pathways of cascading failures can be visualized. Such visualizations help grid operators anticipate impending failures and take proactive measures once initial failures occur on any nodes of the graph. Applications of interaction graphs range from offline cascading studies and development of prevention strategies to online mitigation. As a probabilistic model, an interaction graph built from outage data is characterized by an interaction matrix that estimates the conditional probabilities of child failures when a parent failure occurs [46, 47]. Thus, a random vector on initial failures being repeatedly multiplied by this matrix can produce more samples of cascading failures. In the real-time operating environment, visualizing an interaction graph together with already failed components helps improve situational awareness of the grid operator during a cascading failure and provide decision for mitigation strategies [48–50]. Chapter 3 of the book will provide a more detailed introduction of interaction models and graphs.

An interaction graph can be composed of interconnected multiple layers, which are subgraphs focusing on a specific severity index on consequences such as the number of line outages, the amount of load shedding, and the extent of propagation [15]. Each layer provides distinct statistical insights derived from the same dataset of failures but from different perspectives. Thus, they may not necessarily share the same set of nodes and edges because a critical component whose failure triggers many subsequent failures might not be considered critical in terms of causing significant load loss. The edges connecting nodes across two different layers represent transitions from one type of consequence to another in cascading. For instance, the initial stages of a cascading failure may be evident from local line outages, while intermediate stages may involve longer distance propagation, and later stages could lead to increased load losses. These progression patterns can be captured better by various layers within a multilayer interaction graph.

1.7.4 Physical Models

In this subsection, we utilize the *OPA (ORNL-Pserc-Alaska) model* [33, 34, 51–53] as an example to illustrate critical power system behaviors necessitating modeling and simulation for understanding cascading mechanisms. More accurate and detailed models for simulating multi-timescale dynamics in cascading failures will be discussed in Sect. 1.8.

The OPA model incorporates self-organization processes of a power system, considering engineering reactions to blackouts and long-term economic responses to load growth. This model addresses quasi-steady-state behaviors of a power system and its evolution over two timescales. The module within the short timescale, referred to as "fast dynamics," can simulate failures and controls such as failures, overloading and tripping of lines, generation redispatch, and load shedding. This module can be used to simulate a cascading failure with engineering responses. On the other hand, its long timescale module, referred to as "slow dynamics," can simulate the long-term transmission system upgrades to mitigate the blackout risk arising from load demand growth. This, in turn, aids in determining suitable transmission capacities for the short timescale module to assess overloading of lines.

The OPA model's "fast dynamics" module enables simulation of quasi-steady-state power system behaviors during cascading failures, disregarding the transient dynamics of generators and other dynamic devices. Each simulation using the OPA model begins from a random initial set of line outages, assuming independent failure probability for each line. Whenever a line outage happens, remedial actions are implemented by solving an optimal power-flow problem for the optimal generation redispatch and load shedding. For instance, the original OPA model in [51] solves a standard linear programming problem based on a DC power-flow model of the system. Following these remedial actions, if any line becomes overloaded beyond its transmission capacity, it is tripped, triggering another round of remedial actions. This process of testing for overloads and implementing remedial actions is

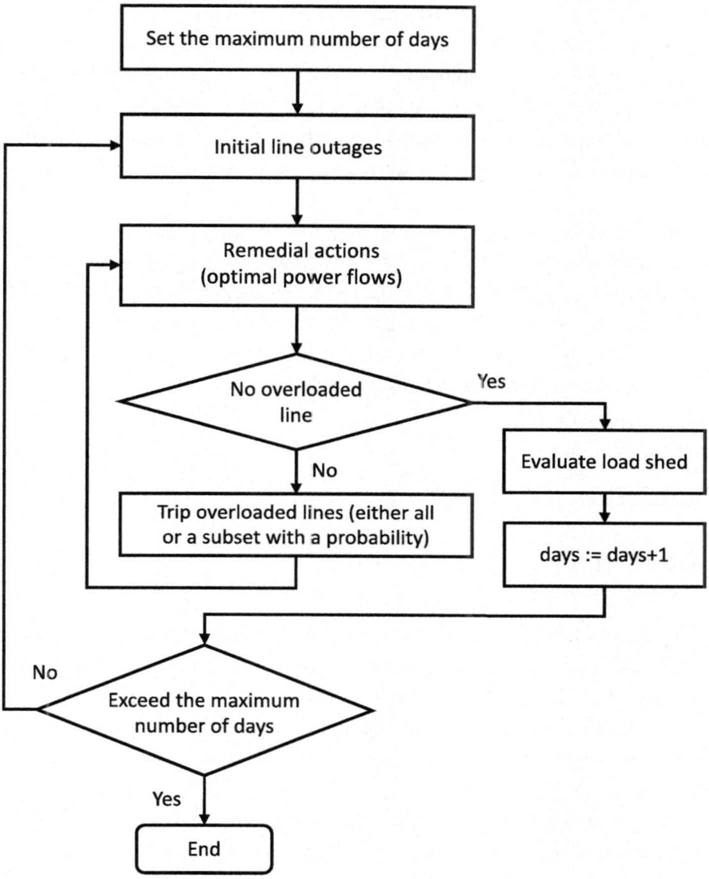

Fig. 1.3 A schematic flow chart of the "fast dynamics" module of the OPA model

iterated until no further line outages occur, effectively generating a cascade sample comprising a sequence of failures.

The "fast dynamics" module of the OPA model is illustrated by a schematic flow chart in Fig. 1.3, in which the maximum number of days defines the total number of cascading scenarios to be generated and simulated, with each day resulting in one scenario. The "slow dynamics" module of the OPA model adds for each day slow growth in generation and load as well as possible line capacity improvements. Thus, when a substantial number of days are simulated, the system undergoes a long-term process of transmission expansion, gradually increasing transmission capacities to mitigate line overloading caused by load growth.

The OPA model can be enhanced to better represent the quasi-steady-state behaviors or even dynamic behaviors of a power system during cascading failures. For instance, substituting the DC power-flow model with an AC power-flow model can allow simulation of voltage collapse and under-voltage load shedding schemes under cascading failures [54, 55]. The incorporation of a frequency regulation model and load frequency characteristics enables simulating frequency drops, frequency protections of generators, and under-frequency load shedding schemes [56]. Moreover, integrating a differential equation solver or a time-domain simulator could assess transient stability and address more protective actions following each failure [57].

Although the OPA model neglects various dynamics and intricate protection and control mechanisms inherent in real-life power systems, it effectively captures critical cascading mechanisms, particularly in the progression and spread of failures. Regarded as an important cascading model, it enables approximate simulations of cascading failures and serves as a foundation for researchers to construct more detailed and precise simulators for cascading failures.

1.8 Simulation

1.8.1 Challenges

A power system manifests various behaviors across very different timescales. For instance, electromagnetic transients (EMTs) in circuits and power electronic controllers occur in the range of microseconds to milliseconds, whereas transient stability and electromechanical oscillations of synchronous generators span milliseconds to seconds. The gradual shifts in loads and adjustments in generation, which lead to equilibrium changes, are often classified as quasi-steady-state dynamics ranging from seconds to minutes. The temperature changes within equipment evolve at an even slower pace of several hours but can contribute to failures of equipment if over-temperature issues arise [58, 59]. The intricate principles and regulations govern a power system's design, planning, and operations and contribute to the extensive range of potential equipment failures and the numerous mechanisms through which failures can interact and propagate.

For the accurate simulation of a power system's response to a contingency, the quality of a power system model is essential. In practical situations where contingency simulations are aimed to assess specific stability concerns, the power system model is often simplified to ignore unrelated dynamics and controls. For instance, when focusing on analyzing the angular stability of synchronous generators during a short post-fault period like 10–30 seconds, the modeling generally concentrates on dynamics in the electromechanical timescale, omitting both the rapid EMT dynamics during and following the fault and its clearance, as well as the slowly drifting equilibrium of the system during this short simulation period.

However, simulating a cascading failure can be very different from simulating a power system contingency. The entire process and impacts of a cascading event can extend over a long period lasting for tens of minutes. This process typically comprises three stages:

1. An early stage begins with the initial equipment failure occurring under system conditions having reduced reliability margins due to factors such as load growth, offline equipment for maintenance, and transmission line overheating. System behaviors in this stage progress gradually and may persist for tens of minutes.
2. A progression stage involves subsequential failures and protections in a relatively localized area of the initial failure. These failures can cause electromechanical oscillations among certain generators, resulting in noticeable frequency responses and voltage variations in this and neighboring regions. A minority of generators may be tripped due to over- or under-excitation protection, but the system can still maintain its stability and integrity. This stage lasts considerably shorter than the early stage and is characterized by faster dynamics.
3. In the final cascade stage, instabilities occur in rotor angles or bus voltages and trigger more protective relays to trip, causing overloading and outages of more equipment. Consequently, failures propagate and spread to a wide area or even the entire system, ultimately reaching a point of no return and resulting in a collapse of the system to lose its stability and integrity. This stage is characterized by greater rapidity, involving even faster dynamics. The system exhibits highly nonlinear behaviors, encompassing a complex combination of continuous and switching dynamics.

It is quite challenging to simulate accurately and entirely all three stages of a cascading process. This needs to address the simulations of all related phenomena including the initial failure and its relay protection, subsequential failures and protective actions, the dynamics of generators and other dynamic devices, fluctuations and controls of frequency and voltages, all encountered instabilities, system collapse and islanding, under-frequency or under-voltage load shedding and other remedial action schemes, and ultimately the blackout of one area or the entire system.

One challenge lies in the requirement of appropriate models for all involved power system components and protection and control schemes as well as the parameters of these models. Moreover, the sensing, measuring, and communication systems involved in the cascading process also need to be modeled to authentically present the mechanism of cascading. This poses a practical challenge in validating and integrating these models for cascading failure simulations. For instance, a conventional objective of the model validation tasks conducted by power system engineers for generators and other dynamic devices is to meet the requirements for transient stability simulations. Thus, the field data used for model validation typically consist of single-fault data obtained from digit fault recorders (DFRs). These recorders are intelligent electronic devices activated upon detecting a fault and record high-resolution data related to the fault for a period until the fault and its associate effects cease. There is no guarantee that a validated generator model based on single-fault data is still accurate in simulating how the generator behaves for an extended period during a cascading failure. Similar issues are also with other dynamic devices. In practice, power system engineers lack field data of cascading events to test the accuracy of their simulation models for extended periods due to the infrequent nature of cascading failures.

Another challenge arises in choosing an effective simulation approach for cascading failures. Modern power system simulators and contingency analysis tools optimize their numerical algorithms and solvers for numerical stability and efficient time performance, focusing on specific timescales. Both the simulation algorithm and model vary based on the dynamics of interests. For instance, EMT models and simulators address fast dynamics in circuits and power electronic controllers; transient stability models and simulators mainly handle electromechanical behaviors of generators and dynamics on a similar timescale; quasi-steady-state simulators address power-flow controls under load or generation changes. For the highest accuracy, a comprehensive EMT grid model that captures intricate power system dynamics may be employed to simulate cascading failures. This is typically practiced on small-scale power systems using commercial real-time digital simulators. However, the computational demands of full EMT simulations can be extremely high for large-scale power grids. Therefore, EMT simulators are typically utilized for specific stages or events tailored from a cascading process, and they are rarely used to simulate the entire cascading process from the initial failure to the final system collapse.

Hence, a compromise for efficient simulation of cascading failures can involve and integrate power system models and simulators in various timescales and adopt a multi-timescale simulation approach [60, 61]. It is important to select appropriate models and the optimal solvers to conduct simulation with the best tradeoff between the computation burden and accuracy. The approach needs to be flexible enough to accommodate switches among models and solvers for computational tasks across different timescales in a cascading process.

1.8.2 Simulation of Multi-timescale Dynamics

Given the primary objective of comprehending cascading mechanisms and developing prevention or mitigation strategies, numerous grid behaviors do not mandate exhaustive simulations of complete dynamics. For instance, scenarios involving line trips that lead to increased loading in remaining lines, or generation adjustments and redispatches for load balancing, might not require in-depth dynamic modeling. Rapidly fading transients following a failure or control action, if not interacting significantly with other dynamic system behaviors, can be adequately captured using simplified dynamic models or power-flow-based quasi-steady-state models. Therefore, for efficient simulations of the multi-timescale dynamics of a large-scale power system during cascading failures, it is viable to utilize a grid model that maintains an acceptable, lowest level of detail while appropriately omitting less important dynamics.

A multi-timescale simulation approach can effectively accelerate simulations of cascading failures by ignoring insignificant dynamics. Instances of such dynamics occur following failures in the early stage of a cascading process. As these failures are temporally spaced, their resultant transients basically fade before the occurrence of the next failure. Subsequentially, as the cascading process advances to the progression stage, the system's dynamic behavior such as oscillation itself could trigger protective actions. This makes dynamic simulations necessary in order to assess stability and dynamic security after each failure. Moreover, as the cascading failure progresses to its the final cascade stage, the system has a massive outbreak of failures with overlapping transients. Consequently, this situation necessitates a fully dynamic simulation across the entire system.

In summary, a multi-timescale simulation approach typically employs a quasi-steady-state simulator in the early stage to primarily compute equilibrium changes following temporally separate failures. For the progression stage, it employs a dynamic simulator, either a transient stability simulator using a phasor-based grid model or an EMT simulator, to determine the system's transient response to each failure and associated protective actions if activated. In the ultimate cascade stage, a thorough dynamic simulation is conducted until the system's instability and collapse become evident.

1.9 Introduction of the Book

This chapter has introduced cascading failures, including the causes, prevention and mitigation of cascading failures, and power system restoration after cascading failures. The chapter has also provided an overview of the risk assessment, modeling, and simulation of cascading failures and discussed their requisites, challenges, and methodologies.

In next decades, the methodologies and tools for grid planning and operations will be reformed with the rapid growth in big data, high-performance computing, and artificial intelligence. The rest of the book will focus on three subjects that have potential for breakthroughs in prevention and mitigation of cascading failures: risk assessment, modeling, and simulation.

Chapter 2 discusses the significance of historical utility data in understanding and quantifying risks of blackouts, highlighting its potential uses in improving contingency lists, creating influence or interaction graphs, sampling cascading outages, and modeling cascading and restoration processes.

Chapter 3 introduces interaction models of cascading failures and discusses an interaction analysis approach that utilizes simulated or real outage data to estimate component interactions during cascading and designs mitigation strategies based on identified critical components.

Chapter 4 focuses on probabilistic analytics of cascading failures using an analytic Markov-based model, leading to a semi-analytic approach for unbiased risk assessment, efficient simulation, and advanced prevention and mitigation strategies.

Chapter 5 discusses how cascading failures in power systems are modeled for computer simulations and focuses on the side-to-side comparison of the statistics and cascading path characteristics from two simulators respectively based on quasi-steady-state models and dynamic models.

Chapter 6 presents a multi-timescale cascading model for quasi-dynamic simulations, along with a Markovian tree-based probabilistic model for identifying high-risk cascading outages and suggesting optimal mitigation strategies.

Chapter 7 introduces a quasi-steady-state approach for simulating cascading failures that incorporates static frequency characteristics of generators and loads, enabling the simulation of frequency-related protection and mitigation actions such as under-frequency load shedding, without the need for a time-domain simulation.

Chapter 8, from an industrial perspective, offers an overview of current practices and criteria against cascading failures, introducing operating state analysis for cascading processes, the standards related to cascading, the approaches for case development, methodologies and tools for cascading analysis, and industrial practices in preventing and mitigating cascading failures.

References

1. Glossary of Terms Used in NERC Reliability Standards, 2023, https://www.nerc.com/pa/Stand/Glossary%20of%20Terms/Glossary_of_Terms.pdf
2. G.S. Vassell, The Northeast blackout of 1965. Publ. Utilities Fortnightly (United States) **126** (1990)
3. G.L. Wilson, P. Zarakas, Anatomy of a blackout: How's and why's of the series of events that led to the shutdown of New York's power in July 1977. IEEE Spectr. **15**(2), 39–49 (1978)
4. 1996 System Disturbances: Review of Selected 1996 Electric System Disturbances in North America, North American Electric Reliability Council, August 2002, https://www.nerc.com/pa/rrm/ea/System%20Disturbance%20Reports%20DL/1996SystemDisturbance.pdf
5. V.X. Filho, L.A.S. Pilotto, N. Martins, A.R.C. Carvalho, A. Bianco, Brazilian defense plan against extreme contingencies. IEEE Power Engineering Society Summer Meeting, Vancouver, BC, Canada,15–19 July 2001
6. U.S.-Canada Power System Outage Task Force, Final Report on the August 14, 2003 Blackout in the United States and Canada: Causes and Recommendations, April 2004
7. A. Berizzi, The Italian 2003 blackout, IEEE Power Engineering Society General Meeting, Denver, CO, USA, 6–10 June 2004
8. Final Report System Disturbance on 4 November 2006, union for the co-ordination of transmission of electricity, January 2007
9. FERC and NERC, Arizona-Southern California Outages on September 8, 2011: Causes and Recommendations, April 2012
10. Report of the Enquiry Committee on Grid Disturbance in Northern Region on 30th July 2012 and in Northern, Eastern & North-Eastern Region on 31st July 2012, August 16th, 2012, New Delhi
11. G. Liang, S.R. Weller, J. Zhao, F. Luo, Z.Y. Dong, The 2015 Ukraine Blackout: Implications for false data injection attacks. IEEE Trans. Power Syst. **32**(4), 3317–3318 (2017)
12. A. Nordrum, Transmission failure causes nationwide blackout in Argentina. IEEE Spectrum: Technology, Engineering, and Science News. 20 June 2019
13. K. Zhou, I. Dobson, Z. Wang, The most frequent N-K line outages occur in motifs that can improve contingency selection. IEEE Trans. Power Syst., Early Access (https://doi.org/10.1109/TPWRS.2023.3249825)
14. J. Qi, K. Sun, S. Mei, An interaction model for simulation and mitigation of cascading failures. IEEE Trans. Power Syst. **30**(2), 804–819 (2015)

15. W. Ju, K. Sun, J. Qi, Multi-layer interaction graph for analysis and mitigation of cascading outages. IEEE J. Emerg. Sel. Top. Circuits Syst **7**(2), 239–249 (2017)
16. P.D. Hines, I. Dobson, P. Rezaei, Cascading power outages propagate locally in an influence graph that is not the actual grid topology. IEEE Trans. Power Syst. (2016)
17. K. Zhou, I. Dobson, Z. Wang, A. Roitershtein, A.P. Ghosh, A Markovian influence graph formed from utility line outage data to mitigate large cascades. IEEE Trans. Power Syst. **35**(4), 3224–3235 (2020)
18. "Remedial Action Scheme" Definition Development: Background and Frequently Asked Questions, NERC, June 2014
19. G. Zhang, K. Sun, H. Chen, R. Carroll, Y. Liu, Application of synchrophasor measurements for improving operator situational awareness. IEEE PES General Meeting, Detroit, MI, 24–29 July 2011
20. K. Sun, Q. Zhou, Y. Liu, A phase locked loop-based approach to real-time modal analysis on synchrophasor measurements. IEEE Trans. Smart Grid **5**(1), 260–269 (2014)
21. J. Qi, K. Sun, W. Kang, Optimal PMU placement for power system dynamic state estimation by using empirical observability Gramian. IEEE Trans. Power Syst. **30**, 2041–2054 (2015)
22. K. Sun, X. Luo, J. Wong, Early warning of wide-area angular stability problems using synchrophasors. IEEE PES General Meeting, 23–26 July 2012, San Diego, 2012
23. F. Hu, K. Sun, A. Del Rosso, E. Farantatos, N. Bhatt, Measurement-based real-time voltage stability monitoring for load areas. IEEE Trans. Power Syst. **31**(4), 2787–2798 (2016)
24. K. Sun, D.-Z. Zheng, Q. Lu, Splitting strategies for islanding operation of large-scale power systems using OBDD-based methods. IEEE Trans. Power Syst. **18**, 912–923 (2003)
25. K. Sun, D.-Z. Zheng, Q. Lu, Searching for feasible splitting strategies of controlled system islanding. IEE Proc. Gener. Transm. Distrib. **153**, 89–98 (2006)
26. K. Sun, K. Hur, P. Zhang, A new unified scheme for controlled power system separation using synchronized phasor measurements. IEEE Trans. Power Syst. **26**(3), 1544–1554 (2011)
27. L.H. Fink, K.L. Liou, C.C. Liu, From generic restoration actions to specific restoration strategies. IEEE Trans. Power Syst. **10**(2), 745–752 (1995)
28. M.M. Adibi, L.H. Fink, Overcoming restoration challenges associated with major power system disturbances – restoration from cascading failures. IEEE Power Energy Mag. **4**(5), 68–77 (2006)
29. Y. Hou, C.-C. Liu, K. Sun, P. Zhang, S. Liu, D. Mizumura, Computation of milestones for decision support during system restoration. IEEE Trans. Power Syst. **26**(3), 1399–1409 (2011)
30. A. Golshani, W. Sun, K. Sun, Real-time optimized load recovery considering frequency constraints. IEEE Trans. Power Syst. **34**(6), 4204–4215 (2019)
31. C. Wang, V. Vittal, K. Sun, OBDD-based sectionalizing strategies for parallel power system restoration. IEEE Trans. Power Syst. **26**(3), 1426–1433 (2011)
32. I. Dobson, Where is the edge for cascading failure? Challenges and opportunities for quantifying black risk. IEEE Power Engineering Society General Meeting, Tampa FL USA, June 2007
33. I. Dobson, B.A. Carreras, V.E. Lynch, D.E. Newman, Complex systems analysis of series of blackouts: cascading failure, critical points, and self-organization. Chaos **17**(2) (2007)
34. H. Ren, I. Dobson, B.A. Carreras, Long-term effect of the n-1 criterion on cascading line outages in an evolving power transmission grid. IEEE Trans. Power Syst. **23**(3), 1217–1225 (2008)
35. S. Cieslewicz, R. Novembri, *Utility Vegetation Management Final Report* (US Federal Energy Regulatory Commission, 2004)
36. IEEE Working Group on understanding, prediction, mitigation and restoration of cascading failures, Benchmarking and validation of cascading failure analysis tools. IEEE Trans. Power Syst. **31**(6), 4887–4900 (2016)
37. IEEE PES CAMS Task Force on Understanding, Prediction, Mitigation and Restoration of Cascading Failures, Initial review of methods for cascading failure analysis in electric power transmission systems. IEEE PES General Meeting, Pittsburgh, PA, 2008, pp. 1–8

38. K.B. Athreya, P.E. Ney, *Branching Processes*, Dover NY 2004 (reprint of Springer-verlag Berlin 1972)
39. T.E. Harris, *Theory of Branching Processes*, Dover NY 1989
40. I. Dobson, Estimating the propagation and extent of cascading line outages from utility data with a branching process. IEEE Trans. Power Syst. **27**(4), 2146–2155 (2012)
41. J. Qi, W. Ju, K. Sun, Estimating the propagation of interdependent cascading outages with multi-type branching processes. IEEE Trans. Power Syst. **32**(2), 1212–1223 (2017)
42. B.A. Carreras, V.E. Lynch, I. Dobson, D.E. Newman, Dynamical and probabilistic approaches to the study of blackout vulnerability of the power transmission grid. 37th HICSS, Hawaii, 2004
43. I. Dobson, B.A. Carreras, D.E. Newman, A loading-dependent model of probabilistic cascading failure. Probab. Eng. Inf. Sci. **19**(1), 15–32 (2005)
44. I. Dobson, B.A. Carreras, D.E. Newman, A branching process approximation to cascading load-dependent system failure. 37th Hawaii International Conference on System Sciences, Hawaii, 2004
45. P.D. Hines, I. Dobson, et al., "Dual Graph" and "Random Chemistry" methods for cascading failure analysis. 46th Hawaii International Conference on System Sciences, HI, January 2013
46. J. Qi, J. Wang, K. Sun, Efficient estimation of component interactions for cascading failure analysis by EM algorithm. IEEE Trans. Power Syst. **33**(3), 3153–3161 (2018)
47. J. Qi, Utility outage data driven interaction networks for cascading failure analysis and mitigation. IEEE Trans. Power Syst. **36**(2), 1409–1418 (2021)
48. W. Ju, J. Qi, K. Sun, Simulation and analysis of cascading failures on an NPCC power system test bed. IEEE Power and Energy Society General Meeting, Denver CO, July 2015
49. C. Chen, W. Ju, K. Sun, S. Ma, Mitigation of cascading outages using a dynamic interaction graph-based optimal power flow model. IEEE Access **7**, 168636–168648 (2019)
50. C. Chen, S. Ma, K. Sun, X. Yang, C. Zheng, X. Tang, Mitigation of cascading outages by breaking inter-regional linkages in the interaction graph. IEEE Trans. Power Syst. **38**(2), 1501–1511 (2023)
51. B.A. Carreras, V.E. Lynch, I. Dobson, D.E. Newman, Critical points and transitions in an electric power transmission model for cascading failure blackouts. Chaos **12**(4), 985–994 (2002)
52. B.A. Carreras, V.E. Lynch, I. Dobson, D.E. Newman, Complex dynamics of blackouts in power transmission systems. Chaos **14**(3), 643–652 (2004)
53. D.E. Newman, B.A. Carreras, V.E. Lynch, I. Dobson, Exploring complex systems aspects of blackout risk and mitigation. IEEE Trans. Reliab. **60**(1), 134–143 (2011)
54. S. Mei, F. He, X. Zhang, et al., An improved OPA model and blackout risk assessment. IEEE Trans. Power Syst. **24**(2), 814–823 (2009)
55. S. Mei, Y. Ni, G. Wang, S. Wu, A study of self-organized criticality of power system under cascading failures based on AC-OPF with voltage stability margin. IEEE Trans. Power Syst. **23**, 1719–1726 (2008)
56. W. Ju, K. Sun, R. Yao, Simulation of cascading outages using a power flow model considering frequency. IEEE Access **6**(1), 37784–37795 (2018)
57. B. Park, X. Su, K. Sun, An enhanced OPA model: Incorporating dynamically induced cascading failures. IEEE Trans. Power Syst. **37**(6), 4962–4965 (2022)
58. R. Yao, K. Sun, F. Liu, S. Mei, Efficient simulation of temperature evolution of overhead transmission lines based on analytical solution and NWP. IEEE Trans. Power Deliv. **33**(4), 1576–1588 (2018)
59. R. Yao, K. Sun, Towards simulation and risk assessment of weather-related outages. IEEE Trans. Smart Grid **10**(4), 4391–4400 (2019)
60. R. Yao, S. Huang, K. Sun, F. Liu, X. Zhang, S. Mei, A multi-timescale quasi-dynamic model for simulation of cascading outages. IEEE Trans. Power Syst. **31**(4), 3189–3201 (2016)
61. R. Yao, S. Huang, K. Sun, F. Liu, X. Zhang, S. Mei, W. Wei, L. Ding, Risk assessment of multi-timescale cascading outages based on Markovian tree search. IEEE Trans. Power Syst. **32**(4), 2887–2900 (2017)

Chapter 2
Analyzing Cascading Failures and Blackouts Using Utility Outage Data

Ian Dobson

2.1 Analysis of Particular Blackouts

An influential and useful approach to large blackouts considers each specific blackout, analyzes the sequence of events and their interrelated causes in detail, and extracts lessons learned that can be implemented to prevent that particular blackout, or a similar blackout, from happening again. Every large blackout has such analysis and reporting, although the public reporting often is of a summary nature with very limited raw data available for independent analysis. Given the large variety of phenomena involved in cascading blackouts, the details of the analysis and the combination of causes vary considerably. Useful examples of these analyses are [1–5], and some of these blackout narratives also describe the mechanisms involved.

At a higher level, the evolution of large blackouts tends to have features in common such as a long complicated chain of unusual events and interactions. Moreover, blackout data shows that there are patterns in blackout statistics that we discuss in the following sections.

2.2 Probability and Risk of Large Blackouts Estimated from Data

Many countries record the size of transmission system blackouts [6–8]. For example, the USA publicly records blackouts above a particular size. The blackout

I. Dobson (✉)
ECpE Department, Iowa State University, Ames, IA, USA
e-mail: dobson@iastate.edu

K. Sun (ed.), *Cascading Failures in Power Grids*, Power Electronics and Power Systems, https://doi.org/10.1007/978-3-031-48000-3_2

size in MW power interrupted as well as the blackout duration and the number of customers affected are often recorded. The blackout duration is less useful than the other measures since blackout duration suffers from a lack of uniform definition, depends heavily on the last few elements restored, and is inherently more variable than the blackout size [9].

The empirical distribution of blackout size is best presented as a survival function on a log–log scale as shown in Fig. 2.1. The survival function is the probability of a blackout exceeding a given size as a function of the size. (The survival function is also called the complementary cumulative distribution function (CCDF) and is one minus the cumulative distribution function (CDF).) The survival function shows the data without the smoothing or binning needed for a probability density function, and the log–log scale shows the large blackouts of small probability and the overall pattern of how the blackout probability decreases as blackout size increases. It is also more useful to consider blackout probability within a given factor (e.g., to within a factor of two or one half) as given by an equal distance above or below the data on the logarithmic vertical axis.

The approximately straight line behavior in the log–log plot of the distribution of blackout size in Fig. 2.1 shows that the distribution of blackout size has a "heavy tail" or power law region, which is influential on blackout probability and risk. (The power law region is of course limited in its highest extent because every grid has a largest possible blackout in which the entire grid blacks out.) The slope of approximately -1 in the survival function in Fig. 2.1 implies that doubling the blackout size only halves the probability that a blackout exceeds that size. The slope of -1 in the survival function corresponds to a slope of -2 in a log–log plot of the probability density function, which implies that a blackout of double the size has only one quarter of the probability. This relationship shows the typical heavy tail dependence that as blackout size increases, probability decreases, but it decreases very slowly. In contrast, for a probability distribution with an exponential tail (such as exponential or normal distributions), doubling the blackout size at most squares

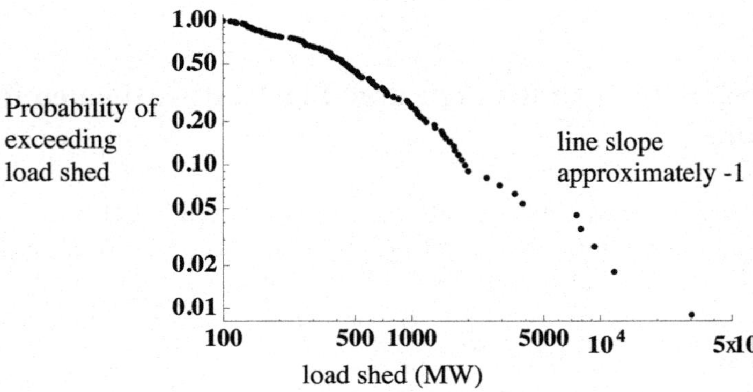

Fig. 2.1 Empirical survival function of blackout size from Western interconnection in the USA from 1984 to 2006

the probability, so that large blackouts are vanishingly unlikely. A probability distribution with heavy tails indicates that blackouts of all sizes, including large blackouts, can occur. Large blackouts are rare, but they are expected to happen occasionally, and they are not "perfect storms." These qualitative remarks are further supported by a more precise description in Sect. 2.2.1.

The approximate power law dependence of blackout size has been confirmed by observed transmission blackout size statistics in many developed countries [6–8]. Cascading failure is a sequence of dependent events that successively weaken the power system. At a given stage in the cascade, the previous events have weakened the power system so that further events are more likely. It is this dependence that makes the long series of cascading events that cause large blackouts likely enough to pose a substantial risk. (If the events were independent, then the probability of a large number of events would be the product of the small probabilities of individual events and would be vanishingly small.)

The power law region can be explained using ideas from complex systems theory. The main idea is that over the long term, the power grid reliability is shaped by the engineering responses to blackouts and the slow load growth and tends to evolve toward the power law distribution of blackout size [6, 10]. It is notable that the power law region appears both in blackouts attributed to weather and blackouts not associated with weather [11].

Blackout risk can be defined as the product of blackout probability and blackout cost. One simple assumption is that blackout cost is roughly proportional to blackout size, although larger blackouts are expected to have costs (especially indirect costs) that increase faster than linearly. In the case of the power law dependence, the larger blackouts can become rarer at a similar rate as costs increase, and then the risk of large blackouts is comparable to or even exceeding the risk of small blackouts [8]. Mitigation of blackout risk should consider both small and large blackouts, because mitigating the small blackouts that are easiest to analyze may inadvertently increase the risk of large blackouts [12].

2.2.1 How the Blackout Probability Decreases as Size Increases

Since the way that blackout probability decreases as blackout size increases is so important to large blackout risk [8] and justifying the study of large cascades, it is useful to describe an idealized form of this relationship more precisely. Suppose that the blackout size above some minimum blackout size $x_{\min}$ MW is the Pareto random variable X with the probability density function

$$f_X(x) = \frac{x_{\min}}{x^2}, \quad x \geq x_{\min} \tag{2.1}$$

Then

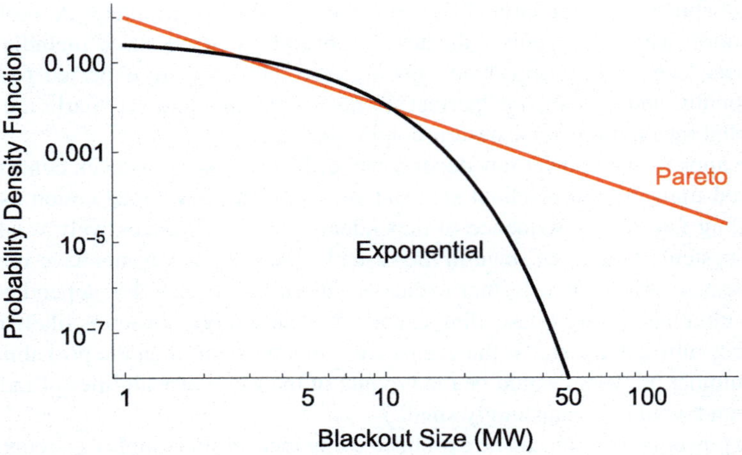

Fig. 2.2 Comparing Pareto and exponential probability density functions on a log–log plot showing that large size events have vanishingly small probability for exponential, whereas the more slowly decreasing Pareto can have rare large events. Slope of Pareto probability density function on this log–log plot is −2. Pareto (Eq. 2.1) has $x_{\min} = 1$ and exponential (Eq. 2.8) has rate $\alpha = \ln\sqrt{2}$ so that both distributions have median 2

$$\log f_X(x) = -2 \log x + \log x_{\min}, \quad x \geq x_{\min} \tag{2.2}$$

so that the plot of $\log f_X(x)$ versus $\log x$ is linear with slope −2 as shown in Fig. 2.2.

Recall, for example, that probability density $f_X(500)$ per MW at $X = 500$ MW means that the probability of a blackout between 500 MW and 501 MW is $f_X(500)$ and the probability of a blackout between 500 MW and 500+dy MW is $f_X(500)dy$. More generally, the probability of a blackout between a MW and b MW is $\int_a^b f_X(y)dy$. The survival function $\overline{F}_X(x)$ is the probability that the blackout size exceeds x so that

$$\overline{F}_X(x) = \int_x^\infty f_X(y)dy = \int_x^\infty \frac{x_{\min}}{y^2} dy = \frac{x_{\min}}{x}, \quad x \geq x_{\min} \tag{2.3}$$

And

$$\log \overline{F}_X(x)) = -\log x + \log x_{\min}, \quad x \geq x_{\min} \tag{2.4}$$

so that the plot of $\log \overline{F}_X(x)$ versus $\log x$ is linear with slope −1.

Then doubling the blackout size from x to $2x$ gives one quarter of the probability density and one half of the survival function:

$$f_X(2x) = \frac{x_{\min}}{(2x)^2} = \frac{1}{4} f_X(x) \tag{2.5}$$

$$\overline{F}_X(2x) = \frac{x_{\min}}{2x} = \frac{1}{2}\overline{F}_X(x) \tag{2.6}$$

It is straightforward to repeat this calculation more generally with a power law x^{-s} of the tail of the probability density of blackout size. Then the survival function has a power law tail x^{1-s}. And on log–log plots the slope of the probability density is $-s$ and the slope of the survival function is $1-s$. Doubling the blackout size gives 2^{-s} times the probability density and 2^{1-s} times the survival function.

In contrast, one can consider a probability distribution with an exponential tail such as

$$f_X(x) = \alpha e^{-\alpha x}, \quad x \geq 0 \tag{2.7}$$

$$\overline{F}_X(x) = e^{-\alpha x}, \quad x \geq 0 \tag{2.8}$$

Then doubling the size from x to $2x$ has an effect proportional to squaring both the probability density and the survival function:

$$f_X(2x) = \alpha e^{-2\alpha x} = \frac{1}{\alpha}(f_X(x))^2 \tag{2.9}$$

$$\overline{F}_X(2x) = e^{-2\alpha x} = (\overline{F}_X(x))^2 \tag{2.10}$$

It follows that if the tail of blackout size were exponential (which it is not), this squaring effect would have made large blackouts vanishingly unlikely as illustrated in Fig. 2.2.

2.3 Processing Utility Outage Data

This section describes detailed utility outage data and how to automatically process a series of outages into events or cascades and further into successive generations of outages within the cascades. The event processing is fundamental for studying cascading and resilience with real data. This differs from the traditional processing for reliability, which tends to study the outages and repairs in a steady state that is averaged over the year.

It is important to note that utility outage data includes events of all sizes and all causes. In particular, large events include those mainly caused by cascading outages within the transmission grid as well as those mainly caused by extreme weather. The outages caused by extreme weather can be identified by their location and timing matching weather records or by cause codes in utility data. Most of the large blackouts are associated with extreme weather [13]. (Cascading effects could also contribute to the weather blackouts, but the extent of this contribution is not known.) The blackouts caused by extreme weather are studied in resilience, whereas the routine outages occurring singly or in small groups are dominant in the study of reliability, partly because small outages are much more numerous. The study of cascading includes large blackouts that can be considered part of resilience and can also include short cascades with a few outages that quickly stop. There are significant overlaps and no clear boundaries between cascading, resilience, and reliability. The important point for data analysis is that if one considers all the automatic outages observed in some real utility data, they encompass and exemplify aspects of cascading, resilience, and reliability, and the differences between these categories mostly depend on the processing and interpretation of the data. Detailed utility outage data are foundational for the study of cascading, resilience, and reliability.

While the focus here is on transmission system outages, it should be noted that Murphy [14, 15] describes correlated generator outages with detailed utility generator outage data from the Generator Availability Data System (GADS) of the North American Electric Reliability Corporation (NERC) in the USA. The increased generator outage rates during extremes of cold or heat and loading have a significant impact on generation adequacy, as, for example, in the February 2021 blackout in Texas. The increased generator failures are correlated by common external weather rather than by cascading via interactions within the power system.

2.3.1 Transmission Utility Outage Data

2.3.1.1 Utility Outage Data

Transmission systems operators and regulators often systematically collect detailed time-stamped outage data for lines, transformers, and other equipment. For example, the North American Electric Reliability Corporation (NERC) has been collecting North American automatic (momentary and sustained) outage data for transmission elements operating at 200 kV and above in its Transmission Availability Data System (TADS) since January 2008 [16]. While the collection of detailed outage data is standard practice in North America and many other countries, there are only a few publicly available sources for this data [17, 18]. The data typically includes the outage start time, end time, bus name(s), rating, cause codes, and whether the outage is automatic (forced) or planned. The processing of the data typically extracts the automatic outages and neglects the planned outages. Some data such as TADS includes only the automatic outages. There may also be an annual inventory of components in the data.

2.3.1.2 Extracting Events from Detailed Outage Data

A key step in processing real data for cascading and resilience is automatically extracting events. The events are often called cascades or resilience events. These events range in size from isolated single outages to events with hundreds of outages.

One approach to defining and extracting events locks at the outage starting times to detect when they bunch up. This approach was developed to study cascading due to all causes in utility data [19]. Note that outages bunching up due to bad weather is well known in reliability [20, 21]. The event definition is simple: if successive outages start within a time threshold such as one hour, they are in the same event. The time threshold of one hour should be adjusted to the data being processed. Although unrelated outages can start within one hour, it is much more likely that the outages are dependent. The processing simply moves through the outage start times in the order of their occurrence and makes a new event when there is a gap between successive outages of more than one hour.

The outages in a cascade event identified by the bunching can be further divided into generations of outages. The initial outages of the cascade are the first generation; these are followed by the second generation of outages, and so on until the cascade ends. The generations of outages within each cascade arise from the fast time scale of less than a minute of the protection system that implements the outages. Multiple automatic outages occurring simultaneously or in very quick succession are grouped together in the same generation, and in particular, successive outages that start within one minute are in the same generation [19].

The described processing of outages into a series of events, each with one or more generations of outages, structures the outage data for cascading analysis. The majority of the events have only one outage and one generation, but there is considerable interest in examining the larger events, which are rarer but more consequential.

Another approach to defining and extracting events considers how outages overlap in time. Two outages overlap in time if the first outage is not restored when the second outage starts. This approach was developed primarily to study weather-related events in utility data [13, 22, 23]. Motivations for considering overlap include the weakening of the grid while an outage persists and considering outages of some positive duration as more significant or impactful than momentary outages. Consider the set union of the durations of the outages as illustrated in Fig. 2.3. This set union of the duration times of the outages is composed of disjoint intervals, and each disjoint interval defines the duration of one event. The outages can be processed into these events using the number of unrestored outages [22], which varies with time as outages occur and are restored. The number of unrestored outages or, more usually, the negative of the number of unrestored outages is familiar in resilience [9, 24, 25] and is sometimes called the resilience performance curve. Under normal conditions the number of unrestored outages stays near zero because outages are generally infrequent and are restored quickly. But under stressed conditions, outages are more frequent and accumulate before they can be restored, and the number of unrestored

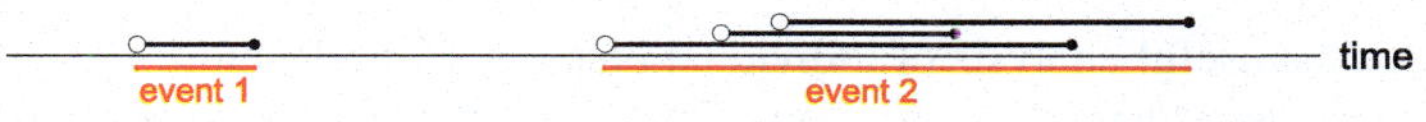

Fig. 2.3 Two events formed by the union of overlapping outages. For each outage duration shown above the axis, the open circle is the outage start time and the dot is the outage restore time

outages has excursions away from zero. These accumulations of outages are the events. The events can be extracted from the data by detecting when the number of unrestored outages passes and returns to zero. In practice, some limitation of the outage duration before processing the data may be necessary to avoid an outage that has an exceptionally long repair time creating an unrealistically long event.

Another approach to find weather events is to identify the time period of bad weather in weather data and then find the outages that occur in that time period [26].

Combinations of bunching and overlap and elaborations restricting the geographic region and handling momentary outages may be applied in practice [9, 13, 23, 27]. For example, in [9, 27], for each interconnection, the automatic outages are grouped together into events based on the bunching and overlaps of their starting times and durations. We quote from [27] the algorithm used: "Every outage in an event has to either start within five minutes of a previous outage in the event or overlap in duration with at least one previous outage in the event that has a difference in starting time not exceeding one hour. In applying this algorithm, repeated momentary outages of the same element are neglected if they occur within 5 minutes of each other." Then events that contain at least one outage with a weather-related initiating or sustained cause code are defined as weather-related.

2.3.2 *Statistical Patterns in Number of Outages and Generations*

The processing of outages into events enables the empirical statistics of the events to be determined, and their statistics augments the heavy tailed behavior of blackout size measured in load shed considered in Sect. 2.2. The probability distribution of the number of outages in events has a heavy tail and a power law character. For example, Fig. 2.4 shows the distribution of the number of outages in events with more than 20 outages in 5 years of North American transmission events [27]. Moreover, as shown in Fig. 2.5, the number of generations in these cascades approximates a Zipf distribution [13, 28], which also has a heavy tail and is a power law distribution. The slope of the dashed line in Fig. 2.5 has been proposed as the SEPSI index of cascading [13, 28], where SEPSI stands for System Event Propagation Slope Index.

2.4 Probabilistic Models Directly Driven by Utility Data

Except for the useful analysis of individual blackouts, almost all of the research on cascading is based on models and simulation. Since there are many and varied mechanisms of cascading, and blackouts typically involve complicated combinations of

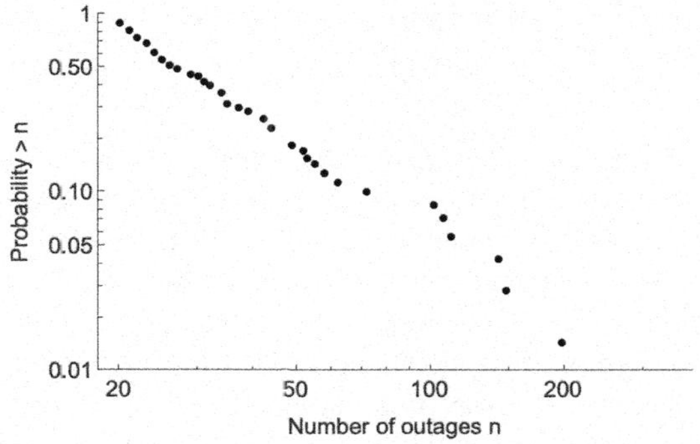

Fig. 2.4 Log–log plot of survival function of the number of outages in North American transmission events with at least 20 outages. Data are from [27]

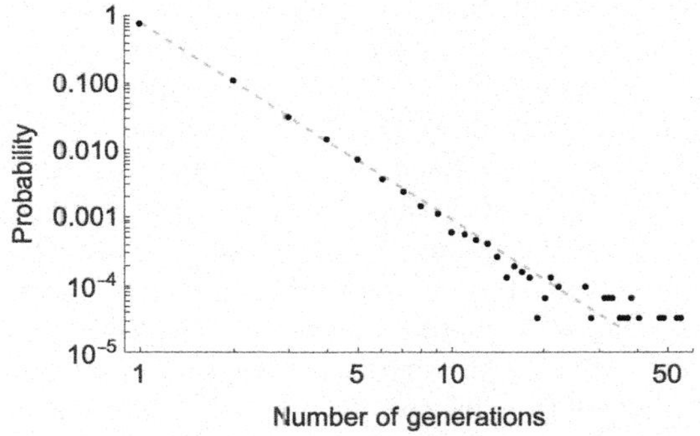

Fig. 2.5 Log–log plot of probability distribution of number of generations for North American transmission events (dots) with dashed gray line showing the slope of the fitted Zipf distribution. Data are from [13]

these mechanisms, cascading is particularly hard to model realistically. There is substantial progress modeling a subset of the mechanisms of cascading to produce cascades that could plausibly occur, but many of the most studied mechanisms of cascading are only those with tractable models, and there is much less work giving definitive validation of the cascading models and simulations [29, 30]. And the restoration from cascading blackouts is not often modeled.

Historical blackout data has an obvious key role to ground cascading models and simulations in reality with calibration and validation. But there are further promising and emerging opportunities to directly exploit historical blackout data. To encourage this direction of research, this section shows how historical outage data can enable better contingency lists, be transformed into influence or interaction graphs, replace simulation by sampling, and describe typical blackouts and their restoration with Poisson processes.

2.4.1 *Contingency Motifs for Multiple Outages Initiating Cascading*

When assessing the risk of cascading outages with a simulation, it is usual to sample multiple initial line outages with some sort of equal probability assumption, such as independent, equal probabilities for the individual line outages that make up each multiple outage [31, 32], or equal probabilities for all $N - 2$ outages [33]. One reason is that networks are designed and operated with the $N - 1$ criterion so that outage of only one element in an idealized simulation does not generally initiate a cascade. Analysis of real utility data shows that these equal probability assumptions are pragmatic but unrealistic, making the cascading simulations start from a set of contingencies that are unlikely in practice and making the resulting cascading risk estimates less credible.

A contingency motif [34] is a spatial pattern or small subgraph of the network of multiple outages that occur much more frequently than corresponding multiple outages chosen randomly with equal probability from the utility network. For example, in real data, the $N - 2$ contingency motif  of two lines with a common bus initially outaging occurs much more frequently in initial outages than a random selection of any two lines in the network. Contingency motifs are quite similar to the motifs conventionally considered in network theory [35], except that the frequency of contingency motifs is high compared to a corresponding random selection from the specific power grid being analyzed, rather than high compared to a corresponding random selection from a random graph with similar characteristics. Ren et al. propose conventional network motifs as an indicator of cascading outage risk [36]. They show that phases of cascading outages as the load level increases correspond to the decrease of the frequency of network motifs. The frequency of motifs reflects the connectivity of the power grid; hence, it can be a warning sign of the cascading outage risk. Other researchers have studied conventional network motifs as an indicator of power grid robustness and reliability using techniques from network science [37, 38].

The initial outages of a cascade can be obtained from real outage data by grouping the outages into cascades and generations within each cascade as in Sect. 2.3.1.2. Then the initiating outages are those in the first generation of each cascade. A utility network on which the outages can be located can be deduced directly from the outage data [39], so that the patterns of the initial outages on the utility network can be detected, classified, and counted. This allows the $N - k$ contingency motifs to be obtained, as those that are, for example, ten times more likely to occur in the initial outages than by random selection of k lines from the network. It turns out for two North American publicly available historical outage data sets that the three contingency motifs  ,  ,  account for most of the probability of multiple line outages [34]. Preferentially including these contingency motifs in a contingency list according to their observed frequencies gives a much more realistic sampling of the initiating contingencies and contributes to a better cascading risk analysis via simulation. Of course, for a complete risk analysis, it is necessary to augment the multiple outages with the more tractable single outages, whose probability can be more accurately determined (at least for lines of a given voltage rating or other class), and whose sampling is straightforward.

2.4.2 Influence/Interaction Graphs Driven by Utility Data

The way in which line outages influence or interact with each other in cascades of outages can be described by graphs (networks) in which the transmission lines are the nodes and the influences or interactions between outages are the links between the nodes. These graphs are called influence graphs or interaction graphs and are obviously different networks than the physical utility network. Influence/interaction graphs were originally developed in [33, 40, 41] and are reviewed in [42, 43]. Influence graphs can be formed from simulated or utility data, and this section only describes some particular issues of influence graphs driven by utility data.

The first issue with processing utility data into an influence graph is that one is limited to the amount of recorded data. Methods to mitigate this include combining together the data for several generations to estimate the transitions for multiple generations at once while exploiting the overall form of cascade propagation and Bayesian methods [43], and exploiting memory of the interactions between generations of the cascades [44].

The second issue with processing utility data into an influence graph is that real data (usually quantized in time to one minute) has multiple outages occurring in the same minute, which is mainly due to the fast time scale of the protection system. (Simulations, at least so far, avoid this by idealizing single outages occurring in sequence.) This can be addressed either by regarding the multiple outages as additional nodes in the influence graph [43] or by running an expectation maximizing algorithm to allocate the influences among the multiple outages [44].

2.4.3 Sampling from Utility Data to Replace Simulation

Cascading failure analysis is dominated by simulations of models of a subset of the mechanisms of cascading. However, there is another way to generate realistic cascades by computer, which is to sample from the observed outage statistics that characterize how outages propagate and spread on the network. This has been done as part of models of transmission system resilience [45, 46] and to estimate how much cascading follows the initial damage from an earthquake [47]. One can either sample directly for the empirical distributions or fit a distribution to the empirical distribution and sample from the fitted distribution. The advantage of using a fitted distribution is that it smooths the rarer data and allows some extrapolation. The sampling based on empirical data only captures the overall statistical form of the cascading, but it requires no modeling assumptions, and is very fast compared with model-based approaches. The sampling approach, while necessarily an approximate description, has some sound basis in reality, especially if the empirical data of the network under study is used.

There are two questions to address with sampling cascade outages: How many outages are in the cascade and where are these outages located on the network? One way to characterize the number of line outages in a cascade uses a branching process statistical model that is estimated from the observed data [19]. The propagation parameter from generation k to generation $k+1$ is the expected number of outages in generation $k+1$ that are produced by each outage in generation k. The propagation parameters for all the generations can be directly calculated from observed outage data grouped into cascades and generations. Then the distribution of the number of line outages can be calculated from the number of initial outages and the propagation parameters [19, 47]. Another way is to empirically estimate and sample from the distribution of the total number of outages in cascades [45, 46]. The form of the distribution of the total number of line outages can have the form of a Zipf distribution for two or more line outages [46]. The sampling approach can extend to other aspects. For example, it is also feasible [46] to sample the number of generator outages from observed generator outage statistics [14, 15] and to sample transmission line restoration times from observed line restoration statistics [48].

The spread of the outages on the network can be statistically described by the empirical distribution of network distance between line outages in cascades [45–47]. To obtain these statistics it is necessary to locate the historical outages on the utility network, and this network can be formed directly from the outage data [39]. After an initial outage or outages are selected on the network, the outages remaining in the total number of outages in the cascade can be located on the network by successively sampling from the network distance distribution. Resampling is used if there are no intact lines available at the sampled network distance.

The sampling approach to cascading is particularly effective when cascading is only one part of the entire calculation and fast samples of cascades are needed. For example, a network model can be used to evaluate the impact of the sampled cascade outages. There is a problem with straightforward sampling of the cascades that the large cascades of the most interest are rare, so that straightforward sampling of cascades is very inefficient in calculating the high impact low frequency cascades. This problem can be solved by stratified sampling with strata of different cascade sizes [46, 49]. Stratified sampling allows sufficient samples at each range of cascade sizes, the results of which are then weighted by the probability of that range of cascade sizes, which in our case is known from the distribution of cascade sizes.

2.4.4 Statistical Models of Outage and Restore Processes

There are several useful ways to describe events by tracking their outages. For each event, the outage process $O(t)$ is the cumulative number of outages at time t and the restore process $R(t)$ is the cumulative number of restores at time t. Both processes

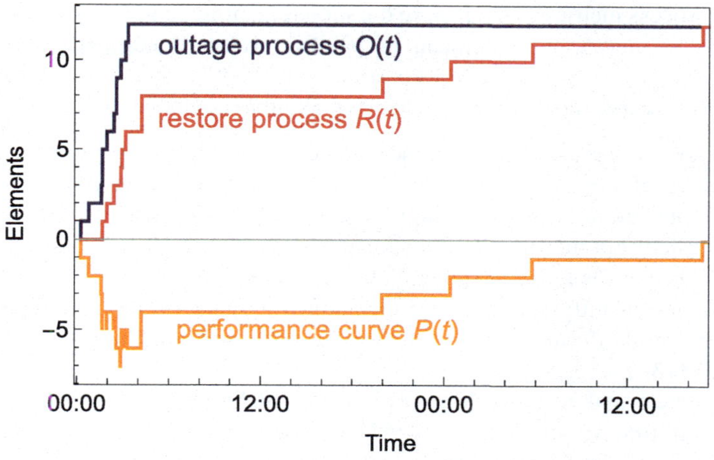

Fig. 2.6 Processes for a transmission system event with 12 outages. Image from [9] is licensed under CC BY 4.0

start at zero at the beginning of the event and increase to the total number of outages n, as shown in the example with $n = 12$ outages in Fig. 2.6. Resilience studies [24, 25, 50, 51] often define for each event a performance (or resilience) curve $P(t)$, which is the negative of the number of *unrestored* outages at time t. The performance curve decrements for each outage and increments for each restore as shown in Fig. 2.6. Indeed, the performance curve is related to the outage and restore processes by

$$P(t) = R(t) - O(t)$$

The performance curve can be uniquely decomposed into its outage and restore processes, and it contains the same information as the outage and restore processes [52].

The outage and restore processes and the performance curve, while straightforward, are fundamental to analyzing events in real outage data. Note that the event analysis is at a systems level and is not focused on tracking individual components: it only counts the number of outages and number of restores and it does not track which outaged component restored when or the order in which components restore. Also, the forms of the outage and restore processes and performance curve readily lead to resilience metrics that describe each process; in particular, it is useful to have separate metrics describing the outage process and the restore process and the performance curve [9, 27]. Examples of the metrics are:

Outage process metrics: number of outages, outage duration, outage rate (can be time dependent)

Restore process metrics:	restore duration (e.g., time to 95% restore, geometric mean of positive restore times [9]), time to first restore, restore rate (can be time dependent)
Performance curve metrics:	area, nadir (maximum components out)

The outage, restore, and performance processes can track quantities other than the number of outages by simply changing the quantity on the vertical axis. For example, the total MVA ratings of the outaged components can be tracked [27, 53]. The list of metrics above assumed that the processes tracked the number of components out. Changing the quantity tracked gives a new set of metrics, except for those metrics quantifying duration. Calculating metrics for utility outage data gives basic and useful information about the various types and magnitudes of outages and blackouts [13, 27, 53, 54].

The outage and restore processes can be stochastically modeled as Poisson processes of time varying rate [9, 55–57]. Consider an outage process of rate $\lambda_O(t)$ that depends on time t. Then, to first order in a small time interval $[t, t+h]$, the probability of a single outage in $[t, t+h] = \lambda_O(t)h$. The restore process can similarly be modeled as a Poisson process of time varying rate $\lambda_R(t)$. The rates of the outage and restore processes can be obtained from utility data. For example, analysis of seven years of North American transmission data shows that the most typical event with n outages has an outage process of constant rate λ_O over a short interval $[0, o_b]$ and a restore process that starts at time r_a after the start of the outages with a rate proportional to a lognormal distribution:

$$\lambda_O(t) = \lambda_O \tag{2.11}$$

$$\lambda_R(t) = \frac{n}{(t-r_a)\sigma\sqrt{2\pi}} \exp[-(\ln(t-r_a)-\mu)^2/(2\sigma^2)], \quad t \geq r_a \tag{2.12}$$

The parameters of these models such as λ_O, r_a, μ, σ can be estimated from utility data [9]. The lognormal distribution of restore rate implies that the rate of restoration typically quickly becomes high and then slows over a long period of time, with the last few restores very delayed. Note that the most typical North American events are dominated by weather events, so it cannot be assumed that the typical cascading event not related to weather is the same as the most typical event. For example, the constant rate outage process of the typical weather event in Eq. 2.11 is probably not appropriate to cascading events since cascading events show an accelerating rate of outages [58].

One can also consider the mean of the outage and restore processes and the resulting performance curve to obtain a typical event for the data that is processed. We denote these mean processes by $\overline{O}(t)$, $\overline{R}(t)$, and $\overline{P}(t)$. Of course, the Poisson processes vary about their mean value, but the mean value is a representative case. For example, Fig. 2.7 shows a typical North American transmission event when all the events over seven years are processed.

The mean processes give a nice formula [57] for the area $\overline{A}$ of the mean performance curve $\overline{P}(t)$ tracking the number of components outages. $\overline{A}$ is also the

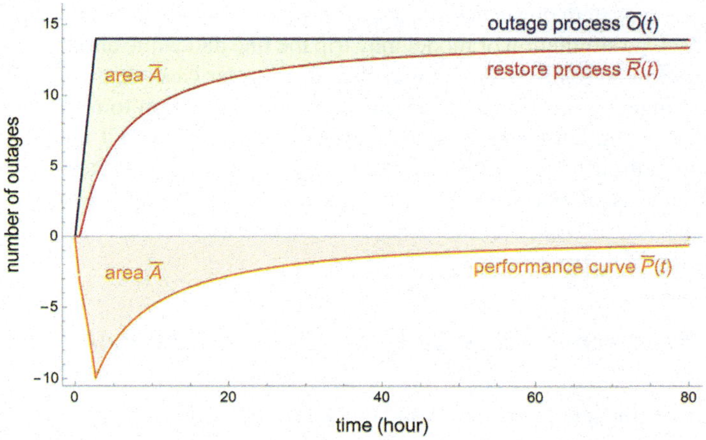

Fig. 2.7 Typical North American transmission system resilience event tracked by number of outages on the vertical scale. The shaded area $\overline{A}$ of $\overline{P}(t)$ is a useful metric and is the same as the shaded area between $\overline{O}(t)$ and $\overline{R}(t)$. Image from [57] is licensed under CC BY 4.0

mean of the area A of the performance curve $P(t)$. Moreover, $\overline{A}$ is also the area between the mean outage and mean restore processes (see Fig. 2.7). Then, writing $\overline{o}$ for the mean outage time and $\overline{r}$ of the mean restore time,

$$\overline{A} = n(\overline{r} - \overline{o}) \tag{2.13}$$

so that (2.13) can be understood as the height n of this area times its average width.

Another way to structure the modeling of events arose in distribution systems [55, 56]. Instead of directly modeling the restore process from data, this approach represents a Poisson outage process arriving at a queue that repairs the outages to produce a restore process. In [56], the response to hurricane Ike is modeled this way using processes that vary in both time and space.

2.4.5 *Validating and Calibrating Simulations with Statistical Data*

The statistical patterns and metrics that can be extracted from utility data are of fundamental importance in calibrating and validating models and simulations to ensure that they can approximate reality well enough. Overall there has been some progress toward calibration and validation for cascading failure [29–31, 59, 60], but much remains to be done.

Since cascading outages encounter many thresholds for discrete actions such as tripping a line or not tripping a line, similar simulations with similar data (or even the real power system on successive days) may behave differently under very similar conditions. One simulation or model may trip the line and while another may not and this can have a large effect on the way that a particular cascade evolves. Therefore, it is too stringent to require an exact match of the simulation to real blackout data. However, one can compare the statistical patterns or the overall metrics produced by the simulation to the observed statistical patterns and metrics. For example, the distributions of blackout sizes or how much the cascades propagate can be compared.

2.4.6 Researcher Access to Utility Data and the Path Forward

There are significant opportunities and much further work to be done processing and analyzing transmission system data. Given the few sources of public detailed outage data [17, 18], one challenge is: How can utilities and regulators be encouraged to share detailed and comprehensive outage data with researchers? One encouraging development is methods to anonymize data. For example, with only some loss of usefulness, bus names can be encrypted and geographic location can be coarse grained. Potential legal liabilities for sharing the data could likely be mitigated by using older data, such as data more than seven years old in the USA. Further collaborative efforts between researchers, regulators, and industry are needed to realize the benefits from the data that has already been collected.

References

1. D. Kosterev, C. Taylor, W. Mittelstadt, Model validation for the August 10, 1996 WSCC system outage. IEEE Trans. Power Syst. **14**, 967–979 (1999)
2. V. Venkatasubramanian, Y. Li, Analysis of 1996 Western American electric blackouts, in *Bulk Power System Dynamics and Control - VI*, Cortina d'Ampezzo, Italy, Aug 2004
3. US-Canada Power System Outage Task Force, Final Report on the August 14, 2003 Blackout in the United States and Canada (2004)
4. Federal Energy Regulatory Commission and the North American Electric Reliability Corporation, Arizona-Southern California Outages on September 8, 2011: Causes and Recommendations (2012)
5. IEEE PES PSDP Task Force on Blackout experience, mitigation, and role of new technologies, blackout experiences and lessons, Best practices for system dynamic performance, and the role of new technologies, IEEE Special Publication 07TP190, July 2007
6. I. Dobson, B.A. Carreras, V.E. Lynch, D.E. Newman, Complex systems analysis of series of blackouts: cascading failure, critical points, and self-organization. Chaos **17**(2), 026103 (2007)
7. P. Hines, J. Apt, S. Talukdar, Large blackouts in North America: historical trends and policy implications. Energy Policy **37**(12), 5249–5259 (2009)
8. B.A. Carreras, D.E. Newman, I. Dobson, North American blackout time series statistics and implications for blackout risk. IEEE Trans. Power Syst. **31**(6), 4406–4414 (2016)
9. I. Dobson, S. Ekisheva, How long is a resilience event in a transmission system? Metrics and models driven by utility data. IEEE Trans. Power Syst. https://doi.org/10.1109/TPWRS.2023.3292328

10. H. Ren, I. Dobson, B.A. Carreras, Long-term effect of the n-1 criterion on cascading line outages in an evolving power transmission grid. IEEE Trans. Power Syst. **23**(3), 1217–1225 (2008)
11. B.A. Carreras, D.E. Newman, I. Dobson, A.B. Poole, Evidence for self-organized criticality in a time series of electric power system blackouts. IEEE Trans. Circuits Syst. Part 1 **51**(9), 1733–1740 (2004)
12. D.E. Newman, B.A. Carreras, V.E. Lynch, I. Dobson, Exploring complex systems aspects of blackout risk and mitigation. IEEE Trans. Reliab. **60**(1), 134–143 (2011)
13. S. Ekisheva, R. Rieder, J. Norris, M. Lauby, I. Dobson, Impact of extreme weather on North American transmission system outages, in *IEEE PES General Meeting*, Washington DC USA, July 2021
14. S. Murphy, J. Apt, J. Moura, F. Sowell, Resource adequacy risks to the bulk power system in North America. Appl. Energy **212**, 1360–1376 (2018). (Also see supplementary information)
15. S. Murphy, F. Sowell, J. Apt, A time-dependent model of generator failures and recoveries captures correlated events and quantifies temperature dependence. Appl. Energy **253**, 113513 (2019). (Also see supplementary information).
16. NERC webpage www.nerc.com/pa/RAPA/tads
17. Bonneville Power Administration transmission services operations & reliability, [Online]. Available: https://transmission.bpa.gov/Business/Operations/Outages/
18. N.K. Carrington, I. Dobson, Z. Wang, Transmission grid outage statistics extracted from a web page logging outages in Northeast America, North American Power Symposium, College Station TX USA, November 2021
19. I. Dobson, Estimating the propagation and extent of cascading line outages from utility data with a branching process. IEEE Trans. Power Syst. **27**(4), 2146–2155 (2012)
20. R. Billinton, G. Singh, Application of adverse and extreme adverse weather: modeling in transmission and distribution system reliability evaluation. IEE Proc.-Gener. Transm. Distrib. **153**(1), 115–120 (2006)
21. R. Billinton, G. Singh, J. Acharya, Failure bunching phenomena in electric power transmission systems. Proc. Inst. Mech. Eng. Part O J. Risk Reliab. **220**(1), (2006). https://doi.org/10.1243/1748006XJR
22. N.K. Carrington, S. Ma, I. Dobson, Z. Wang, Extracting resilience statistics from utility data in distribution grids, in *IEEE PES General Meeting*, Montreal, Canada, Aug. 2020
23. M. Papic, S. Ekisheva, E. Cotilla-Sanchez, A risk-based approach to assess the operational resilience of transmission grids. Appl. Sci. **10**(14), 4761 (2020)
24. M. Panteli, D.N. Trakas, P. Mancarella, N.D. Hatziargyriou, Power systems resilience assessment: hardening and smart operational enhancement. Proc. IEEE **105**(7), 1202–1213 (2017)
25. A. Stankovic et al., Methods for analysis and quantification of power system resilience. IEEE Trans. Power Systems. **38**(5), 4774–4787 (2023). https://doi.org/10.1109/TPWRS.2022.3212688
26. E.A. Morris, K.R. Bell, I.M. Elders, Spatial and temporal clustering of fault events on the GB transmission network, in *Probabilistic Methods Applied to Power Systems Conference*, Beijing China, October 2016
27. S. Ekisheva, I. Dobson, J. Norris, R. Rieder, Assessing transmission resilience during extreme weather with outage and restore processes, in *Probabilistic Methods Applied to Power Systems Conference*, Manchester UK, June 2022
28. I. Dobson, Finding a Zipf distribution and cascading propagation metric in utility line outage data. Preprint (2018). arXiv:1808.08434 [physics.soc-ph]
29. IEEE Working Group on understanding, prediction, mitigation and restoration of cascading failures, Benchmarking and validation of cascading failure analysis tools. IEEE Trans. Power Syst. **31**(6), 4887–4900 (2016)
30. P. Henneaux, E. Ciapessoni, D. Cirio, E. Cotilla-Sanchez, R. Diao, I. Dobson, A. Gaikwad, S. Miller, M. Papic, A. Pitto, J. Qi, N. Samaan, G. Sansavini, S. Uppalapati, R. Yao, Benchmarking quasi-steady state cascading outage analysis methodologies, in *Probability Methods Applied to Power Systems*, Boise, Idaho, USA June 2018

31. B.A. Carreras, D.E. Newman, I. Dobson, N.S. Degala, Validating OPA with WECC data, in *Proc. 46th Hawaii Int. Conf. Syst. Sci. (HICSS)*, Maui, HI, USA, Jan. 2013, pp. 2197–2204
32. J. Qi, I. Dobson, S. Mei, Towards estimating the statistics of simulated cascades of outages with branching processes. IEEE Trans. Power Syst. **28**(3), 3410–3419 (2013)
33. P.D.H. Hines, I. Dobson, P. Rezaei, Cascading power outages propagate locally in an influence graph that is not the actual grid topology. IEEE Trans. Power Syst. **32**(2), 958–967 (2017)
34. K. Zhou, I. Dobson, Z. Wang, The most frequent N-k line outages occur in motifs that can improve contingency selection. IEEE Trans. Power Syst. (2023). https://doi.org/10.1109/TPWRS.2023.3249825
35. R. Milo et al., Network motifs: simple building blocks of complex networks. Science **298**(5594), 824–827 (2002)
36. Q. Chen, H. Ren, C. Sun, Z. Mi, D. Watts, Network motif as an indicator for cascading outages due to the decrease of connectivity, in *IEEE PES General Meeting*, Chicago, Illinois, USA, Jul. 2017
37. A.K. Dey, Y.R. Gel, H.V. Poor, What network motifs tell us about resilience and reliability of complex networks. Proc. Natl. Acad. Sci. **116**(39), 19368–19373 (2019)
38. A. Tajer, S.M. Perlaza, H.V. Poor, *Advanced Data Analytics for Power Systems* (Cambridge University Press, Cambridge, 2021)
39. I. Dobson, B.A. Carreras, D.E. Newman, J.M. Reynolds-Barredo, Obtaining statistics of cascading line outages spreading in an electric transmission network from standard utility data. IEEE Trans. Power Syst. **31**(6), 4831–4841 (2016)
40. P.D.H. Hines, I. Dobson, E. Cotilla-Sanchez, M. Eppstein, "Dual graph" and "random chemistry" methods for cascading failure analysis, in *Proc. 46th Hawaii Intl. Conf. Syst. Sci.*, Maui, HI, USA, Jan. 2013, pp. 2141–2150
41. J. Qi, K. Sun, S. Mei, An interaction model for simulation and mitigation of cascading failures. IEEE Trans. Power Syst. **30**(2), 804–819 (2015)
42. U. Nakarmi, M. Rahnamay-Naeini, M.J. Hossain, M.A. Hasnat, Interaction graphs for reliability analysis of power grids: A survey. Energies **13**(9), 2219 (2020). https://doi.org/10.3390/en13092219
43. K. Zhou, I. Dobson, Z. Wang, A. Roitershtein, A.P. Ghosh A Markovian influence graph formed from utility line outage data to mitigate large cascades. IEEE Trans. Power Syst. **35**(4), 3224–3235 (2020)
44. J. Qi, Utility outage data driven interaction networks for cascading failure analysis and mitigation. IEEE Trans. Power Syst. **36**(2), 1409–1418 (2021)
45. M.R. Kelly-Gorham, P.D.H. Hines, K. Zhou, I. Dobson, Using utility outage statistics to quantify improvements in bulk power system resilience, in *Power Systems Computation Conference*, Porto, Portugal, June 2020 and *Electric Power Systems Research*, vol 189, 106676, December 2020
46. M.R. Kelly-Gorham, P.D.H. Hines, I. Dobson, Ranking the impact of interdependencies on power system resilience using stratified sampling of utility data. IEEE Trans. Power Syst. https://doi.org/10.1109/TPWRS.2023.3260119
47. B. Cheng, L. Nozick, I. Dobson, Investment planning for earthquake-resilient electric power systems considering cascading outages. Earthquake Spectra, **38**(3), 1734–1760 (2022)
48. S. Kancherla, I. Dobson, Heavy-tailed transmission line restoration times observed in utility data. IEEE Trans. Power Syst. **33**(1), 1145–1147 (2018)
49. F. Faghihi, P. Henneaux, P.E. Labeau, M. Panteli, An efficient probabilistic approach to dynamic resilience assessment of power systems, Congrès Lambda Mu 22 "Les risques au coeur des transitions, (e-congrès)-22e Congrès Maîtrise des Risques Sûreté Fonctionnement, 2020
50. C. Nan, G. Sansavini, A quantitative method for assessing resilience of interdependent infrastructures. Reliab. Eng. Syst. Safety **157**, 35–53 (2017)
51. S. Poudel, A. Dubey, A. Bose, Risk-based probabilistic quantification of power distribution system operational resilience. IEEE Syst. J. **14**(3), 3506–3517 (2020)
52. N.K. Carrington, I. Dobson, Z. Wang, Extracting resilience metrics from distribution utility data using outage and restore process statistics. IEEE Trans. Power Syst. **36**(2), 5814–5823 (2021)

53. NERC, 2022 State of reliability, An assessment of 2021 bulk power system performance, July 2022. Available: www.nerc.com
54. M. Barkakati, A. Pal, A comprehensive data driven outage analysis for assessing reliability of the bulk power system, in *IEEE PES General Meeting*, Atlanta, GA, USA, 2019
55. C.J. Zapata, S.C. Silva, H.I. Gonzalez, O.L. Burbano, J.A. Hernandez, Modeling the repair process of a power distribution system, in *IEEE/PES T&D Conf. & Exp.*: Latin America, Bogota, Columbia, 2008
56. Y. Wei, C. Ji, F. Galvan, S. Couvillon, G. Orellana, J. Momoh, Non-stationary random process for large-scale failure and recovery of power distribution. Appl. Math. **7**(3), 233–249 (2016)
57. I. Dobson, Models, metrics, and their formulas for typical electric power system resilience events. IEEE Trans. Power Syst. https://doi.org/10.1109/TPWRS.2023.3300125
58. M. Noebels, I. Dobson, M. Panteli, Observed acceleration of cascading outages. IEEE Trans. Power Syst. **36**(4), 3821–3823 (2021)
59. B.A. Carreras, J. M. Reynolds Barredo, I. Dobson, D.E. Newman, Validating the OPA cascading blackout model on a 19402 bus transmission network with both mesh and tree structures, in *Fifty-Second Hawaii International Conference on System Sciences*, Maui, HI, January 2019
60. M. Noebels, R Preece, M. Panteli, AC cascading failure model for resilience analysis in power networks. IEEE Syst. J. **16**(1), 374–385 (2022)

Chapter 3
Interaction Models for Analysis and Mitigation of Cascading Failures

Shuchen Huang, Junjian Qi, and Kai Sun

3.1 Cascading Failure Interaction Analysis Approach

Cascading failure is a common phenomenon in both natural and engineered systems, such as electric power systems [1, 2], natural gas systems [3], transportation networks [4], disease transmission networks [5], and interdependent networks [6]. For example, there have been several large-scale blackouts, such as the 2003 U.S.-Canadian blackout [7], the 2011 Arizona-Southern California blackout [8], and the 2012 Indian blackout [9], which have led to many component failures, extensive outage propagation, and significant economic losses and social impacts. It is thus critical to understand why and how cascading blackouts happen and to further propose effective prevention and mitigation measures to greatly reduce the cascading risk and enhance the power grid resilience. In [10], many quasi-steady-state cascading outage analysis methodologies are reviewed and compared on the standard RTS-96 3-area system model [11].

In order to simulate and analyze cascading failures, many models with different levels of details have been developed [1], such as Manchester model [12], hidden failure model [13, 14], CASCADE model [15], OPA model [16–18], AC OPA model [19], OPA model considering slow process [17], PRA model [20], dynamic model [21], cascading failure model with detailed protection systems [22], and sandpile model [23]. In [24], a cascading failure model is presented to explicitly consider

S. Huang · J. Qi (✉)
Department of Electrical Engineering and Computer Science, South Dakota State University, Brookings, SD, USA
e-mail: Shuchen.Huang@jacks.sdstate.edu; Junjian.Qi@sdstate.edu

K. Sun
Department of Electrical Engineering and Computer Science, University of Tennessee, Knoxville, TN, USA
e-mail: kaisun@utk.edu

K. Sun (ed.), *Cascading Failures in Power Grids*, Power Electronics and Power Systems, https://doi.org/10.1007/978-3-031-48000-3_3

ambient temperature disturbances and the subsequent demand and dynamic line rating changes to investigate the correlations among different events such as line outage, generator tripping, and undervoltage of load buses.

Cascading failure simulations based on various models can produce many samples of cascades. The branching process can extract high-level statistical information from these cascades and quantify the extent of outage propagation by a simple parameter called average propagation [25, 26]. The interdependencies between different types of outages or different critical infrastructure systems can also be analyzed by the multi-type branching process [27]. However, the branching process cannot be used to study how the outages propagate in the system from one component to another component in detail, because it does not retain any information about the network topology or the power flow.

Therefore, a more detailed interaction analysis that retains more information about how components in a system interact during a cascading failure is thus more useful and is able to provide another more useful way to extract propagation patterns in the original cascades. A brief summary of the evolution of the interaction analysis approach over the past 10 years is provided below:

- In 2013, initial ideas were presented to use a "dual graph" to describe the line interactions in [28]. The dual graph has lines as nodes and the links between nodes are decided either using the power system topology or through the $N-1-1$, $N-2$, and $N-3$ contingency analysis. The analysis is mostly for a simple case in which one line outage is followed by another line outage. Tentative ideas are discussed for the case in which multiple line outages are grouped in one generation.
- In 2015, an interaction network approach [29] was proposed for many cascades that are simulated from detailed cascading failure simulations. Each cascade can have multiple generations/stages, in each of which there could be multiple line outages. An interaction matrix that contains the empirical probabilities of one component failure following another component failure is calculated based on a simple assumption that the component j failure in generation $g+1$ is caused only by the component in generation g that is in the previous generation of component j among all cascades for the largest number of times. The corresponding interaction network is built to describe the outage propagation patterns. A highly probabilistic interaction model is proposed to efficiently simulate many cascades and to further evaluate the effect of mitigation strategies, utilizing the distribution of initial outages and the estimated component failure interactions. Critical links are identified and a mitigation strategy is developed based on the identified critical links and a relay blocking mechanism. The same method is also applied to the Northeast Power Coordinating Council (NPCC) power system in [30].
- In 2017, an influence graph was proposed in [31]. This influence graph is built based on the estimation of two types of probabilities: the probability of having k outages in generation $g+1$ given a single outage of component i in generation g and the probability that component j fails in generation $g+1$ given a single outage of component i in generation g and one outage in generation $g+1$. An

efficient simulation of cascades is performed using the two probabilities. Critical components are identified using an equation with which one can quickly identify modifications to a system that will substantially reduce cascade propagation.

- In 2017, a multi-layer interaction graph was proposed in [32] as an extension of the single-layer interaction network in [29]. In addition to line outages, the load shed and electrical distances are also analyzed. Each layer focuses on one of several aspects that are critical for the system operators' decision support, such as the number of line outages, the amount of load shed, and the electrical distance of the outage propagation. Although the load shed is considered as one layer, the cause of the load shed at buses is equally distributed to the failed lines.
- In 2018, the component failure interactions were estimated by the expectation maximization (EM) algorithm [33]. Bayes' theorem is integrated into the EM algorithm, making the interaction estimation more mathematically rigid than the method in [29] and greatly improving the estimation accuracy and efficiency. The EM-based method can accurately estimate the interactions and identify the key links and key components only using a small number of the original cascades from a detailed cascading blackout model, which is critical for online cascading failure analysis and decision-making.
- In 2020, a Markovian influence graph was formed for utility line outage data in [34]. The influence graph represents cascading as a Markov chain and the Markov chain state can include multiple lines. The transition probabilities in the Markov chain transition matrix are estimated using the utility outage data. One challenge is that there are not enough data to accurately estimate the Markov chain transition matrix that describes the outage propagation between two consecutive generations. To address this problem, an estimation method is developed by grouping together data for higher generations and using empirical Bayesian methods to improve the required estimates of cascade stopping probabilities.
- In 2021, an interaction network approach was developed in [35] specially for utility outage data, addressing several unique challenges. To accurately estimate the interactions between component outages, two mechanisms are introduced: the evolution of interactions over generations and the memory between consecutive generations. Critical components are identified based on the estimated interaction networks by solving a set of carefully formulated linear equations, considering loops due to the complex component interactions. A generation-dependent interaction model is developed to efficiently simulate cascades that well capture the properties of the original outage data.
- In 2021, a coupled interaction model was proposed in [36] to explicitly model the interdependency between two types of outages: line outages and the load shed. The interaction matrix in [29, 33, 35] is generalized to the coupled interaction matrix that can describe the interactions between line outages, between line outages and the load shed, and between the load shed. The coupled interaction matrix is effectively estimated by the EM algorithm. A coupled interaction model is further proposed to efficiently generate cascades with both line outages and the load shed based on the coupled interaction matrix and the distribution of initial outages. Critical links are identified based on the coupled interaction matrix by

calculating a comprehensive severity index that considers the consequences of both line outages and the load shed.

- In 2023, a critical component identification method considering spatial propagation was proposed for utility outage data in [37]. The spatial distance between two generations of outages, the total spatial distance between the outages in a cascade, and the average spatial propagation velocity for two consecutive generations are defined. Critical components are identified based on a metric that combines the information of the expected number of outages and that of the spatial distance.

The rest of this chapter is mainly focused on the interaction analysis approaches developed in [29, 33, 35–37]. The other approaches are equally important for the establishment of this research direction.

3.2 Cascading Failure Data Sources for Interaction Analysis

3.2.1 Simulated Data

Traditionally simulation-based approaches dominate cascading failure study, partly due to the scarcity of real outage data for the very rare cascading events. Many models have been developed to simulate cascading failures.

Interaction analysis has been performed using the data generated by simulation models. In [29, 33], the simulated data on the IEEE 118-bus system from the AC OPA model are used to quantify the interactions among transmission line outages. In [30], the simulated data are from simulations on the NPCC system, which represents the northeastern region of the Eastern Interconnection (EI) system. Furthermore, a multi-layer interaction graph [32] is built based on the simulated data on the NPCC system using the DC OPA model.

There are major challenges for using simulated cascading data for interaction analysis. First, the cascading failure simulation models are difficult to benchmark or validate [10, 38]. Although there could be many mechanisms in a cascading failure, the simulation models can only select a few mechanisms and it is usually not clear how realistic the simulated cascades are compared with what has happened in real systems and what will most probably happen in the future. Second, sufficient simulated data may not be readily available, as the computational complexity is rapidly increasing with the increase of the system size and the number of mechanisms to be considered.

3.2.2 Utility Outage Data

A different approach directly analyzes real outage data. In [39], the Bonneville Power Administration (BPA) outage data in the Transmission Availability Data System (TADS) of North American Electric Reliability Corporation (NERC) are analyzed, and the branching process estimation of the distribution of the total number of line outages matches well with the empirical distribution. In [40], 22 years of the online records of blackout size and duration published by NERC's Disturbance Analysis Working Group are analyzed for investigating the blackout size distributions, the statistics of waiting times, and the long-term correlations between blackouts. In [41], the power network topology for the same data used in [39] is formed, and the overall spatial spreading is analyzed.

The recorded outages by BPA for 14 years since January 1999 include 44,593 automatic and planned transmission line outages [41, 42]. After preprocessing the outage data by deleting the outages that are remote from the main system and for lines with rated voltage below 68 kV or without bus names, adjusting bus names to eliminate duplication, and combining buses in the same or adjacent substation, there are 42,561 automatic and planned line outages [41]. Usually only the 10,942 automatic outages are used to analyze cascading outage propagation, mainly because the cascading failure analysis focuses on uncontrolled outages, as in NERC's definition of "cascading" as "the uncontrolled successive loss of system elements triggered by an incident at any location" [43].

In order to analyze the interactions between component outages, we need to group the outages into different cascades and generations. One cascade corresponds to one cascading failure sample, while one generation corresponds to one stage in a cascade. Each cascade starts with initial outages in generation 0 followed by further outages grouped into generations 1, 2, $\cdots$ until the cascade stops. This can be done according to the gaps in start time between successive outages [39]. If successive outages have a gap of more than one hour (operator actions can usually be completed in one hour), the outage after the gap starts a new cascade. In each cascade, if successive outages have a gap of more than one minute (fast transients and protection actions are completed within one minute), the outage after the gap starts a new generation of the cascade.

This procedure is applied to the 10,942 automatic outages. In some generations the same outage appears more than once, probably due to the reclosing of the protective relay within a very short time. In this case we only keep one outage for that particular component in that generation. After this we have 6,687 cascades with 10,779 automatic outages for which $n = 582$ lines (components) are involved. The number of line outages and generations in each cascade varies significantly. 84% of the cascades only have one generation of outages with an average number of line outages as 1.16, while there is one cascade that has 109 generations and 143 line outages.

How to group the cascading sequences into cascades and generations is still an open question. In addition to the very simple method that is only based on the gap between the outages, an optimal decomposition approach is proposed in [44] by approximating the temporal pattern of the outage sequences, in hope of revealing more failure interactions and better mitigating the heterogeneity in cascades.

3.2.3 Data Format

Either the data comes from simulations or utility outage records, it can be processed into a uniform data format to facilitate further interaction analysis. Two typical data formats are presented below: one for line outages and the other for both line outages and the load shed.

3.2.3.1 Data Format for Line Outages

In power systems, the components can be branches such as the transmission lines or transformers. In order to estimate the interaction between the component failures, we need to have a large amount of data that record the processes of cascading failures. The data, which come either from utility outage records or from cascading failure simulation models, can be grouped into different cascades and generations. Assume we have a total of M cascades as

	Generation 0	Generation 1	Generation 2	$\cdots$
Cascade 1	$\mathcal{F}_0^{(1)}$	$\mathcal{F}_1^{(1)}$	$\mathcal{F}_2^{(1)}$	$\cdots$
Cascade 2	$\mathcal{F}_0^{(2)}$	$\mathcal{F}_1^{(2)}$	$\mathcal{F}_2^{(2)}$	$\cdots$
$\vdots$	$\vdots$	$\vdots$	$\vdots$	$\vdots$
Cascade M	$\mathcal{F}_0^{(m)}$	$\mathcal{F}_1^{(m)}$	$\mathcal{F}_2^{(m)}$	$\cdots$

Here $\mathcal{F}_g^{(m)}$ is the set of the failed components in generation g of cascade m.

3.2.3.2 Data Format for Line Outages and the Load Shed

Although the high-level statistical models such as the multi-type branching process in [27] provide useful tools to capture the pattern of outage propagation, it is still very challenging to analyze the detailed interdependency between line outages and the load shed, especially with limited samples. To address this challenge, a coupled interaction analysis is developed in [36] to describe the interaction/interdependency between line outages and the load shed at buses.

Cascading failures in power systems involve successive line outages and load shed at buses. According to the outage sequence from either cascading failure simulations or utility outage record, a cascade can be divided into different generations. Assume there are N_l lines and N_b buses with load in the system. The index numbers of lines and load buses are converted so that the set of lines is $\{1, 2, \cdots, N_l\}$ and the set of buses with load is $\{1, 2, \cdots, N_b\}$. Assume there are M cascades in the dataset:

	Generation 0	Generation 1	Generation 2	$\ldots$
Cascade 1	$\mathcal{F}^{1,0}$	$\mathcal{F}^{1,1}$	$\mathcal{F}^{1,2}$	$\ldots$
Cascade 2	$\mathcal{F}^{2,0}$	$\mathcal{F}^{2,1}$	$\mathcal{F}^{2,2}$	$\vdots$
$\vdots$	$\vdots$	$\vdots$	$\vdots$	$\vdots$
Cascade M	$\mathcal{F}^{M,0}$	$\mathcal{F}^{M,1}$	$\mathcal{F}^{M,2}$	$\ldots$

Here $\mathcal{F}^{m,g} = \{\mathcal{S}_{\mathsf{L}}^{m,g}, \mathcal{S}_{\mathsf{B}}^{m,g}, \boldsymbol{Z}^{m,g}\}$, $\mathcal{S}_{\mathsf{L}}^{m,g}$ and $\mathcal{S}_{\mathsf{B}}^{m,g}$ are, respectively, the set of lines that outage and the set of buses with load shed in generation g of cascade m, and $\boldsymbol{Z}^{m,g} \in \mathbb{Z}^{N_b}$ is the vector for the amount of discretized load shed at each bus in generation g of cascade m.

In both utility outage data and cascading failure simulations, the load shed is usually recorded in MW. Let $X_v^{m,g}$ be the load shed at bus v in MW in generation g of cascade m. Discretization should be first performed on the load shed data by the technique in [26, 27]. If Δ_v MW is the chosen unit of discretization for bus v, an integer multiple of Δ_v MW can be obtained for the load shed at bus v as

$$Z_v^{m,g} = \text{int}\left[\frac{X_v^{m,g}}{\Delta_v} + 0.5\right], \tag{3.1}$$

where int$[x]$ is the integer part of x. A systematic approach for choosing Δ_v will be discussed in Sect. 3.5.3.

3.3 Formulation of Component Failure Interactions

3.3.1 Interaction Matrix for Line Outages

After obtaining the original cascades for line outages, we can estimate the component interactions, which are defined as the *interaction matrix* $\boldsymbol{B} \in \mathbb{R}^{n\times n}$, where n is the number of components in the system. The element of $\boldsymbol{B}$, b_{ij}, is the empirical probability that component i failure causes component j failure (the empirical probability that component j fails following component i failure).

In order to estimate $\boldsymbol{B}$, an auxiliary matrix $\boldsymbol{A} \in \mathbb{R}^{n\times n}$ is needed whose element a_{ij} is the expected number of component j failure caused by component i failure among all successive generations of all cascades. There is

$$b_{ij} = \frac{a_{ij}}{N_i}, \tag{3.2}$$

where N_i is the number of times that component i fails. The interaction matrix $\boldsymbol{B}$ describes how the components in the system interact with each other. Its nonzero

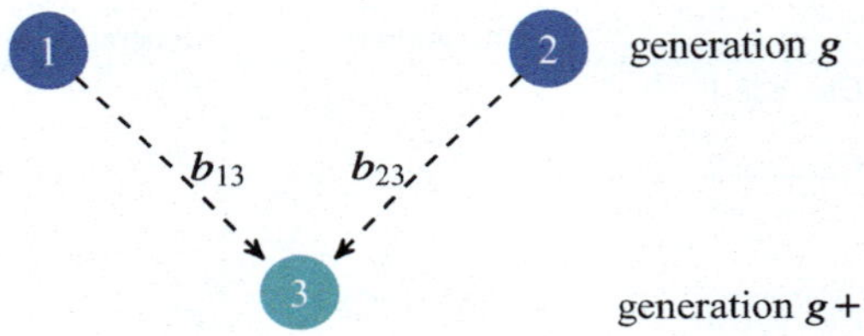

Fig. 3.1 Illustration for finding the cause of a component failure

elements are called *links*. For example, for a nonzero element b_{ij}, there is a link $l : i \rightarrow j$, representing that the source component i failure causes the destination component j failure with a positive probability. All of the links form a directed *interaction network* $\mathcal{G}(C, \mathcal{L})$ with the set of vertices C and the set of links $\mathcal{L}$.

If we know matrix $\boldsymbol{A}$, then matrix $\boldsymbol{B}$ can be very easily estimated by (3.2). If the interaction matrix $\boldsymbol{B}$ is known, for two consecutive generations in each cascade we can infer how probable one component failure causes another component failure, based on which we can get matrix $\boldsymbol{A}$. For the very simple example shown in Fig. 3.1, if $b_{13} \approx 1$ while $b_{23} \approx 0$, it is much more probable that the failure of component 3 is caused by component 1 failure. In these two consecutive generations, the probability that component 1 failure causes component 3 failure is approximately 1.0, while that for component 2 failure causing component 3 failure is around 0.

However, neither $\boldsymbol{A}$ nor $\boldsymbol{B}$ is initially known. Therefore, the estimation of the interactions between component failures is actually a typical parameter estimation problem with incomplete data.

In [29], the interaction matrix $\boldsymbol{B}$ is directly estimated from an approximated matrix $\boldsymbol{A}$ that is not statistically inferred from $\boldsymbol{B}$ but is only estimated based on some simple assumptions.

First, it is assumed that there are interactions between any failed component in the last generation and that in this generation to guarantee that no interactions will be ignored. Based on this assumption, a matrix $\boldsymbol{A}^0 \in \mathbb{Z}^{n\times n}$ can be constructed, whose entry a_{ij}^0 is the number of times that component i fails in one generation before the failure of component j among all original cascades. Since $\boldsymbol{A}^0$ is obtained by using all M_u cascades, it does not depend on the order that the cascades are processed.

The assumption based on which $\boldsymbol{A}^0$ is obtained may exaggerate the interactions between component failures since a component failing in the last generation of another component failure does not necessarily mean these components interact with each other. Therefore, for each failed component in generation 1 and the following generations, the failed component that most probably causes it should be determined.

Specifically, for any two consecutive generations k and $k+1$ of any cascade m, the failure of component j in generation $k+1$ is considered to be caused by a set of failed components in generation k that appear in the previous generation of component j among all cascades for the largest number of times, which can be described as

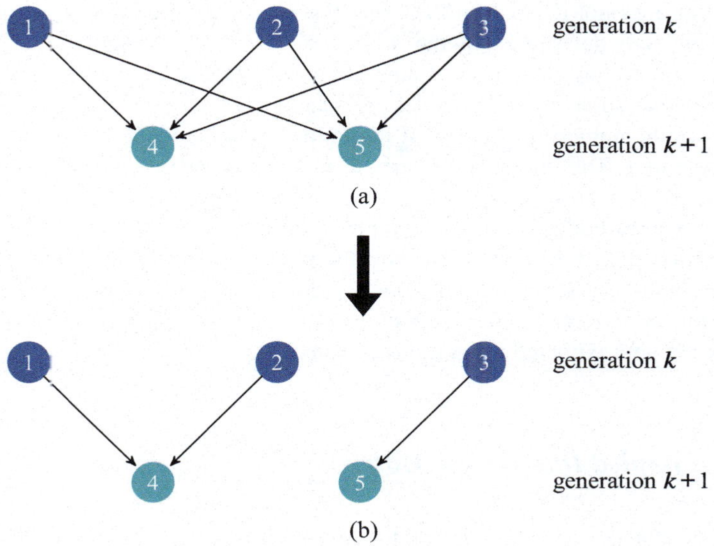

Fig. 3.2 Illustration for determining the cause of component failures

$$\{i_c | i_c \in F_k^{(m)} \text{ and } a^0_{i_c j} = \max_{i \in F_k^{(m)}} a^0_{ij}\}. \tag{3.3}$$

Note that it is possible that two or more components in generation k are considered as the cause of the failure of component j. When the number of cascades used for estimation is not large enough, this will be more possible because in this case no component has much greater a^0_{ij} than the others. In the extreme case in which the a^0_{ij}'s for $i \in F_k^{(m)}$ are all the same, it will become impossible to determine which component more possibly causes the failure of component j, and thus all components will be considered as the cause.

An illustration of two consecutive generations of a cascade is shown in Fig. 3.2. If we assume that

$$a^0_{14} = a^0_{24} = \max_{i \in \{1,2,3\}} a^0_{i4} \tag{3.4}$$

$$a^0_{35} = \max_{i \in \{1,2,3\}} a^0_{i5}, \tag{3.5}$$

we can determine the cause of component 4 failure as components 1 and 2 and the cause of component 5 failure as component 3.

What should be emphasized is that $\boldsymbol{A}^0$ stays unchanged when determining the most possible causes for failed components in generation 1 and the following generations. Therefore, the determination of the component that causes a failed component does not depend on the order that the original cascades are processed but is completely determined by the $\boldsymbol{A}^0$ matrix.

After determining the cause of any component failure in generation 1 and the following generations for all cascades, $\boldsymbol{A}^0$ can be corrected to be $\boldsymbol{A} \in \mathbb{Z}^{n \times n}$, whose entry a_{ij} is the number of times that the failure of component i causes the failure of component j. The interaction matrix $\boldsymbol{B} \in \mathbb{R}^{n \times n}$ can then be calculated from $\boldsymbol{A}$ based on (3.2).

The above method is based on a simple assumption that the component j failure in generation $g + 1$ is caused only by the components in generation g that are in the previous generation of component j for the largest number of times. This assumption may overestimate or ignore/underestimate some interactions and thus can only get approximated estimations of $\boldsymbol{A}$ and $\boldsymbol{B}$.

3.3.2 *Coupled Interaction Matrix*

Interaction analysis can be performed not only for line outages but also by considering the interdependency between line outages and the load shed at different buses. This is useful because the load shed corresponds to more direct economic losses and is of more interest by the utilities. The analysis of interdependency between line outages and the load shed at buses enables better capturing of the cascading failure propagation patterns and can help develop more effective mitigation strategies.

In [32], a multi-layer interaction graph is proposed as an extension of a single-layer interaction network. In this multi-layer graph, line outages, the load shed, and the electrical distances are analyzed. Each layer focuses on one of several aspects that are critical for the system operators' decision support, such as the number of line outages, the amount of load shedding, and the electrical distance of the outage propagation. Although the load shed is considered as one layer, the cause of the load shed at buses is equally distributed to the failed lines, which may not always be reasonable as the contribution of line outages to the load shed can be significantly different.

In [36], in order to describe the interdependency between line outages and the load shed, a *coupled interaction matrix* $\boldsymbol{B} \in \mathbb{R}^{(N_l+N_b)\times(N_l+N_b)}$ is defined as

$$\boldsymbol{B} = \begin{bmatrix} \boldsymbol{B}^{\mathsf{LL}} & \boldsymbol{B}^{\mathsf{LB}} \\ \boldsymbol{B}^{\mathsf{BL}} & \boldsymbol{B}^{\mathsf{BB}} \end{bmatrix},$$

where $\boldsymbol{B}^{\mathsf{LL}}$, $\boldsymbol{B}^{\mathsf{LB}}$, $\boldsymbol{B}^{\mathsf{BL}}$, and $\boldsymbol{B}^{\mathsf{BB}}$ are defined as follows. Assume $M_{\mathrm{u}} \leq M$ cascades are used for estimating the coupled interaction matrix.

- $\boldsymbol{B}^{\mathsf{LL}} \in \mathbb{R}^{N_l \times N_l}$ captures the interactions between lines outages, which is the same as the interaction matrix in Sect. 3.3.1. Its entry in the ith row and jth column, b_{ij}^{LL}, is the empirical probability that line j fails following the outage of line i.

- $\boldsymbol{B}^{\mathsf{LB}} \in \mathbb{R}^{N_l \times N_b}$ captures the interactions from line outages to the load shed at buses. As has been adopted and verified in [25–27], the Poisson distribution can capture the statistical properties for offspring outages being selected from a large number of possible outages that have small probability and are approximately independent. Therefore, we use the Poisson distribution to approximate the distribution of the discretized load shed at buses following line outages. For example, to capture the discretized load shed at bus v following the outage of line i, the mean of the Poisson distribution is recorded as b_{iv}^{LB}, which is the entry in the ith row and vth column of $\boldsymbol{B}^{\mathsf{LB}}$. If line i fails in generation g, k_v units of discretized load shed at bus v will occur in generation $g+1$ with probability

$$p_{iv}^{\mathsf{LB}} = \frac{\left(b_{iv}^{\mathsf{LB}}\right)^{k_v}}{k_v!} e^{-b_{iv}^{\mathsf{LB}}}. \tag{3.6}$$

- $\boldsymbol{B}^{\mathsf{BL}} \in \mathbb{R}^{N_b \times N_l}$ captures the interactions from the load shed to line outages. Note that the interactions between the load shed at buses and the further line outages are not necessarily causal but actually capture what might consequently happen in the next generation following the load shedding event in the current generation. When load shedding cannot eliminate all line overloading or other technical constraint violations [45, 46], line outages can still be observed following the load shedding event. As has been verified in [27], line outages following load shedding are rare. Therefore, we consider a simplified case in which line outages are not sensitive to the amount of load shed at buses. When load shedding occurs at bus u, line j will fail in the next generation with a constant probability b_{uj}^{BL}, which is the entry in the uth row and jth column of $\boldsymbol{B}^{\mathsf{BL}}$.
- $\boldsymbol{B}^{\mathsf{BB}} \in \mathbb{R}^{N_b \times N_b}$ captures the interactions between the load shed at buses. Similar to $\boldsymbol{B}^{\mathsf{BL}}$, $\boldsymbol{B}^{\mathsf{BB}}$ captures the successive load shedding in consecutive generations that is also not necessarily causal. As the system has been significantly weakened when the cascading outage evolves, successive load shedding may occur at different buses or at the same bus more than once. We assume each unit of the discretized load shed at bus u independently generates the discretized load shed at bus v by a Poisson distribution with mean b_{uv}^{BB}. Therefore, if k_u units of load are shed at bus u in generation g, k_v units of discretized load shed at bus v will occur in generation $g+1$ with the following probability:

$$p_{uv}^{\mathsf{BB}} = \frac{\left(k_u b_{uv}^{\mathsf{BB}}\right)^{k_v}}{k_v!} e^{-k_u b_{uv}^{\mathsf{BB}}}. \tag{3.7}$$

The matrix $\boldsymbol{B}$ determines how components, either lines or buses, interact with each other. The nonzero entries of $\boldsymbol{B}$ are called *links*. Link $l : i \to j$ corresponds to $\boldsymbol{B}$'s nonzero entry in the ith row and jth column. By putting all links together, a directed network $\mathcal{G}(C, \mathcal{L})$ called *coupled interaction network* can be obtained: the vertices C are components, including both lines and buses, and the directed link $l \in$

$\mathcal{L}$ represents that the destination vertex component fails following the source vertex component outage with probability greater than 0. Different from the interaction network in Sect. 3.3.1, there are four different types of links: $\mathsf{L} \rightarrow \mathsf{L}$ links, $\mathsf{L} \rightarrow \mathsf{B}$ links, $\mathsf{B} \rightarrow \mathsf{L}$ links, and $\mathsf{B} \rightarrow \mathsf{B}$ links.

To estimate the coupled interaction matrix $\boldsymbol{B}$, the following auxiliary matrix $\boldsymbol{A} \in \mathbb{R}^{(N_l+N_b)\times(N_l+N_b)}$ is needed:

$$\boldsymbol{A} = \begin{bmatrix} \boldsymbol{A}^{\mathsf{LL}} & \boldsymbol{A}^{\mathsf{LB}} \\ \boldsymbol{A}^{\mathsf{BL}} & \boldsymbol{A}^{\mathsf{BB}} \end{bmatrix},$$

whose four sub-matrices are defined below:

- $\boldsymbol{A}^{\mathsf{LL}} \in \mathbb{R}^{N_l \times N_l}$ has entry a_{ij}^{LL} as the total number of times that line j fails following the outage of line i in the dataset. Based on a_{ij}^{LL}, b_{ij}^{LL} can be estimated as

$$b_{ij}^{\mathsf{LL}} = \frac{a_{ij}^{\mathsf{LL}}}{N_i^{\mathsf{L}}}, \tag{3.8}$$

  where N_i^{L} is the total number of outages of line i in the M_{u} cascades.
- $\boldsymbol{A}^{\mathsf{LB}} \in \mathbb{R}^{N_l \times N_b}$ has entry a_{iv}^{LB} as the total discretized amount of load shed at bus v following the outage of line i in the M_{u} cascades. Then b_{iv}^{LB} can be estimated by

$$b_{iv}^{\mathsf{LB}} = \frac{a_{iv}^{\mathsf{LB}}}{N_i^{\mathsf{L}}}. \tag{3.9}$$

- $\boldsymbol{A}^{\mathsf{BL}} \in \mathbb{R}^{N_b \times N_l}$ has entry a_{uj}^{BL} as the total number of outages of line j following the load shed at bus u, from which b_{uj}^{BL} is estimated as

$$b_{uj}^{\mathsf{BL}} = \frac{a_{uj}^{\mathsf{BL}}}{N_u^{\mathsf{B}}}, \tag{3.10}$$

  where N_u^{B} is the total number of times that bus u has nonzero discretized load shed in the M_{u} cascades.
- $\boldsymbol{A}^{\mathsf{BB}} \in \mathbb{R}^{N_b \times N_b}$ has entry a_{uv}^{BB} as the total discretized amount of load shed at bus v following the load shed at bus u. As the load shed at bus v after each unit of load shed at bus u independently follows Poisson distribution, whose mean b_{uv}^{BB} can be estimated by

$$b_{uv}^{\mathrm{BB}} = \frac{a_{uv}^{\mathrm{BB}}}{\sum_{m=1}^{M_{\mathrm{u}}} \sum_{g=0}^{G^m-1} Z_u^{m\ g}}, \tag{3.11}$$

where G^m is the number of generations in cascade m and the denominator is the total discretized amount of load shed at bus u in the M_{u} cascades.

3.4 EM Algorithm

In [33], the EM algorithm [47, 48], a simple yet effective method for performing maximum likelihood estimation of parameters when there are incomplete data, is applied to deal with incomplete data and more accurately estimate the interaction matrix. The corresponding maximum likelihood estimation problem is to estimate the parameters $\boldsymbol{B}$ in order to maximize $\log \mathrm{P}(\boldsymbol{A}, \boldsymbol{y}; \boldsymbol{B})$, which is the logarithm of the joint probability of having the specific interactions between components in any two successive generations among all used cascades that are represented in $\boldsymbol{A}$ and the observed result $\boldsymbol{y}$ as the M original cascades [33]. This section will provide a brief introduction to the EM algorithm.

3.4.1 A Coin-Flipping Example

The coin-flipping experiment in [48] is adapted to illustrate how the EM algorithm works. Assume there are n coins denoted by $c_1, \cdots, c_n$, and coin c_i lands on heads and tails with probability θ_i and $1 - \theta_i$, respectively. The following experiments are performed for m times:

1. Exactly one coin is selected from the n coins and each coin is selected by the same probability.
2. A total of K tosses are independently performed for the chosen coin.

Based on the results of these experiments, we want to estimate $\boldsymbol{\theta} = (\theta_1, \cdots, \theta_n)$. During the experiments we record $\boldsymbol{x} = [x_1, \cdots, x_m]$ where $x_j = i$ if c_i is chosen in the jth set of tosses and $\boldsymbol{y} = \left[\boldsymbol{y}_1^\top, \cdots, \boldsymbol{y}_m^\top\right]^\top$ where $\boldsymbol{y}_j = [y_{j1}, \cdots, y_{jK}]$ and $y_{jk} = 1$ if the selected coin in the jth set of tosses lands on heads for the kth toss and $y_{jk} = 0$ otherwise.

Note that in this parameter estimation problem there are *complete data*. This is because both the type of the coin used for each toss and the result of each coin toss are known. The parameters $\boldsymbol{\theta}$ can be estimated by maximum likelihood estimation:

$$\hat{\theta}_i = \frac{\sum_{j|x_j=i} \sum_{k=1}^{K} y_{jk}}{K \sum_{j=1}^{m} I(x_j = i)}, \tag{3.12}$$

where $j|x_j = i$ indicates the set of tosses with coin c_i and I equals one if $x_j = 1$ and zero otherwise. The estimated parameters $\hat{\boldsymbol{\theta}} = (\hat{\theta}_1, \cdots, \hat{\theta}_n)$ from (3.12) maximize $\log P(\boldsymbol{x}, \boldsymbol{y}; \boldsymbol{\theta})$, which is the logarithm of the joint probability of having the coin types $\boldsymbol{x}$ and the observed result $\boldsymbol{y}$.

Then we change the problem settings by only recording the result of the coin-flipping $\boldsymbol{y}$ but not the types of the coins $\boldsymbol{x}$. We refer to $\boldsymbol{x}$ as hidden variables. Because we do not have the data for the types of the coins, in this parameter estimation problem there are *incomplete data*.

The EM algorithm can perform parameter estimation under this new setting. Specifically, starting from some initial parameters $\hat{\boldsymbol{\theta}}^{(0)} = (\hat{\theta}_1^{(0)}, \cdots, \hat{\theta}_n^{(0)})$, the EM algorithm uses the parameters in this iteration to calculate the probabilities for each possible case of the incomplete data. Then the classic maximum likelihood estimation method is modified to be able to consider these probabilities, based on which the updated parameter estimates $\hat{\boldsymbol{\theta}}^{(t+1)}$ can be obtained. The EM algorithm iterates between the E-step and M-step as follows until convergence:

- **E-step:** Estimate a probability distribution of the incomplete data based on the parameter in this iteration.
- **M-step:** Estimate the parameters using the completions in the E-step.

For the above coin-flipping example, when we consider the incomplete data case the EM algorithm can be formulated as follows:

- **E-step:** For the jth set of tosses, the number of heads is $n_j^{\text{head}} = \sum_{k=1}^{K} y_{jk}$ and that for tails is $n_j^{\text{tail}} = K - n_j^{\text{head}}$. Then the probability distribution of x_j is

$$P(x_j = i) = \frac{\left(\hat{\theta}_i^{(t)}\right)^{n_j^{\text{head}}} \left(1 - \hat{\theta}_i^{(t)}\right)^{n_j^{\text{tail}}}}{\sum_{l=1}^{n} \left(\hat{\theta}_l^{(t)}\right)^{n_j^{\text{head}}} \left(1 - \hat{\theta}_l^{(t)}\right)^{n_j^{\text{tail}}}}, \tag{3.13}$$

and the corresponding expected number of heads for coin c_i is

$$E_i^{\text{head}} = n_j^{\text{head}} P(x_j = i). \tag{3.14}$$

- **M-step:** The parameters can be updated as

Table 3.1 Quantities in E-step

j	n_j^{head}	$P(x_j = 1)$	$P(x_j = 0)$	E_1^{head}	E_2^{head}
1	3	0.266	0.734	0.798	2.202
2	4	0.352	0.648	1.409	2.591
3	4	0.352	0.648	1.409	2.591
4	8	0.733	0.267	5.868	2.132
5	7	0.647	0.353	4.531	2.470
Total	–	–	–	14.014	11.987

$$\hat{\theta}_i^{(t+1)} = \frac{\sum_{j=1}^{m} n_j^{\text{head}} P(x_j = i)}{K \sum_{j=1}^{m} P(x_j = i)}. \tag{3.15}$$

For example, we assume there are two coins, c_1 and c_2, with $\theta_1 = 0.7$ and $\theta_2 = 0.4$. We set $m = 5$ and $K = 10$ and get the following data:

$$\boldsymbol{x} = [1 \quad 0 \quad 0 \quad 1 \quad 1],$$

$$\boldsymbol{y} = \begin{bmatrix} 0\,1\,0\,0\,0\,1\,0\,0\,1\,0 \\ 0\,0\,1\,1\,0\,1\,0\,0\,0\,1 \\ 0\,1\,1\,1\,0\,0\,0\,1\,0\,0 \\ 1\,0\,1\,1\,1\,1\,0\,1\,1\,1 \\ 1\,1\,1\,0\,1\,1\,0\,1\,1\,0 \end{bmatrix}.$$

For the EM algorithm, the initial parameter is assumed to be $\hat{\boldsymbol{\theta}}^{(0)} = (0.6, 0.5)$. Then in E-step the number of heads, the probability that a specific coin is chosen, and the expected numbers of heads for each coin in each set of the tosses are listed in Table 3.1. In M-step we update the parameters as $\hat{\theta}_1^{(1)} = 0.596$ and $\hat{\theta}_2^{(1)} = 0.453$. After nine iterations we get $\hat{\theta}_1 = 0.702$ and $\hat{\theta}_2 = 0.383$, both of which are very close to the real parameters.

3.4.2 *Mathematical Foundation*

With complete data, the objective function of the maximum likelihood estimation ($\log P(\boldsymbol{x}, \boldsymbol{y}; \boldsymbol{\theta})$) usually only has one global optimum, which can often be obtained in a closed form, such as by (3.12) in the coin-flipping example [48]. However, with incomplete data, the modified maximum likelihood estimation has to find $\hat{\boldsymbol{\theta}}$ that maximizes $\log P(\boldsymbol{y}; \boldsymbol{\theta})$, which usually has multiple local optima. If there are several local optima, there might be only one global optimum.

In order to solve this problem, the EM algorithm converts the one single optimization problem of $\log P(\boldsymbol{y}; \boldsymbol{\theta})$ into a series of subproblems, each of which has an objective function with a unique global optimum. In the E-step, it chooses a function g_t that lower bounds $\log P(\boldsymbol{y}; \boldsymbol{\theta})$ and satisfies $g_t\left(\hat{\boldsymbol{\theta}}^{(t)}\right) = \log P\left(\boldsymbol{y}; \hat{\boldsymbol{\theta}}^{(t)}\right)$. In the M-step, it determines the updated parameter $\hat{\boldsymbol{\theta}}^{(t+1)}$ that maximizes g_t. Since g_t matches $\log P(\boldsymbol{y}; \boldsymbol{\theta})$ at $\hat{\boldsymbol{\theta}}^{(t)}$, there is $\log P\left(\boldsymbol{y}; \hat{\boldsymbol{\theta}}^{(t)}\right) = g_t\left(\hat{\boldsymbol{\theta}}^{(t)}\right) \leq g_t\left(\hat{\boldsymbol{\theta}}^{(t+1)}\right) = \log P\left(\boldsymbol{y}; \hat{\boldsymbol{\theta}}^{(t+1)}\right)$, meaning that the objective function monotonically increases during each iteration [48].

The estimated parameter increases the likelihood function after each iteration until a local maximum is achieved. Similar to most optimization methods for non-convex functions, there is no guarantee that the EM algorithm will converge to a global maximum. Starting from different initial parameters may get different solutions. Running the algorithm for multiple times by using different initial parameters may help to get the solution with global optimum. Although other numerical optimization methods, such as gradient descent or Newton's method [49], can in theory be used to solve the optimization problem, the EM algorithm provides a simple, robust, and easy-to-implement tool for parameter estimation in models with incomplete data [48].

3.5 Estimating Component Failure Interactions

3.5.1 Interaction Estimation for Simulated Line Outage Data

Assume $M_{\mathrm{u}} \leq M$ original cascades are utilized to estimate the interactions between component failures. Here we discuss how to estimate the interaction matrix by applying the EM algorithm introduced in Sect. 3.4. Specifically, the EM algorithm can be implemented in the following four steps.

1. **Initialization:** In order to avoid ignoring any useful information, we assume that any failed component in generation g is the cause of the component failures in generation $g + 1$. Then the initial matrix $\boldsymbol{A}^{(0)} \in \mathbb{Z}^{n\times n}$ can be obtained from all original cascades and the initial interaction matrix $\boldsymbol{B}^{(0)}$ can be calculated from $\boldsymbol{A}^{(0)}$ by using (3.2).

   The assumption here tends to overestimate the component interactions. This is because a component that fails before the failure of another component may not be the cause of that particular failure. However, it is appropriate to use $\boldsymbol{A}^{(0)}$ to get the initial guess for $\boldsymbol{B}$ since in this way we will not miss any interaction between component failures. As mentioned in Sect. 3.4.2, running the algorithm for multiple times by using different initial parameters may help to get the solution with global optimum. Based on numerical experiments, this is not

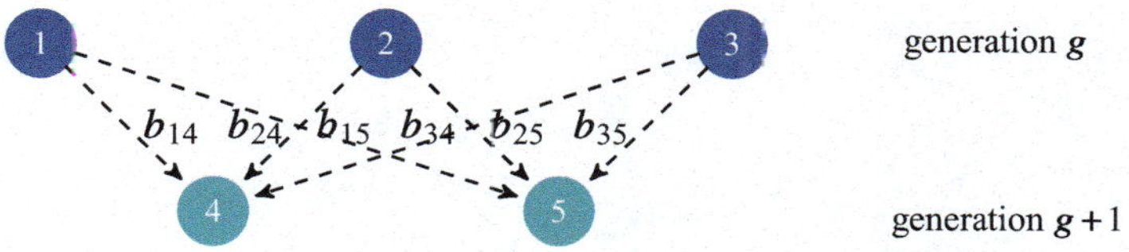

Fig. 3.3 Illustration for inferring the probability of one component failure in generation $g + 1$ caused by a specific component failure in generation g

necessary because the chosen $\boldsymbol{A}^{(0)}$ has a clear physical meaning and provides a good initial guess.

2. **E-step:** Estimate $\boldsymbol{A}^{(k+1)}$ based on $\boldsymbol{B}^{(k)}$.
For any two successive nonzero generations g and $g+1$ of any cascade m, under the condition that component j has failed, the component j failure in generation $g+1$ is caused by component $i \in F_g^{(m)}$ in generation g by probability

$$p_{ij}^{(k+1)m,g} = \frac{b_{ij}^{(k)}}{1 - \prod\limits_{l \in F_g^{(m)}} \left(1 - b_{lj}^{(k)}\right)}. \tag{3.16}$$

If $i \notin F_g^{(m)}$, $p_{ij}^{(k+1)m,g} = 0$. For example, for the two consecutive generations shown in Fig. 3.3, there is

$$p_{14} = \frac{b_{14}}{1 - (1 - b_{14})(1 - b_{24})(1 - b_{34})}, \tag{3.17}$$

and p_{15}, p_{24}, p_{25}, p_{34}, and p_{35} can also be written in a similar manner.
The updated entry of $\boldsymbol{A}^{(k+1)}$ can be obtained as the summation over all consecutive nonzero generations for all cascades

$$a_{ij}^{(k+1)} = \sum_{m=1}^{M_{\mathrm{u}}} \sum_{g=0}^{G^m - 2} p_{ij}^{(k+1)m,g}, \tag{3.18}$$

where G^m is the number of generations with nonzero number of outages in cascade m.

3. **M-step:** Estimate $\boldsymbol{B}^{(k+1)}$ based on $\boldsymbol{A}^{(k+1)}$.
After $\boldsymbol{A}^{(k+1)}$ is obtained, the updated interaction matrix $\boldsymbol{B}^{(k+1)}$ can be calculated by using (3.2).
4. **End:** Iterate the E-step and M-step until

$$\frac{\left\| \boldsymbol{B}_{ij}^{(k+1)} - \boldsymbol{B}_{ij}^{(k)} \right\|_{\mathrm{F}}}{\sqrt{N}} < \epsilon, \tag{3.19}$$

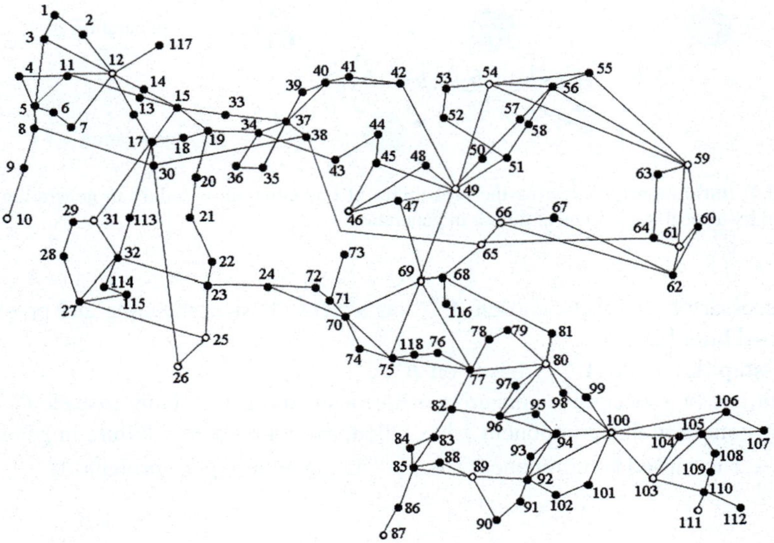

Fig. 3.4 IEEE 118-bus system

where $||\boldsymbol{X}||_\mathrm{F}$ is the Frobenius norm of a $u \times v$ matrix $\boldsymbol{X}$ defined as

$$||\boldsymbol{X}||_\mathrm{F} = \sqrt{\sum_{i=1}^{u}\sum_{j=1}^{v} |X_{ij}|^2}, \tag{3.20}$$

$N = N_{\neq 0}$ if $N_{\neq 0}$, which is the number of nonzero elements in $\boldsymbol{B}^{(k+1)} - \boldsymbol{B}^{(k)}$, is greater than 0, otherwise $N = 1$, ϵ is the tolerance, and the $\boldsymbol{B}_{ij}^{(k+1)}$ that satisfies (3.19) will be the estimated interaction matrix.

A total of $M = 50{,}000$ cascades are obtained by open-loop AC OPA simulation [19] on the IEEE 118-bus system [50] shown in Fig. 3.4. AC OPA is a variant of the basic OPA [18, 51]. Different from the basic OPA that uses DC optimal power flow (OPF), AC OPA uses AC OPF and thus can consider reactive power and voltage.

The component interactions (links) are estimated by the EM-based method in this section and the method in Sect. 3.3.1 based on M_u cascades. The EM algorithm converges after five iterations. In Table 3.2, we present the number of identified links (card($\mathcal{L}$)). The ratio of nonzero elements in the interaction matrix $\boldsymbol{B}$ is defined as $r = \mathrm{card}(\mathcal{L})/n^2$, where n is the number of components. The very small value of r suggests that $\boldsymbol{B}$ is a sparse matrix and only a very small number of components have interactions in terms of cascading outage propagation.

Based on the interaction matrix, the interaction networks are constructed which provide a graphical representation of the interactions between the components. Figures 3.5, 3.6, 3.7, 3.8 show the interaction networks obtained by using different

Table 3.2 Number of estimated links

Method	M_u	n	card($\mathcal{L}$)	r
EM method	41,000	186	715	0.0207
EM method	400	186	170	0.0049
Method in Sect. 3.3.1	41,000	186	343	0.0099
Method in Sect. 3.3.1	400	186	77	0.0022

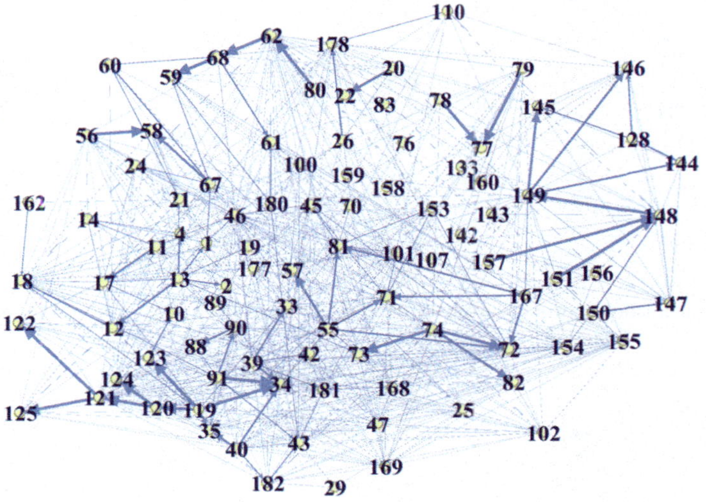

Fig. 3.5 Interaction network estimated from EM algorithm using $M_u = 41{,}000$ cascades

estimation methods and a different number of original cascades. Note that only the components with links are shown and the line thickness indicates the weight.

When using the EM algorithm most major component interactions can still be estimated even when only 400 cascades are used. By contrast, when using the method in Sect. 3.3.1, the interaction estimation using only 400 cascades is not effective.

3.5.2 *Interaction Estimation for Utility Line Outage Data*

In Sect. 3.5.1, when estimating the interaction network using simulated cascades, the data across all generations are used, without considering the evolution of cascading behavior during different stages. The empirical probabilities that one component outage happens following another component outage are averaged over all generations. The overall outage propagation may be greatly underestimated, because as the operating conditions become more abnormal in later stages, these empirical probabilities capturing the component interactions may become more

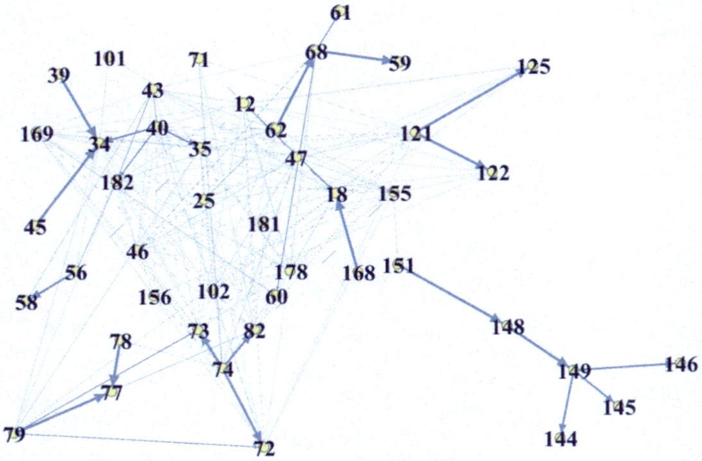

Fig. 3.6 Interaction network estimated from EM algorithm using $M_u = 400$ cascades

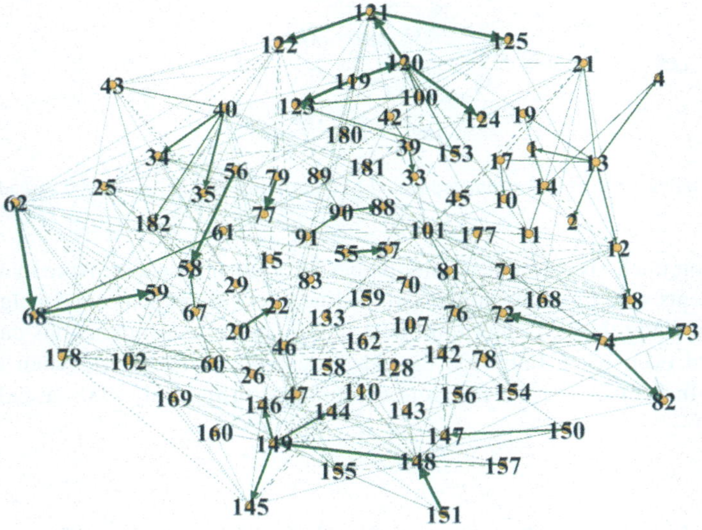

Fig. 3.7 Interaction network estimated from the method in Sect. 3.3.1 using $M_u = 41{,}000$ cascades

significant, although the number of components involved in cascading may become smaller.

For estimating interaction networks from real outage data, it is much more challenging due to many reasons, including (1) more obvious evolution among generations/stages, (2) high heterogeneity among cascades (both very short and very

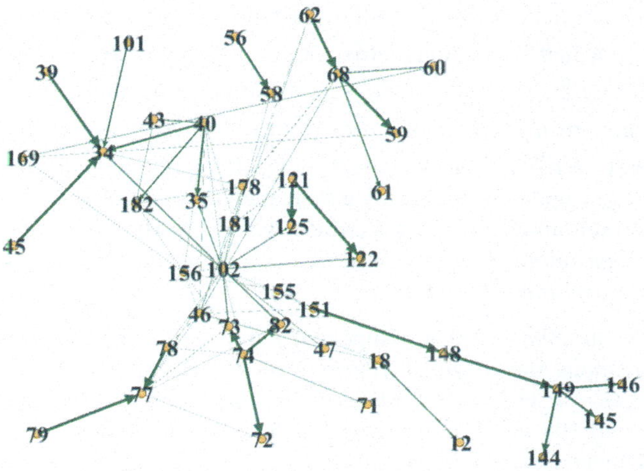

Fig. 3.8 Interaction network estimated from the method in Sect. 3.3.1 using $M_u = 400$ cascades

long cascades occur; in the BPA outage data, one cascade has 109 generations and 143 outages), and (3) data scarcity.

The data scarcity challenge will be addressed in Sect. 3.8.2 by generating more cascades using a highly probabilistic model. Furthermore, in [52] a deep convolutional generative adversarial network (DCGAN)-based method is proposed to learn the implicit features for component failure interactions and a systematic method is developed to evaluate the performance of the learning method on missing interaction recovery and new interaction discovery.

In [35], the first two challenges that are unique for real outage data are addressed by the following two mechanisms:

1. **Estimating interaction matrices for different generations**: To capture the evolution of interactions over different generations, separate failure interactions are estimated for different generations rather than performing estimation for the entire data. Specifically, instead of estimating one interaction matrix for all data, an individual interaction matrix $\boldsymbol{B}_{g-1}^0$ is estimated for any two consecutive generations $g-1$ and g using the EM algorithm in Sect. 3.4 based only on the data for generations $g-1$ and g. Assuming the largest generation number is G, $\boldsymbol{B}_g^0$ for $g = 0, \cdots, G-1$ can be obtained.
2. **Memory of interactions in previous generations**: In order to get sufficient propagation capacity and generate very long cascades, the interaction networks are assumed to be correlated in the sense that the interaction network in the current generation shares part of the interactions in a few previous generations to allow for memory. Specifically, the failure interactions in $\boldsymbol{B}_g^0$ are assumed to last $\bar{g} > 1$ generations unless the corresponding elements are updated in a future failure interaction matrix. For example, as a special case of $\bar{g} = 2$, the interaction

matrix has memory of its last failure interaction matrix $\boldsymbol{B}^0_{g-1}$. Specifically, we first set the actual interaction matrix for generation g as $\boldsymbol{B}_g = \boldsymbol{B}^0_g$. For the zero elements in $\boldsymbol{B}^0_g$, $g = 1, \cdots, G-1$, if the corresponding elements in $\boldsymbol{B}^0_{g-1}$ are nonzero, then in $\boldsymbol{B}_g$ those zero elements in $\boldsymbol{B}^0_g$ will be updated to their counterparts in $\boldsymbol{B}^0_{g-1}$. Note that not $\boldsymbol{B}_{g-1}$ but $\boldsymbol{B}^0_{g-1}$ is used to update $\boldsymbol{B}_g$, because using $\boldsymbol{B}_{g-1}$ to update would involve more and more components in cascading as the generation number increases, which would greatly overestimate cascading failure propagation. Based on numerical experiments, $\bar{g} = 2$ appears to be a good choice for the BPA dataset.

Since it is usually rare that component i fails following the outage of itself, in our estimations the diagonal elements of $\boldsymbol{B}_g$ are strictly less than one for $g = 0, \cdots, G-1$. For the BPA dataset, the maximum value among all diagonal elements for all 108 interaction matrices is 0.5. This is reasonable because if any diagonal element is equal to one, the cascade would not be able to stop once the corresponding component fails, which would conflict with either engineering experience or the original BPA dataset where all cascades eventually stop.

The interaction matrices $\boldsymbol{B}_g$'s describe how the components interact with each other in different generations. The nonzero (or to be more exact positive) elements are called *links*. For a nonzero element b_{ij} of $\boldsymbol{B}_g$, there is a link $l : i \rightarrow j$, representing that there is a positive probability for the destination component j outage in generation g following source component i outage in generation $g-1$. All links of $\boldsymbol{B}_g$ form a directed *interaction network* $\mathcal{G}^g(\mathcal{C}^g, \mathcal{L}^g)$ with the set of vertices $\mathcal{C}^g$ and the set of links $\mathcal{L}^g$.

Figures 3.9, 3.10, 3.11 show the interaction networks from $\boldsymbol{B}_0$, $\boldsymbol{B}_4$, and $\boldsymbol{B}_9$, respectively. In these figures, only the components with links are displayed and the line thickness indicates the weight. From the estimated interaction networks, it is found that with the increase of the generation number, the number of links quickly decreases while the average probability increases. For example, the complementary cumulative distributions (CCDs) of the nonzero elements are shown in Figs. 3.12, 3.13, 3.14. Note that the elements in $\boldsymbol{B}_g$'s that are less than 10^{-6} are ignored. The probability that the nonzero (positive) elements in $\boldsymbol{B}_0$ are equal to one is 0.0042, while for $\boldsymbol{B}_4$ and $\boldsymbol{B}_9$ it increases to 0.34 and 0.61. This may be a consequence of sampling from cascades. Since the later generations have much fewer samples, they will reveal correspondingly fewer interactions. Moreover, with the decreasing number of samples, the average probability of each of these revealed interactions will increase because there are fewer of them. Besides, to some extent Figs. 3.12, 3.13, 3.14 seem to show a decreasing number of samples from the same distribution.

3.5.3 Interaction Estimation for Coupled Interaction Matrix

When there are multiple line outages and the load shed at multiple buses in two successive generations, it is challenging to decide the actual interactions between outages, either line outages or the load shed, and further determine $\boldsymbol{A}$ and $\boldsymbol{B}$. To address this issue, the EM algorithm is adopted to infer the interactions between outages [27, 33]. The EM algorithm iterates over E-step and M-step to maximize the likelihood estimation of the coupled interaction matrix $\boldsymbol{B}$ until convergence. As the capability of Poisson distribution to capture the propagation of load shed can be improved by the choice of the discretization unit Δ_v [26], the discretization units for load buses are adaptively updated with the EM algorithm. The detailed procedure is described as follows:

1. **Initialization of discretization units**: An initial discretization unit $\Delta_v^{(0)} = 50$ MW is chosen for each load bus v. The load shed at buses in the original M_u cascades is processed by (3.1) with the initial discretization units.
2. **Initialization of $\boldsymbol{A}$ and $\boldsymbol{B}$**: Each component in generation $g+1$ is initially assumed to follow each outage in generation g. An initial matrix $\boldsymbol{A}^{(0)}$ can thus be constructed by processing the original M_u cascades, based on which the initial $\boldsymbol{B}^{(0)}$ can be calculated by (3.8)–(3.11).
3. **E-step**: Update $\boldsymbol{A}^{(k+1)}$ based on $\boldsymbol{B}^{(k)}$

   In the $(k+1)$th iteration, $p_{ij}^{m,g(k+1)}$ and $p_{uj}^{m,g(k+1)}$, the probability of line j outage in generation $g+1$ following line i outage and that following the load shed at bus u in generation g of cascade m, are estimated as

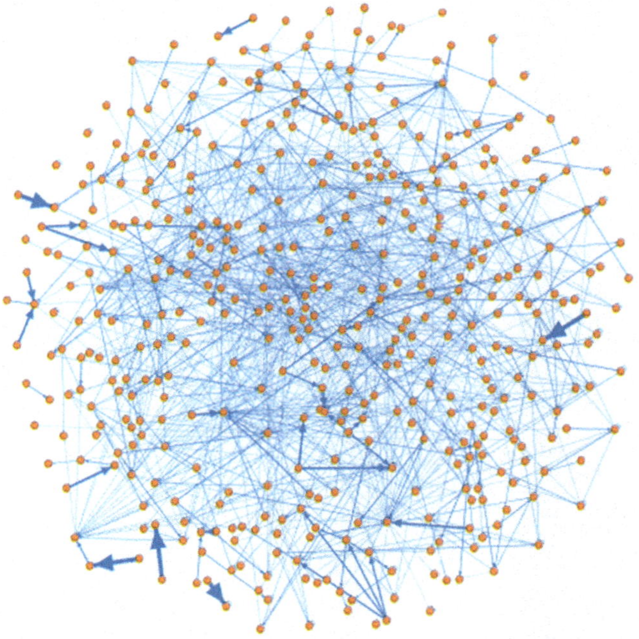

Fig. 3.9 Interaction network from $\boldsymbol{B}_0$

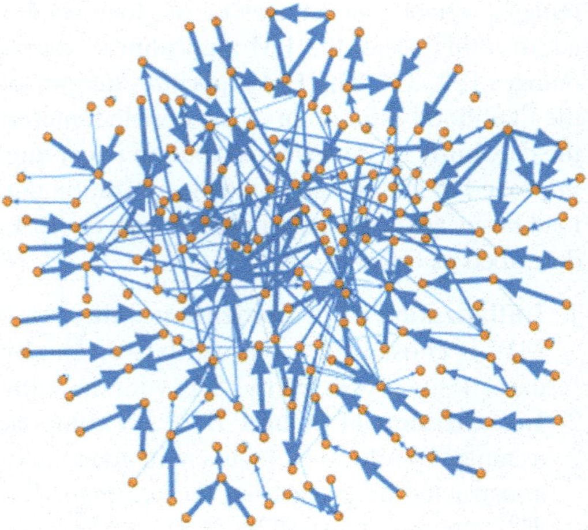

Fig. 3.10 Interaction network from $\boldsymbol{B}_4$

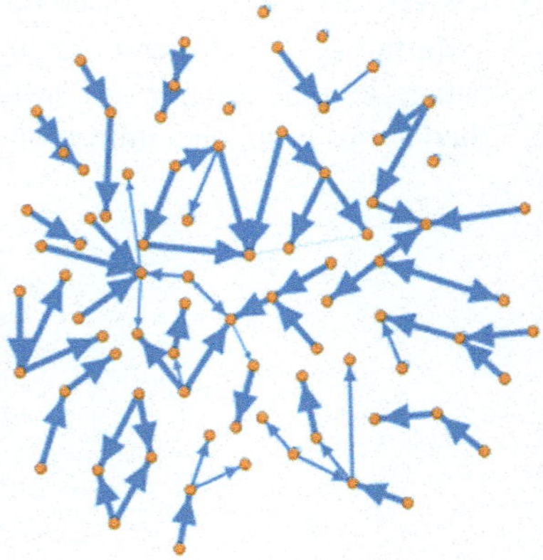

Fig. 3.11 Interaction network from $\boldsymbol{B}_9$

$$p_{ij}^{m,g(k+1)} = \frac{b_{ij}^{\mathsf{LL}(k)}}{1 - \prod_{c \in \mathcal{S}_{\mathsf{L}}^{m,g}} \left(1 - b_{cj}^{\mathsf{LL}(k)}\right) \prod_{c \in \mathcal{S}_{\mathsf{B}}^{m,g}} \left(1 - b_{cj}^{\mathsf{BL}(k)}\right)}$$

$$p_{uj}^{m,g(k+1)} = \frac{b_{uj}^{\mathsf{BL}(k)}}{1 - \prod_{c \in \mathcal{S}_{\mathsf{L}}^{m,g}} \left(1 - b_{cj}^{\mathsf{LL}(k)}\right) \prod_{c \in \mathcal{S}_{\mathsf{B}}^{m,g}} \left(1 - b_{cj}^{\mathsf{BL}(k)}\right)}.$$

Similarly, $p_{iv}^{m,g(k+1)}$ and $p_{uv}^{m,g(k+1)}$, the probability of having $Z_v^{m,g+1}$ units of discretized load shed at bus v in generation $g+1$ following the outage of line i and that following the load shed at bus u in generation g of cascade m, are estimated as

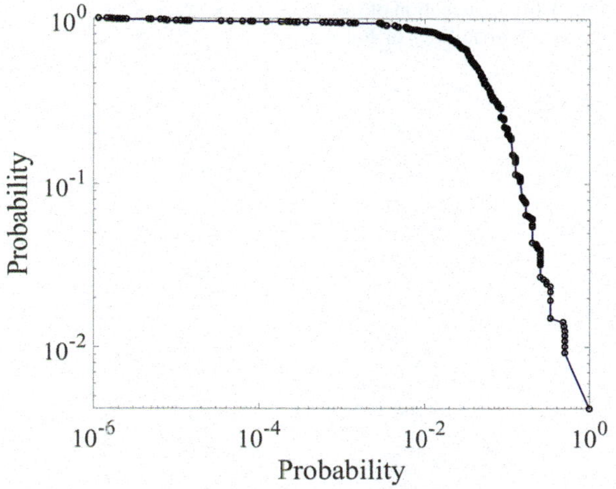

Fig. 3.12 Distribution of the nonzero probabilities in $\boldsymbol{B}_0$

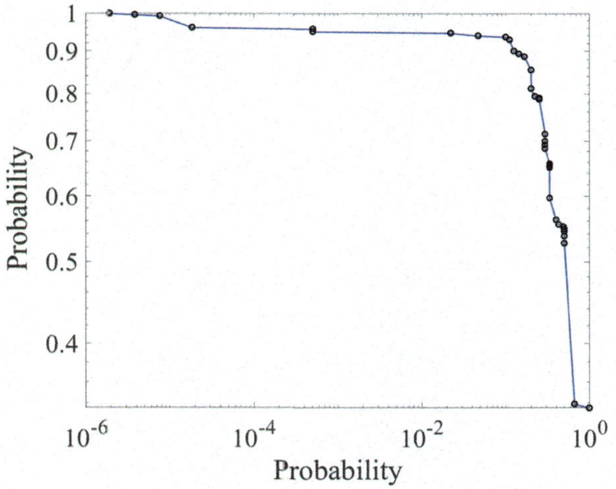

Fig. 3.13 Distribution of the nonzero probabilities in $\boldsymbol{B}_4$

$$p_{iv}^{m,g(k+1)} = \frac{p_{iv}^{\mathsf{LB}(k)}}{1 - \prod\limits_{c \in \mathcal{S}_{\mathsf{L}}^{m,g}} \left(1 - p_{cv}^{\mathsf{LB}(k)}\right) \prod\limits_{c \in \mathcal{S}_{\mathsf{B}}^{m,g}} \left(1 - p_{cv}^{\mathsf{BB}(k)}\right)}$$

$$p_{uv}^{m,g(k+1)} = \frac{p_{uv}^{\mathsf{BB}(k)}}{1 - \prod\limits_{c \in \mathcal{S}_{\mathsf{L}}^{m,g}} \left(1 - p_{cv}^{\mathsf{LB}(k)}\right) \prod\limits_{c \in \mathcal{S}_{\mathsf{B}}^{m,g}} \left(1 - p_{cv}^{\mathsf{BB}(k)}\right)},$$

where $p_{iv}^{\mathsf{LB}(k)}$ and $p_{uv}^{\mathsf{BB}(k)}$ can be calculated by (3.6)–(3.7) based on $\boldsymbol{B}^{(k)}$. The entries of $\boldsymbol{A}^{\mathsf{LL}}$ and $\boldsymbol{A}^{\mathsf{BL}}$ are updated as

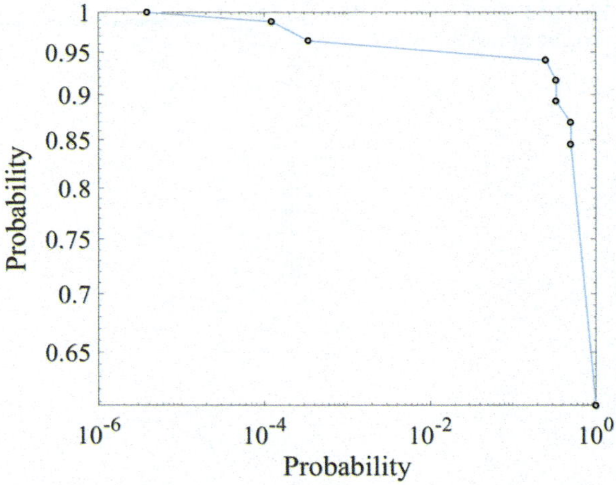

Fig. 3.14 Distribution of the nonzero probabilities in $\boldsymbol{B}_9$

$$a_{ij}^{\mathsf{LL}(k+1)} = \sum_{m=1}^{M_\mathrm{u}} \sum_{g=0}^{G^m-2} p_{ij}^{m,g(k+1)} \tag{3.21}$$

$$a_{uj}^{\mathsf{BL}(k+1)} = \sum_{m=1}^{M_\mathrm{u}} \sum_{g=0}^{G^m-2} p_{uj}^{m,g(k+1)}, \tag{3.22}$$

where $p_{ij}^{m,g(k+1)}$ is zero if line $i \notin \mathcal{S}_{\mathsf{L}}^{m,g}$ or line $j \notin \mathcal{S}_{\mathsf{L}}^{m,g+1}$. And $p_{uj}^{m,g(k+1)}$ is zero if bus $u \notin \mathcal{S}_{\mathsf{B}}^{m,g}$ or line $j \notin \mathcal{S}_{\mathsf{L}}^{m,g+1}$.

The entries of $\boldsymbol{A}^{\mathsf{LB}}$ and $\boldsymbol{A}^{\mathsf{BB}}$ are updated as

$$a_{iv}^{\mathsf{LB}(k+1)} = \sum_{m=1}^{M_\mathrm{u}} \sum_{g=0}^{G^m-2} Z_v^{m,g+1} p_{iv}^{m,g(k+1)} \tag{3.23}$$

$$a_{uv}^{\mathsf{BB}(k+1)} = \sum_{m=1}^{M_\mathrm{u}} \sum_{g=0}^{G^m-2} Z_v^{m,g+1} p_{uv}^{m,g(k+1)}, \tag{3.24}$$

where $p_{iv}^{m,g(k+1)}$ is zero if line $i \notin \mathcal{S}_{\mathsf{L}}^{m,g}$ or bus $v \notin \mathcal{S}_{\mathsf{B}}^{m,g+1}$. And $p_{uv}^{m,g(k+1)}$ is zero if bus $u \notin \mathcal{S}_{\mathsf{B}}^{m,g}$ or bus $v \notin \mathcal{S}_{\mathsf{B}}^{m,g+1}$.

The interactions of the load shed following line outages are also recorded to update the discretization units. Take the load shed at bus v following the outages of line i as an example. First let U_v denote the maximum amount of discretized load shed at bus v. Let the vector $\boldsymbol{C}_{iv}^{\mathsf{LB}(k+1)} \in \mathbb{R}^{U_v+1}$ record the number of cumulative times of the load shed at bus v following the N_i^{L} times of the outage of line i in M_u

cascades. For the nth outage of line i, if there is no discretized load shed at bus v, $C_{iv}^{\mathsf{LB}(k+1)}(0)$ increases by 1. Otherwise, if $Z_v^{m,g+1}$ units of discretized load shed at bus v follow the nth outage of line i with a probability of $p_{iv}^{m,g(k+1)}$, $C_{iv}^{\mathsf{LB}(k+1)}(0)$ and $C_{iv}^{\mathsf{LB}(k+1)}(Z_v^{m,g+1})$ increase by $1-p_{iv}^{m,g(k+1)}$ and $p_{iv}^{m,g(k+1)}$, respectively.

4. **M-step**: Update $\boldsymbol{B}^{(k+1)}$ based on $\boldsymbol{A}^{(k+1)}$

(4.1) The estimation of interaction matrix $\tilde{\boldsymbol{B}}^{(k+1)} \in \mathbb{R}^{(N_l+N_b)\times(N_l+N_b)}$ is updated with $\boldsymbol{A}^{(k+1)}$ according to (3.8)–(3.11).

(4.2) The sample variance of the discretized load shed at bus v following the outage of line i is calculated as

$$S_{iv}^{2(k+1)} = \frac{\sum_{l=0}^{U_v} C_{iv}^{\mathsf{LB}(k+1)}(l)\left(l-\tilde{b}_{iv}^{\mathsf{LB}(k+1)}\right)^2}{N_i^{\mathsf{L}}-1}. \tag{3.25}$$

(4.3) For each load bus v, Δ_v is updated as

$$\Delta_v^{(k+1)} = \Delta_v^{(k)} \sqrt{\frac{\sum_{i\in\mathcal{S}_v^{\mathsf{LB}}} N_i^{\mathsf{L}} \frac{S_{iv}^{2(k+1)}}{\tilde{b}_{iv}^{\mathsf{LB}(k+1)}}}{\sum_{i\in\mathcal{S}_v^{\mathsf{LB}}} N_i^{\mathsf{L}} \frac{\tilde{b}_{iv}^{\mathsf{LB}(k+1)}}{S_{iv}^{2(k+1)}}}}, \tag{3.26}$$

where $\mathcal{S}_v^{\mathsf{LB}}$ is the set of row indices at which the vth column of $\tilde{\boldsymbol{B}}^{\mathsf{LB}(k+1)}$ has nonzero entries. A more detailed explanation of (3.26) can be found in the Appendix.

(4.4) Using the updated discretization units, the load shed at buses in the original M_{u} cascades is reprocessed by (3.1). And the entries in $\boldsymbol{B}^{\mathsf{LB}(k+1)}$ and $\boldsymbol{B}^{\mathsf{BB}(k+1)}$ are updated from $\tilde{\boldsymbol{B}}^{\mathsf{LB}(k+1)}$ and $\tilde{\boldsymbol{B}}^{\mathsf{BB}(k+1)}$ as

$$b_{iv}^{\mathsf{LB}(k+1)} = \frac{\Delta_v^{(k)}}{\Delta_v^{(k+1)}}\tilde{b}_{iv}^{\mathsf{LB}(k+1)} \tag{3.27}$$

$$b_{uv}^{\mathsf{BB}(k+1)} = \frac{\Delta_u^{(k+1)}\Delta_v^{(k)}}{\Delta_u^{(k)}\Delta_v^{(k+1)}}\tilde{b}_{uv}^{\mathsf{BB}(k+1)}. \tag{3.28}$$

5. **End**: Steps 3–5 are repeated until the following condition is satisfied:

$$\sigma_{\boldsymbol{B}} = \sqrt{\frac{\sum_{i=1}^{N_l+N_b}\sum_{j=1}^{N_l+N_b}\left(b_{ij}^{(k+1)}-b_{ij}^{(k)}\right)^2}{N}} \le \varepsilon, \tag{3.29}$$

where $N = N_{\neq 0}$ if $N_{\neq 0}$, which is the number of nonzero entries in $\boldsymbol{B}^{(k+1)} - \boldsymbol{B}^{(k)}$, is greater than 0, otherwise $N = 1$, and ε is the preset threshold.

After the convergence of the EM algorithm, the estimated matrix $\boldsymbol{B}$ and the discretization units in the last iteration will be used for the cascading outage simulation and mitigation.

The open-loop OPA [51] is applied on the IEEE 300-bus system [50] to generate 30,000 original cascades. This system has 191 buses with load and 411 lines. The total load and the total generation capacity are, respectively, 23,848 MW and 32,678 MW. The parameters of the OPA model in [51] are set as $\gamma = 2$, $\alpha = 0.95$, $\beta = 0.30$, and $p_0 = 0.001$. The initialization of the discretization units in the EM algorithm should consider the amount of load at different buses. A too large discretization unit can lead to a lot of the load shed at some buses to zero. The initial discretization units for all load buses are chosen as 50 MW. In the original cascades for the IEEE 300-bus system, the load shed at 88% of the buses is not less than 25 MW and thus is discretized as nonzero integers according to (3.1). The load shed that is greater than 25 MW accounts for 99.21% of the total amount of load shed.

Only 10% of the original cascades ($M_{\mathrm{u}} = 3000$) are randomly selected to estimate the coupled interaction matrix $\boldsymbol{B}$. The ε in (3.29) is set as 0.01. Figure 3.15 shows the corresponding coupled interaction network. The orange and red vertices, respectively, denote lines and buses with load. The green, blue, purple, and cyan

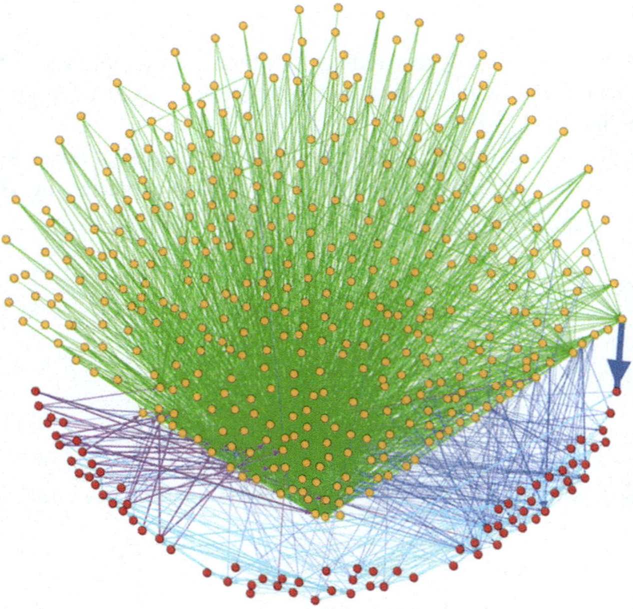

Fig. 3.15 Coupled interaction network for IEEE 300-bus system

arrows are the links in $\boldsymbol{B}^{\mathrm{LL}}$, $\boldsymbol{B}^{\mathrm{LB}}$, $\boldsymbol{B}^{\mathrm{BL}}$, and $\boldsymbol{B}^{\mathrm{BB}}$, respectively. Most links are green, indicating that line outages are the dominant factors in cascading failure propagation. This is because a more dramatic redistribution of power flow and more heavy-loaded lines tend to be observed after the outages of some critical lines rather than after load shedding at some buses. This is consistent with the conclusion in [27]. The line width in Fig. 3.15 indicates the weight of the corresponding link, which is the value in the $\boldsymbol{B}$ matrix. For example, each blue arrow shows the interaction parameter b_{uv}^{BB}. The blue arrow at the lower right corner has a large weight value of 5.0.

The entry b_{iv}^{LB} of the submatrix $\boldsymbol{B}^{\mathrm{LB}}$ captures the interaction that the load shed at bus v follows the outage of line i. The interactions can be analyzed by the connection of line i and bus v in the transmission system. It is found that bus v is connected with line i only for 9.76% of the 246 interactions in $\boldsymbol{B}^{\mathrm{LB}}$. In 16.67% of the interactions, line i is connected with generation buses rather than load buses. This indicates that a line outage may not only impact the load supply at the buses of the outaged line. This can be explained by various complicated causes of the load shed in cascading outages, including but not limited to the isolation of load buses, disconnection of generation buses, and the transmission capacity limits.

3.6 Identifying Components Critical for Outage Propagation

As an intuitive way of identifying the critical components, the number of outages of a component r, denoted by $I'(r)$, can be used to indicate how critical a component is in cascading failures. The set of critical components identified by $I'(r)$ is denoted by C_1. For the BPA outage data in Sect. 3.2.2, C_1 could be identified as the top ten components that fail the most times as $\{2, 8, 92, 24, 42, 41, 101, 17, 237, 234\}$. The corresponding numbers of times they fail are 354, 334, 240, 191, 173, 163, 146, 143, 117, and 108. However, even a component fails for many times, it is not necessarily true that many components will consequently fail. To address this problem, a better metric, the *expected number of outages* following each component outage, can be calculated based on the interaction networks.

3.6.1 Expected Number of Outages Following a Component Outage in Generation g

For each interaction network a subgraph that starts with a component r with at least one outgoing link can be obtained, as illustrated in Fig. 3.16. This subgraph provides useful information about the outage propagation pattern starting with the root component: the probability of each link indicates how possible one component

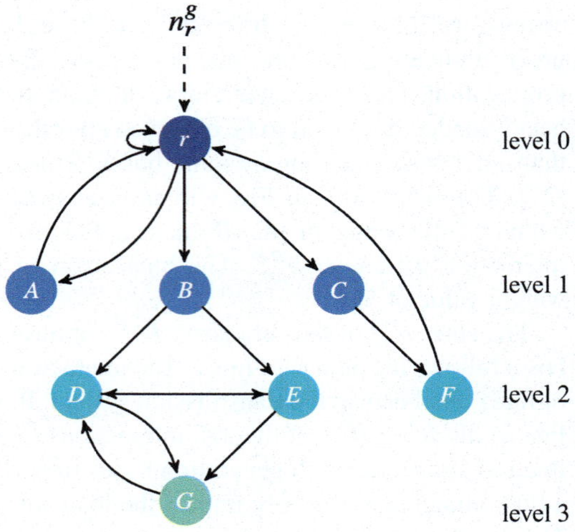

Fig. 3.16 Illustration for a subgraph starting with r

outage happens following another component outage; the expected number of outages of each component indicates the risk level of that particular component.

One major challenge for calculating the expected number of outages following a component is that there may be loops (directed circles or self-loops) in the subgraph, due to complicated interactions among components. In [29] some links are removed in order to obtain a directed acyclic subgraph which, however, will inevitably lose useful information. In [35] a better approach is developed in which the original subgraph structure is retained and a set of carefully formulated linear equations related to the corresponding subgraph is solved. Below is a description of this approach.

For any node r that has outaged in generation g at least once, from $\boldsymbol{B}_g$ there is a subgraph starting with node r (called root node), denoted by $\mathcal{G}_r^g(\mathcal{C}_r^g, \mathcal{L}_r^g)$ with $\text{card}(\mathcal{C}_r^g) = N_r^g$, where card$(\cdot)$ is the cardinality of a set. It is obvious that there is at least one path from r to any node $c \in \mathcal{C}_r^g$ and any node in $\mathcal{C}_r^g$ has at least one incoming link. The root node is at level 0 and any other node that can be reached from the root node after a minimum k hops is at level k. The node numbers in the subgraph are re-ranked from 1 which corresponds to the root node r. For the new node number j, we denote j^0 as its old number before re-ranking. An adjacency matrix $\boldsymbol{C}^g = [c_{uv}^g] \in \mathbb{R}^{N_r^g \times N_r^g}$ is built, for which $c_{uv}^g = b_{u^0 v^0}^g$ if there is a link from node u^0 to node v^0 in $\boldsymbol{B}_g$; otherwise $c_{uv}^g = 0$.

For $\forall j \in \mathcal{C}_r^g$, its expected number of outages, e_j^g, can be easily obtained from $\mathcal{G}_r^g$ as

$$e_j^g = \sum_{i=1}^{N_r^g} c_{ij}^g e_i^g + d_j^g, \tag{3.30}$$

where the first term on the right-hand side is the expected number of outages caused by the nodes in $\mathcal{G}_r^g$ (including j itself) and d_j^g is the expected number of outages of node j due to any factor outside $\mathcal{G}_r^g$. Since the objective is to calculate the expected number of outages following node j outage in generation g, only when $j = 1$ (j is the root node) is $d_j^g \in \mathbb{Z}_+$ the number of times that the root node fails in generation g (denoted by n_r^g), while for $\forall j \in C_r^g \backslash \{1\}$ there is $d_j^g = 0$.

Let the vector of the expected number of outages for the nodes be $\boldsymbol{e}^g \in \mathbb{R}^{N_r^g \times 1} = [e_1^g \; e_2^g \cdots e_{N_r^g}^g]^\top$. Then $\boldsymbol{e}^g$ can be obtained by solving the following linear equations:

$$\boldsymbol{D}^g \boldsymbol{e}^g = \boldsymbol{d}^g, \tag{3.31}$$

where $\boldsymbol{D}^g = \boldsymbol{I}_{N_r^g} - (\boldsymbol{C}^g)^\top$, $\boldsymbol{I}_{N_r^g}$ is an identity matrix with dimension N_r^g, and $\boldsymbol{d}^g = [n_r^g \; 0 \; \cdots \; 0]^\top$. Since $\mathcal{G}_r^g$ is weakly connected, there does not exist any row of $\boldsymbol{D}^g$ whose elements are all zeros.

3.6.2 *Existence of Unique Positive Solution for (3.31)*

Theorem 3.1 *$\boldsymbol{D}^g$ is singular if and only if there exist two nodes $i, j \in C_r^g$ for which*

1. *There are links both from i to j and from j to i.*
2. *At least one of them has a self-loop.*
3. *There are no incoming links to node i or j from any node other than i or j.*
4. *There is*

$$(1 - c_{ii}^g)(1 - c_{jj}^g) = c_{ij}^g c_{ji}^g.$$

The proof of Theorem 3.1 is straightforward and is thus omitted due to space limit. It is based on linear dependency of two rows of $\boldsymbol{D}^g$ and the fact that the diagonal element $0 < 1 - c_{ii}^g \leq 1$ while the nonzero off-diagonal element $-c_{ij}^g < 0$. Here we only show how condition 4 is derived when conditions 1–3 are satisfied. Assume $i, j \in C_r^g$ satisfy conditions 1–3. The submatrix of the ith and jth rows of $\boldsymbol{D}^g$ that has nonzero elements is

$$\begin{array}{c} \\ i \\ j \end{array} \begin{array}{c} \begin{array}{cc} \quad i & \quad\quad j \end{array} \\ \begin{bmatrix} 1 - c_{ii}^g & -c_{ji}^g \\ -c_{ij}^g & 1 - c_{jj}^g \end{bmatrix} \end{array}, \tag{3.32}$$

where $c_{ii}^g c_{jj}^g \neq 0$ due to condition 2 in Theorem 3.1. It is obvious that only when condition 4 is satisfied will rows i and j be linearly dependent and $\boldsymbol{D}^g$ be singular.

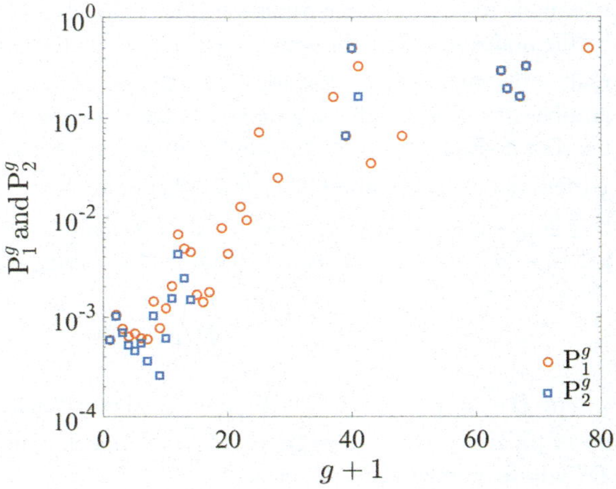

Fig. 3.17 P_1^g and P_2^g for different generations

Instead of calculating the probability of satisfying conditions 1–3 for each subgraph $\mathcal{G}_r^g$, we directly evaluate the whole interaction network $\mathcal{G}^g$. It is obvious that if conditions 1–3 cannot be satisfied by $\mathcal{G}^g$, they will not be satisfied by any of its subgraph $\mathcal{G}_r^g$. Let $P_1^g = P(\text{condition 1})$, $P_2^g = P(\text{conditions 1–2})$, and $P_3^g = P(\text{conditions 1–3})$ be the probabilities for any pair of nodes in $\mathcal{G}^g$ to satisfy condition 1, conditions 1–2, and conditions 1–3, respectively. For the estimated $\boldsymbol{B}_g$'s from BPA data, the nonzero P_1^g's and P_2^g's for $g = 0, \cdots, 107$ are shown in Fig. 3.17 and $P_3^g = 0$ for all g's. For any g that does not have a data point for P_1^g or P_2^g in Fig. 3.17, there is $P_1^g = 0$ or $P_2^g = 0$, Therefore, conditions 1–3 cannot be satisfied by any two nodes of $\mathcal{G}^g$. Furthermore, for $\mathcal{G}_r^g$ starting with any root node r that has outaged in generation $g-1$, not any two nodes in $\mathcal{G}_r^g$ could satisfy conditions 1–3 at the same time, and thus $\boldsymbol{D}^g$ is always invertible, indicating the existence of a unique solution for $\boldsymbol{e}^g$ in (3.31).

Remark 1 Even though conditions 1–3 in Theorem 3.1 are indeed satisfied for $\mathcal{G}_r^g$, since $c_{ii}^g, c_{jj}^g, c_{ij}^g$, and c_{ji}^g are usually small it is safe to have $(1-c_{ii}^g)(1-c_{jj}^g) \gg c_{ij}^g c_{ji}^g$, and the chance to satisfy condition 4 is very low.

For the estimated $\boldsymbol{B}_g$'s from BPA data, for all pairs of nodes in $\mathcal{G}^g$, $g = 1, \cdots, 108$ that satisfy conditions 1–2 (note that no node pair satisfies conditions 1–3), the conditional probability for satisfying condition 4 is as low as 0.69%, while the conditional probability for $(1-c_{ii}^g)(1-c_{jj}^g)/c_{ij}^g c_{ji}^g \geq 2$ is as high as 98.27%. The average and maximum values of $(1-c_{ii}^g)(1-c_{jj}^g)/c_{ij}^g c_{ji}^g$ among all ij pairs that satisfy conditions 1–2 are, respectively, 4.51×10^{38} and 6.50×10^{40}.

Next we show that this unique solution is positive. Note that e_j^g is the expected number of outages and has a clear physical meaning. For $j = 1$, since c_{ij}^g is small and $e_i^g \ll e_1^g$ for $\forall i \neq j$, the right-hand side of (3.30) is dominant by

$d_1^g > 0$. Therefore, it is safe and reasonable to assume that $e_1^g > 0$. In reality, in the calculations for the OPA dataset, there is always $e_1^g \approx d_1^g > 0$.

Lemma 1 *If there exists any $j \neq 1$ for which $e_j^g \leq 0$, the assumption $e_1^g > 0$ would not hold.*

Proof of Lemma 1 Assume there exists $j \neq 1$ at level $k > 0$ for which $e_j^g \leq 0$. Since $j \neq 1$ there is $d_j^g = 0$. From (3.30), it is obvious that at level $k - 1$ either $e_i^g = 0$ for any node i that has a link to j or at least one of the nodes that has a link to node j, such as l, should have $e_l^g < 0$. This derivation can be continued until $k = 0$ for which there will be $e_1^g \leq 0$, which conflicts with the assumption $e_1^g > 0$.

From Lemma 1, it is easy to obtain the following theorem.

Theorem 3.2 *Under the assumption that $e_1^g > 0$, the unique solution of the linear equations in (3.31) is positive.*

The proof of Theorem 3.2 is straightforward based on Lemma 1 and is thus omitted.

3.6.3 Metric Based on Expected Number of Outages

Based on Theorems 1 and 2, the linear equations in (3.31) can be readily solved to obtain a unique positive solution. Then the expected number of outages following n_r^g node r outages in generation g can be calculated as

$$I_g(r) = \sum_{j=1}^{N_r^g} e_j^g(r) - n_r^g, \tag{3.33}$$

where n_r^g included in e_1^g is subtracted. Furthermore, the total expected number of outages following one particular node r outage in all generations can be calculated as

$$I(r) = \sum_{g=0}^{G-1} I_g(r), \tag{3.34}$$

which is used as a metric to identify critical components that play important roles in cascading outage propagation. The set of critical components identified from $I(r)$ is denoted by C_2.

We compare the identified 20 most critical components in BPA outage data by using $I'(r)$ and $I(r)$ in Table 3.3. When calculating $I(r)$, the elements in $\boldsymbol{B}_g$'s that are less than 10^{-6} are ignored. Component 83 is the most critical component identified by the interaction network-based approach but is only ranked as the 12th

Table 3.3 Identified key components

Rank	$C_1(I')$	$C_2(I)$	Rank	$C_1(I')$	$C_2(I)$
1	2 (354)	83 (136.0)	11	37 (98)	201 (50.8)
2	8 (334)	17 (101.0)	12	83 (93)	61 (48.7)
3	92 (240)	234 (86.3)	13	446 (91)	56 (43.2)
4	24 (191)	24 (74.4)	14	5 (83)	59 (42.1)
5	42 (173)	76 (65.3)	15	97 (81)	126 (41.2)
6	41 (163)	2 (64.2)	16	19 (79)	26 (40.6)
7	101 (146)	85 (62.2)	17	148 (76)	92 (39.8)
8	17 (143)	8 (54.0)	18	29 (75)	73 (39.8)
9	237 (117)	101 (53.6)	19	72 (75)	308 (39.5)
10	234 (108)	187 (53.6)	20	128 (75)	42 (39.3)

most critical component by the intuitive approach. Component 83 only fails for 93 times and appears in 40 cascades, compared with the outage of 354 times for component 2. However, among these 40 cascades, in 22 cascades as many as 341 outages happen after the first outage of component 83 (in one cascade there can be more than one outage of the same component). Therefore, component 83 is indeed very critical in terms of cascading outage propagation.

On the other hand, some very critical components identified by the intuitive approach have much lower ranking in the interaction network-based approach. For example, component 92 fails for 240 times and thus is considered as the third most critical component by the intuitive approach. But it is only considered as the 17th most critical component by the interaction network-based approach using $I(r)$. Although it fails for 240 times, for most times the cascading stops after its outage and there are only 29 times for which there are other outages after the outage of component 92. Only as few as 49 outages happen after the first outage of component 92. Therefore, it is not that critical for propagation of outages.

The expected numbers of outages in each generation starting with component 83 and component 92, as shown in Fig. 3.18, can be used to indicate their importance in outage propagation over generations. Although I_1 of component 92 is greater than that of component 83, component 83 has almost the same I_2 as component 92 and significantly greater I_g for $g \geq 3$. Overall $I(83) = 136.0$ is greater than $I(92) = 39.8$, and thus component 83 is much more critical than component 92. As shown in Fig. 3.19, in the subgraph of $\boldsymbol{B}_3$ starting with component 83 there are 236 links, while in that of component 92 there are only 4 links (including 1 self-loop at the root component 92) with a very simple topology.

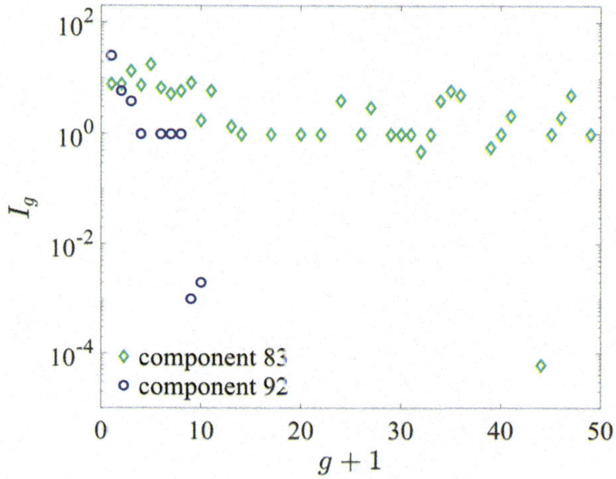

Fig. 3.18 I_g for the subgraphs starting with components 83 and 92

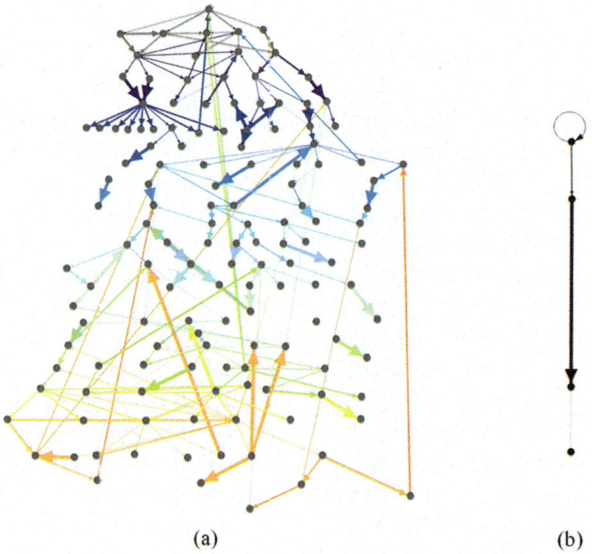

Fig. 3.19 Subgraph in $\boldsymbol{B}_3$ starting with component 83 (**a**) and component 92 (**b**). Line width indicates the value of the $\boldsymbol{B}_3$ entry. The different colors of the links in (**a**) indicate the levels of cascading propagation

3.7 Identifying Critical Components Considering Spatial Propagation

In Sect. 3.6, critical components that play the most important roles in outage propagation are identified. However, only focusing on the number of outages cannot provide a detailed description of the cascading failure impacts. How long time a cascading failure lasts and how wide an area a cascading failure can spread to are also critical factors that need to be considered. The failure propagation paths and

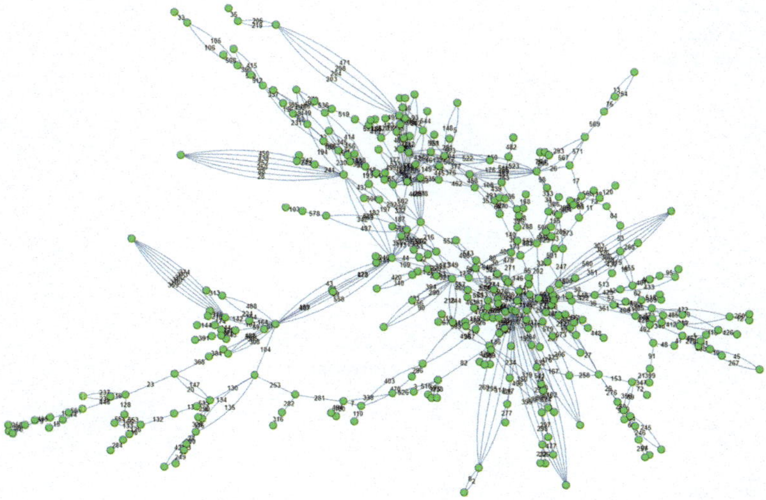

Fig. 3.20 System topology based on the BPA outage data

spatial distances between outages need to be explicitly considered in the mitigation strategy, helping prevent large-scale blackouts in a wide area.

In [32], the propagation path is described by electrical distance, which is the equivalent impedance between two components. The component interaction, the number of outages, and the amount of load shedding are combined in the mitigation strategy design. However, using electrical distance to measure spatial distance could lose the topology information such as the actual geographical lengths of the components, leading to ineffective spatial propagation mitigation. The topology information of the BPA outage data is studied in [41] based on cascade spreading statistics, providing a new direction for mitigation. However, critical components are not identified by directly considering spatial propagation.

In this section, the cascading failure spatial propagation analysis method developed in [37] is discussed. It is applied to the 14-year BPA utility outage data described in Sect. 3.2.2. The spatial distance between two generations of outages, the total spatial distance between the outages in a cascade, and also the average spatial propagation velocity for two consecutive generations are defined. Furthermore, a critical component identification method based on a new metric called the total spatial distance which combines the information of the expected number of outages and that of the spatial distance is also discussed.

The power system topology of the transmission lines (components) involved in the BPA outage data in Sect. 3.2.2 is shown in Fig. 3.20. There are 346 buses and $n = 582$ components. The green dots are buses and the lines are components. Lines with different voltage levels between two buses are drawn as parallel lines. Note that though the figure shows the transmission line length between the components, it is not the actual geographical topology. In power systems, the topological distance $d^{\mathrm{t}}_{i_1 i_2}$ between buses i_1 and i_2 can be defined as the number of components along the shortest path between these two buses. The geographical distance $d^{g}_{i_1 i_2}$ can be defined as the length of the components on the shortest path. Since we want to keep the information of component length for practical considerations, the distance between component $i : i_1 \rightarrow i_2$ and component $j : j_1 \rightarrow j_2$ is defined as

$$d_{ij} = \min\{d^{g}_{i_1 j_1}, d^{g}_{i_1 j_2}, d^{g}_{i_2 j_1}, d^{g}_{i_2 j_2}\}, \tag{3.35}$$

which is the shortest geographical distance among all possible geographical distances between one bus of component i and one bus of component j. Note that $d_{ij} = d_{ji}$ and $d_{ij} = 0$ when components i and j share at least one bus.

Then the spatial distance between the outage components in two successive generations g and $g + 1$ of cascade m can be calculated based on the final $p^{m,g}_{ij}$ obtained in the interaction matrix estimation as

$$d^{(m)}_{g\rightarrow g+1} = \sum_{j\in\mathcal{F}^{(m)}_{g+1}} \sum_{i\in\mathcal{F}^{(m)}_{g}} \frac{p^{m,g}_{ij}}{\sum_{i\in\mathcal{F}^{(m)}_{g}} p^{m,g}_{ij}} d_{ij}. \tag{3.36}$$

Note that a weighted averaging is performed for the component pairs in successive generations based on the final $p^{m,g}_{ij}$ that indicates the dependencies between components i and j from Sect. 3.5.

The *total spatial distance* between the outages in cascade m can be further calculated as

$$d^{(m)}_{\text{total}} = \sum_{g=0}^{G-1} d^{(m)}_{g\rightarrow g+1}. \tag{3.37}$$

Investigating the relationship between the total number of outages of the cascades, generations, and their $d^{(m)}_{\text{total}}$ could help reveal the detailed features in spatial propagation.

Then the *average spatial propagation velocity* from generation g to generation $g + 1$ can be calculated as

$$\bar{v}_{g\rightarrow g+1} = \frac{1}{M_{g\rightarrow g+1}} \sum_{m=1}^{M_{g\rightarrow g+1}} d^{(m)}_{g\rightarrow g+1}, \tag{3.38}$$

where $M_{g\rightarrow g+1}$ is the number of cascades that contains generations g and $g + 1$. Here one generation is considered as one discrete time step.

In [53], based on very simple cascading overload simulations, it is found that the propagation velocity of a Euclidean distance of the failures from the center of the initial failure is approximately constant and is similar for different networks. However, in many real cascading blackouts such as the infamous 2003 U.S.-

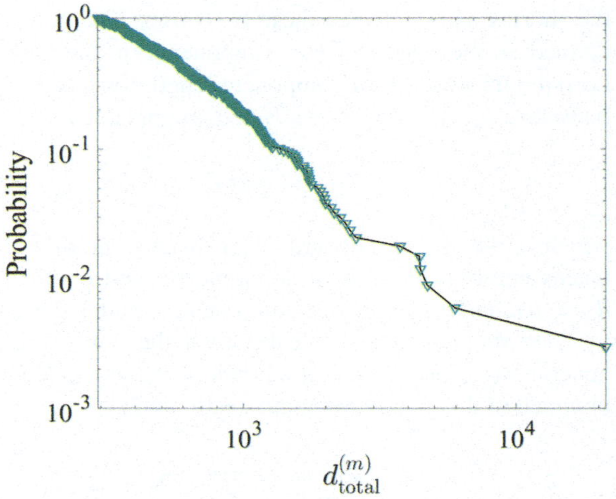

Fig. 3.21 CCD of large $d_{\text{total}}^{(m)}$'s

Canadian blackout the spatial propagation actually accelerated significantly as cascading progressed.

For the BPA dataset, most cascades have one generation and only 752 cascades have positive $d_{\text{total}}^{(m)}$. Figure 3.21 shows the CCD of the relatively large $d_{\text{total}}^{(m)}$'s (i.e., $d_{\text{total}}^{(m)} > 10^{-2}$ which is around the 90th percentile of $d_{\text{total}}^{(m)}$ for all cascades), indicating that there is a dramatic difference between different cascades. The maximum $d_{\text{total}}^{(m)}$ is 2.13×10^4 for $m = 4005$ that has the largest number of generations.

The spatial propagation velocity $\bar{v}_{g \to g+1}$ is shown in Fig. 3.22. When $g = 97$, $\bar{v}_{g \to g+1}$ has the highest value which is 1, 700. When $g = 39, 53, 66, 82, 100, 105, 106, 107$, $\bar{v}_{g \to g+1} = 0$, which is due to the components in these generations sharing the same buses. Based on (3.36), if two components i and j connect to at least one common bus, the responding d_{ij} is zero, which further leads to zero $\bar{v}_{g \to g+1}$.

The $\bar{v}_{g \to g+1}$ for different generations is significantly different. Meanwhile, the number of outages is 2658 for $g = 0$, while it decreases sharply in further generations. To better reveal the cascading failure spatial propagation properties, the outages from successive generations are grouped. Specifically, the generation grouping rules are listed below:

1. Starting with generation 0, a generation group combines several consecutive generations until the total number of outages in this group is larger than a predetermined threshold. This is repeated for the remaining generations. In a special case, if the number of outages of a single generation meets the condition, it is a group.
2. Before generation $g_e = 41$, each generation has hundreds or even thousands of outages, while after generation g_e each generation only has less than ten outages. Therefore, we choose two different thresholds for the number of outages: for a group whose first generation is before generation g_e, the threshold is chosen as

Fig. 3.22 $\bar{v}_{g\to g+1}$ without mitigation

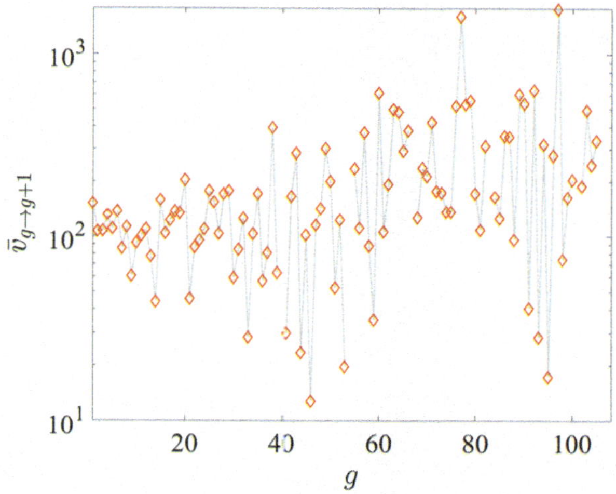

$O_1 = 129$ (the 95th percentile of the number of outages in all generations); for the remaining generation groups, the threshold is set as $O_2 = 65$, which is about half of O_1.

3. If the remaining ungrouped generations do not have greater than or equal to O_2 number of outages, a few generations in the previous generation group will be combined with the remaining ungrouped generations until the number of outages in the last generation group is greater than or equal to O_2.

Then $\bar{v}_{g\to g+1}$ is calculated for the grouped generations and is shown in Fig. 3.23, where a clear increasing tendency of the spatial propagation velocity is revealed, indicating that the extent of spatial propagation is increased in later generations. This is consistent with the previous blackouts such as the 2003 U.S.-Canadian blackout in which the spatial propagation accelerated significantly as the cascading progressed.

In Sect. 3.6, critical components are identified by calculating the expected number of outages starting with each component, and the larger the expected number of outages is, the more critical the corresponding component is in cascading failure propagation. A mitigation strategy is also developed based on the identified critical components. However, this mitigation does not consider component spatial distances and thus cannot guarantee to suppress cascading failure spatial propagation.

To develop a more comprehensive mitigation strategy, the expected number of outages and the spatial distance between components are combined to create a new metric for the identification of critical components, reducing both the number of outages and spatial propagation.

Following Sect. 3.6.1, for every node $u^0 \in C_r^g \backslash \{r\}$ in the failure interaction network from $\boldsymbol{B}_g$, if it is in the subgraph $\mathcal{G}_r^g$ staring with node r and its re-ranked node number is u, the outage measure from component r to component u^0 will

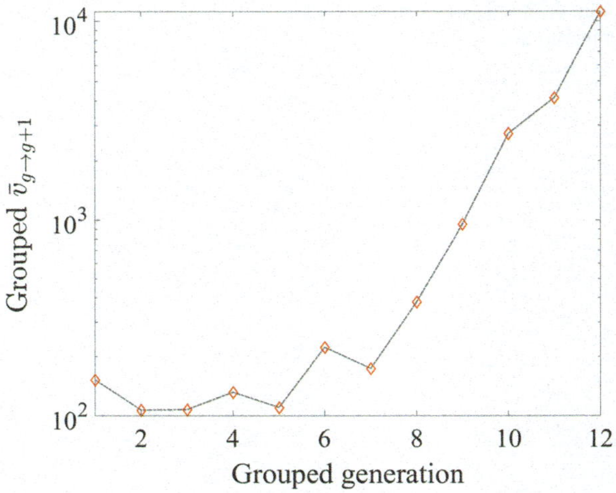

Fig. 3.23 $\bar{v}_{g \to g+1}$ for grouped generations

be $s^g_{r,u^0} = e^g_u$. If node u^0 is node r, then $s^g_{r,r} = e^g_1 - c^{g-1}_r$. After completing all calculations for each node in each failure interaction matrix, we combine the expected number of outages and the spatial distances to calculate a new metric for each node in C^g_r, the *expected spatial propagation* $I^g_d(r)$:

$$I^g_d(r) = \sum_{i=1}^{N^g_r} s^g_{r,i}\, d_{ri}. \tag{3.39}$$

Furthermore, the *total spatial propagation*, $I_d(r)$, over all generations is obtained as

$$I_d(r) = \sum_{g=0}^{G-1} I^g_d(r). \tag{3.40}$$

The metric in (3.40) considers not only the number of outages following a component failure but also the extent of spatial propagation after that component failure. A component is considered critical if its failure leads to many outages with extensive spatial propagation.

3.8 Interaction Model

As large blackouts are usually rare, simulating large blackouts with physical cascading failure models is rather inefficient, while the number of available utility outage data samples is limited. Highly probabilistic interaction models [29, 33–35] can efficiently generate a large number of cascades to better estimate the blackout size [54]. Moreover, the effect of mitigation schemes can also be efficiently tested by modifying the parameters of highly probabilistic models.

3.8.1 Basic Interaction Model

Based on the tripping probability of each component in generation 0 and the interactions between component failures, a cascading failure model called interaction model can very efficiently simulate a large number of cascades whose statistical properties match those of the original data.

Assume there are a total of M original cascades available. Note that we do not necessarily need to use all the M cascades but only M_u of them to generate the tripping probability of each component in generation 0 and the interaction matrix since a smaller number of cascades can capture how frequently the components fail in generation 0 and how the component failures interact with each other, especially when any one of the original cascades can be considered as an independent and identical realization of an underlying process.

It is assumed that all components are initially unfailed and each component fails with a small probability. The component failures in the same generation cause other component failures independently. The flow chart of the interaction model is shown in Fig. 3.24, in which $M_{\max}$ is the number of cascades to be simulated. The model contains two loops, and in each outer loop a cascade is simulated. Specifically, the model is implemented in the following three steps.

- **Step 1: Generate initial outages**
  In the kth outer iteration, each component i randomly fails with probability τ_i to simulate accidental faults and the failed components form generation 0 (initial outages) of the simulated cascade. The probability that a component i fails as initial outages can be estimated by using the generation 0 component failures of the M_u original cascades as

$$\tau_i = \frac{f_0^i}{M_u}, \tag{3.41}$$

  where f_0^i is the number of cascades for which component i fails in generation 0.
- **Step 2: Corresponding columns of *B* are set to zero**
  The columns of $\boldsymbol{B}$ corresponding to the component failures are set zero since in this model once a component fails it will remain that way until the end of the simulation.
- **Step 3: Failed components cause other component failures**
  The component failures in one generation independently generate other component failures. Specifically, if component i fails in this generation, it will cause the failure of any other component j with probability b_{ij}. Once it causes the failures of some components, these newly caused component failures will

Fig. 3.24 Flow chart of the interaction model

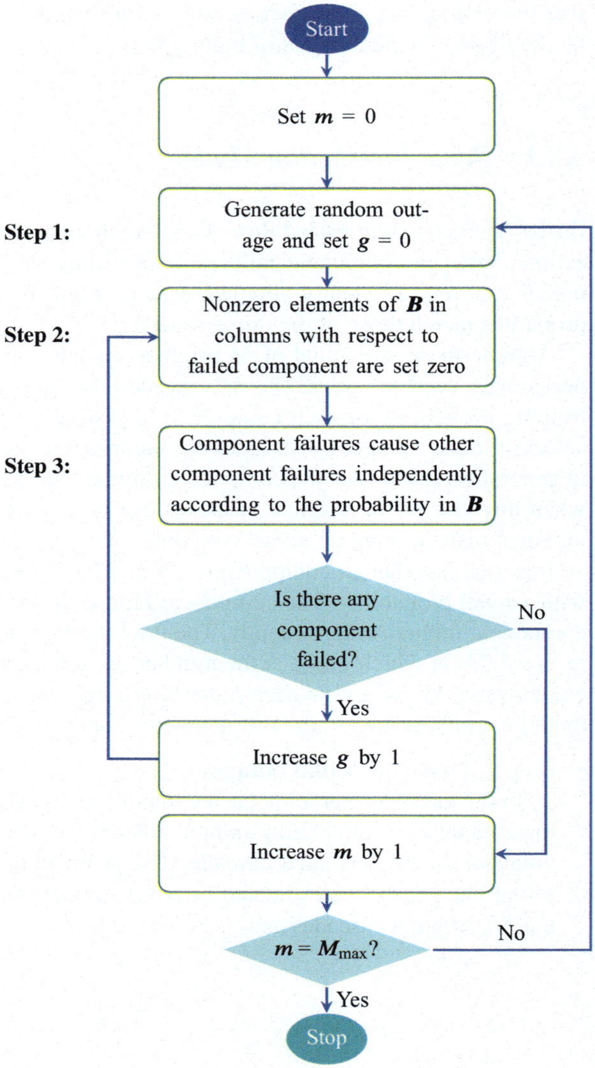

comprise the next generation; then go back to step 2. If no component failure is caused, the inner loop stops.

By using the interaction model, we can simulate as many cascades as possible (greater than M_u or M). Although the simulated cascades are generated by utilizing the information of the initial outages and the interactions contained in the original cascades, they can reveal rare new events due to the high-level probabilistic property of the interaction model, thus helping recover the missing information due to using fewer original cascades. Therefore, as long as the cascades from the interaction

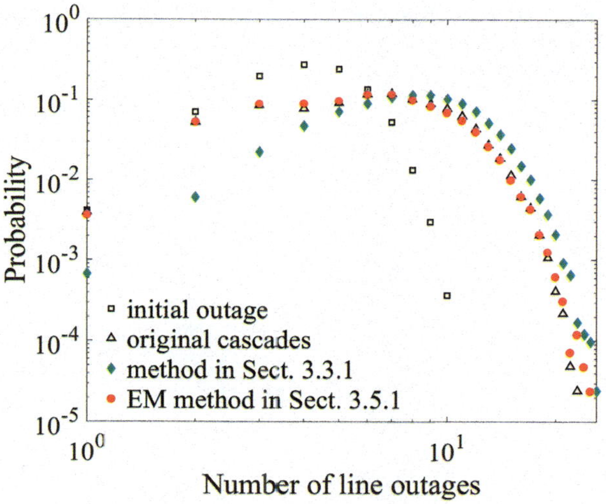

Fig. 3.25 Probability distributions of the line outages. *open square* and *open triangle* denote initial outages and total outages of the original cascades, and *green filled diamond* and *red filled circle* denote total outages of the simulated cascades using 400 cascades, respectively, for the method in Sect. 3.3.1 and the EM method in Sect. 3.5.1

model are well validated, it can be much more time-efficient to first quantify the interactions between the component failures with fewer original cascades from a more detailed cascading failure model and then perform the interaction model simulation than it is to directly simulate a large number of cascades with a more detailed model.

Figure 3.25 shows the probability distributions of the line outages for the 41,000 original cascades and the 41,000 cascades obtained by the basic interaction model based on the interactions estimated from the method in Sect. 3.3.1 and the EM-based method in Sect. 3.5.1. Note that the interaction model uses the distribution of the initial outages and the interactions obtained from 400 original cascades to generate 41,000 cascades. The simulated cascades by using the interactions obtained from 400 original cascades with the EM method have very similar statistical features to the original 41,000 cascades. By contrast, the distribution of the simulated 41,000 cascades based on the interactions obtained from the same number of cascades using the estimation method in Sect. 3.3.1 is very different from the original cascades, indicating that the EM method can more accurately estimate the interactions of component failures.

The interaction model can not only be used for offline study of cascading failures but can also be used for online decision-making support. The interaction matrix can be obtained offline from statistical utility data or simulations of more detailed cascading failure models. It contains important information about the interactions between component failures. By utilizing this information the interaction model has the potential to predict the consequences of events. If something unusual happens in the system, the operators can apply the interaction model to quickly find out which components or which areas of the system will most probably be affected so that a fast response can be performed to pull the system back to normal conditions and to avoid or at least reduce the economic and social losses.

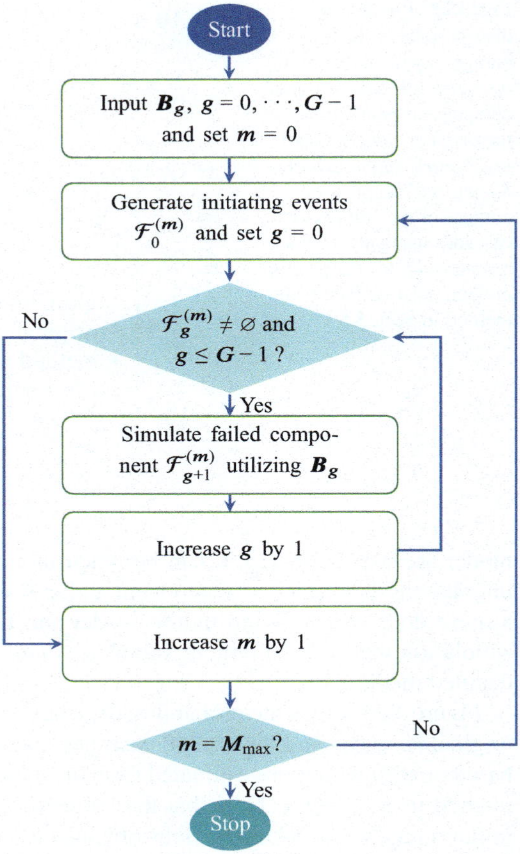

Fig. 3.26 Flow chart of the generation-dependent interaction model

3.8.2 Generation-Dependent Interaction Model

From the perspective of complex systems, the system-level failures are not caused by any specific event but by the property that the components in the system are tightly coupled and interdependent [29, 33]. With the estimated component interactions, a highly probabilistic generation-dependent interaction model can generate cascades that could capture and extend what have been observed in real outage data, validate the estimated component interactions, and further evaluate mitigation strategies.

The generation-dependent interaction model is illustrated in Fig. 3.26, in which the interaction matrix B_g changes with the number of generations.

1. The same initiating events, i.e., the outages in generation zero of the real data, are used for the interaction model simulation. In order to consider the randomness and obtain statistically reliable results, we simulate for each initiating event in

the real data for more than once. Alternatively, we can also simulate cascades with other initiating events in order to reveal their consequences.
2. The outages in generation g are generated independently by outages in generation $g-1$ using interaction matrix $\boldsymbol{B}_{g-1}$. If one component is to fail in generation g for more than once, only one outage is kept.
3. The columns of $\boldsymbol{B}_g$ corresponding to the component outages are not set to be zero to take into account the fact that a component may fail more than once partly due to the operation and reclosing of the protective relays.

The generation-dependent interaction model is validated below using the BPA outage data in Sect. 3.2.2 by comparing the distribution of the number of line outages and the offspring mean of branching process calculated using the original cascades and the generated cascades from the generation-dependent interaction model.

3.8.2.1 Comparison of Distribution of the Number of Line Outages

We simulate ten times of the number of cascades as in the utility outage data by using the generation-dependent interaction model. The initial outages for the interaction model simulation are the same as those in the utility outage data. Figure 3.27 shows a comparison between the CCD of the number of line outages for the original BPA data and those for the simulated cascades from different interaction matrix estimation methods. Each CCD for the simulated cascades is the averaged result over $K = 40$ estimations, each of which is performed using $10M$ cascades generated from the generation-dependent interaction model. Here we choose $K = 40$ because the results for more estimations are similar. The gray lines next to the CCD estimation in Fig. 3.27 are the $C = 95\%$ confidence interval of the estimation [55], i.e., $\mathrm{P}(\bar{p} - t^* \frac{s}{\sqrt{K}} \le p \le \bar{p} + t^* \frac{s}{\sqrt{K}}) = 95\%$, where $\bar{p}$ is the sample mean, s is the sample standard error, and t^* is the upper $(1-C)/2$ critical value for the t-distribution with $K-1$ degrees of freedom. It is seen that the CCD of the generated cascades using the interaction matrix obtained based on the two mechanisms in Sect. 3.5.2 matches very well with that of the original utility outage data and greatly extends it, while using one single interaction matrix estimated for the entire data as in [29, 33] or the generation-dependent interaction matrix without memory of interactions in previous generations greatly underestimates the risk of large cascades. Therefore, considering the evolution of interactions over generations and enabling memory between consecutive generations are important for effectively capturing the outage interactions in the original utility outage data and further the statistical properties of the line outage distribution.

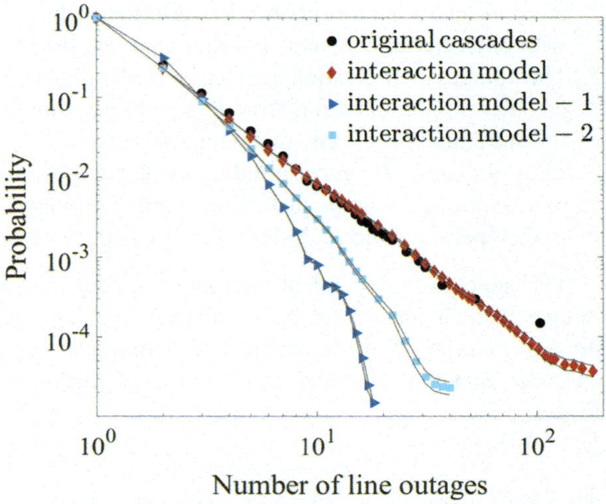

Fig. 3.27 Comparison of the CCDs of the number of line outages from original and simulated cascades. The result for "interaction model" is from using the $\boldsymbol{B}_g$'s estimated based on the two mechanisms in Sect. 3.5.2, while those for "interaction model – 1" and "interaction model – 2" are from using the estimated interaction matrix for the data with all generations and without memory, respectively. The gray lines indicate the 95% confidence interval

3.8.2.2 Comparison of Offspring Mean of Branching Process

The overall offspring mean of branching process (the mean number of child outages generated by each parent outage), $\hat{\lambda}$, can be estimated as [26, 56]

$$\hat{\lambda} = \frac{\sum_{g=1}^{G} Z_g}{\sum_{g=0}^{G} Z_g}, \tag{3.42}$$

where Z_g is the number of outages in generation g. The $\hat{\lambda}$ estimated from the generated $10M$ cascades by the interaction model is 0.252, which is close to that estimated from the original BPA outage data, 0.273.

The mean number of child outages in generation $g+1$ for each parent in generation g is offspring mean λ_g of the branching process. For $g \leq 9$, λ_g is estimated using generations g and $g+1$ as $\hat{\lambda}_g = Z_{g+1}/Z_g$. Similar to [39], for $g \geq 10$ one single offspring mean is estimated as

$$\hat{\lambda}_{10+} = \frac{\sum_{g=11}^{G} Z_g}{\sum_{g=10}^{G-1} Z_g}. \tag{3.43}$$

In Fig. 3.28 it is shown that the estimated $\hat{\lambda}_g$ from the simulated $10M$ cascades using the generation-dependent interaction model matches well with that from the original cascades. Besides, $\hat{\lambda}_g$ increases with the increase of g as the cascade proceeds, which is consistent with the results in [39].

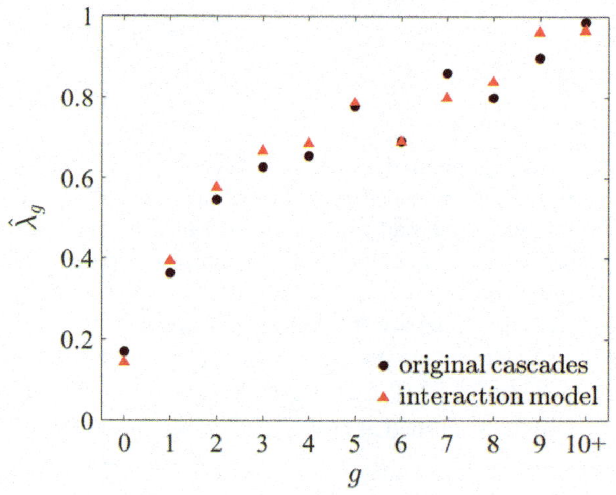

Fig. 3.28 Comparison of the offspring mean of branching process $\hat{\lambda}_g$ estimated from original and simulated cascades

3.8.3 *Coupled Interaction Model*

In [36], a coupled interaction model that utilizes the coupled interaction matrix and the distribution of initial outages is presented to efficiently simulate cascading failures with both line outages and the load shed at buses. It can efficiently generate a large number of cascades to better estimate the blackout size [54]. Moreover, the effect of mitigation schemes can also be efficiently tested by modifying the parameters of the coupled interaction model. The coupled interaction model has the following four steps to generate one cascade.

- **Step 1: Generate initial outages**

  Set $g = 0$. The line outages in generation 0 are independently generated according to their occurrence frequency in generation 0 of the original cascades. Specifically, line i fails in generation 0 by probability

$$\pi_i^{\mathsf{L}} = \frac{\sum_{m=1}^{M_{\mathrm{u}}} \mathbf{1}[i \in \mathcal{S}_{\mathsf{L}}^{m,0}]}{M_{\mathrm{u}}}, \tag{3.44}$$

  where $\mathbf{1}$[event] is an indicator function that evaluates to one if the event happens and evaluates to zero when the event does not happen.

  Similarly, bus u has k units of discretized load shed in generation 0 by probability:

$$\pi_{u,k}^{\mathsf{B}} = \frac{\sum_{m=1}^{M_{\mathrm{u}}} \mathbf{1}[Z_u^{m,0} = k]}{M_{\mathrm{u}}}. \tag{3.45}$$

All generated outages comprise $\mathcal{S}_{\mathsf{L}}^{m,0}$, $\mathcal{S}_{\mathsf{B}}^{m,0}$, and $\boldsymbol{Z}^{m,0}$. Considering the timescale of cascading failures, most cascading failure simulation models do not include any repair process of the failed lines. As each line fails at most once in a simulation, once line i fails the entries in the ith column of $\boldsymbol{B}^{\mathsf{LL}}$ and $\boldsymbol{B}^{\mathsf{BL}}$ are set to be zero.

- **Step 2: Generate further line outages**
  Each outage in $\mathcal{S}_{\mathsf{L}}^{m,g}$ and $\mathcal{S}_{\mathsf{B}}^{m,g}$ independently generates line outages in generation $g+1$. Line j fails in generation $g+1$ following line i outage in $\mathcal{S}_{\mathsf{L}}^{m,g}$ and the load shed at bus u in $\mathcal{S}_{\mathsf{B}}^{m,g}$ with probability b_{ij}^{LL} and b_{uj}^{BL}, respectively. All sampled line outages comprise $\mathcal{S}_{\mathsf{L}}^{m,g+1}$. The columns of $\boldsymbol{B}^{\mathsf{LL}}$ and $\boldsymbol{B}^{\mathsf{BL}}$ that are corresponding to the failed lines are set to be zero.
- **Step 3: Generate further load shed**
  For line $i \in \mathcal{S}_{\mathsf{L}}^{m,g}$, the discretized load shed at bus v follows Poisson distribution with mean b_{iv}^{LB}. For bus $u \in \mathcal{S}_{\mathsf{B}}^{m,g}$, the discretized load shed at bus v follows Poisson distribution with mean $Z_u^{m,g} b_{uv}^{\mathsf{BB}}$. The load shed at bus v are independently sampled for each outage in $\mathcal{S}_{\mathsf{L}}^{m,g}$ and $\mathcal{S}_{\mathsf{B}}^{m,g}$. Since the total load shed at bus v from generation 0 to $g+1$ cannot exceed its total discretized load $\mathrm{Z}_v^{\mathsf{t}}$, the load shed at bus v in generation $g+1$ is recorded as $Z_v^{m,g+1} = \min\{\mathrm{Z}_v^{\mathsf{t}} - \sum_{l=0}^{g} Z_v^{m,l}, \sum_{i \in \mathcal{S}_{\mathsf{L}}^{m,g}} Z_{v\leftarrow i}^{m,g+1} + \sum_{u \in \mathcal{S}_{\mathsf{B}}^{m,g}} Z_{v\leftarrow u}^{m,g+1}\}$, where $Z_{v\leftarrow i}^{m,g+1}$ and $Z_{v\leftarrow u}^{m,g+1}$ are the load shed at bus v in generation $g+1$ generated from line outage i and the load shed at bus u, respectively. All buses with load shed comprise $\mathcal{S}_{\mathsf{B}}^{m,g+1}$ and the corresponding discretized load shed is recorded in $\boldsymbol{Z}^{m,g+1}$.

  The simulation ends if both $\mathcal{S}_{\mathsf{L}}^{m,g+1}$ and $\mathcal{S}_{\mathsf{B}}^{m,g+1}$ are empty. Otherwise, increase g by one and go back to Step 2.

The above steps can be repeated to simulate many cascades for better understanding and mitigating cascading failures. The flow chart is shown in Fig. 3.29 in which $M_{\max}$ cascades are generated.

Compared with detailed cascading failure models, the coupled interaction model is highly probabilistic and is thus much more time-efficient. It has been shown in [36] that a speedup of 9.03 can be achieved by the coupled interaction model for simulating 30,000 cascades.

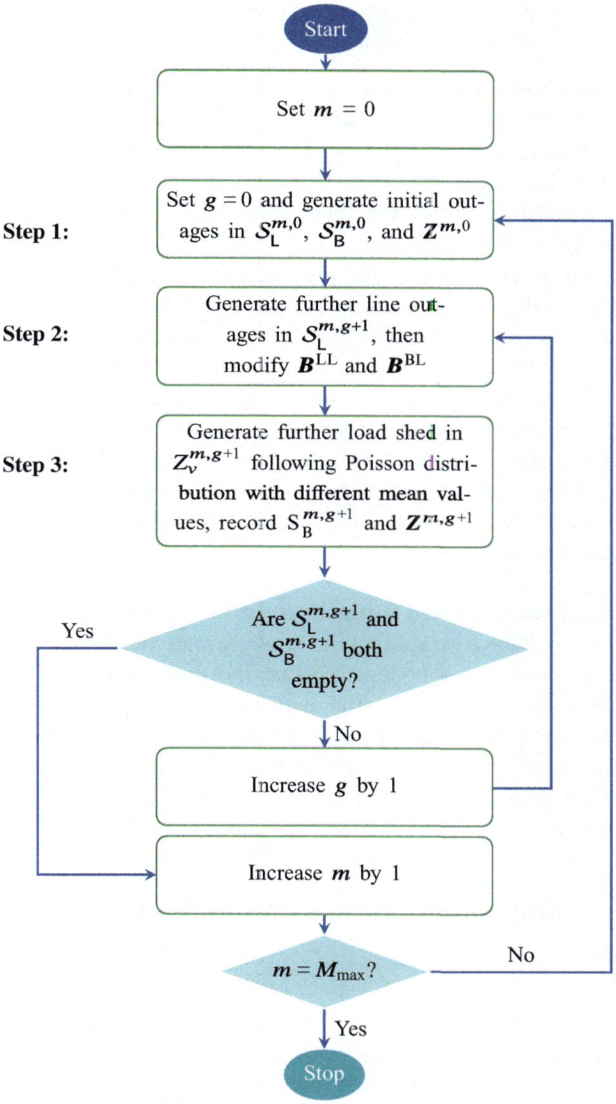

Fig. 3.29 Flow chart of the coupled interaction model

3.9 Cascading Failure Mitigation

Cascading failure mitigation strategies can be designed based on the estimated failure interactions to help reduce the risk of cascading. There are mainly two types of mitigation strategies that have been explored: critical links-based mitigation and critical components-based mitigation.

- **Critical links-based mitigation**: In [29, 33], critical links are identified based on which a mitigation strategy is designed by blocking some specific protective relays. For example, a Zone 3 relay blocking method called adaptive distance relay scheme has been discussed in [57]. The relays are blocked under the condition of the tripping of the lines corresponding to the source vertices of the critical links. Since the critical links can cause a large number of failures and thus play crucial roles in the propagation of cascading failures, it should be beneficial to the overall security of the system to stop the propagation from the source vertices of critical links to the destination vertices by blocking the operation of the relay of the destination vertices, thus securing time for the operators to take remedial actions, such as re-dispatching the generation or even shedding some loads, and finally helping mitigate catastrophic failures. Further in [36] critical links are identified based on the coupled interaction matrix by calculating a comprehensive severity index that considers the consequences of both line outages and the load shed. The critical links-based mitigation strategy can be used for online operation utilizing the identified critical links through offline study.
- **Critical components-based mitigation**: In [35] critical components are identified based on which a mitigation strategy can be designed by upgrading the identified critical components so that the probability that they will fail is significantly reduced. In [37] spatial distance information is further incorporated in the critical component identification to mitigate not only the number of outages but also the spatial propagation of the outages. The critical components-based mitigation can be implemented in the planning stage.

3.9.1 Cascading Failure Mitigation for Utility Line Outage Data

After useful information is extracted from utility outage data, we can utilize it to mitigate cascading failure risks, such as by upgrading the identified critical components so that the probability that they will fail is significantly reduced.

In the generation-dependent interaction model simulations, cascading outage propagation can be mitigated by reducing the elements of the columns of $\boldsymbol{B}_g$, $g = 0, \cdots, G-1$ corresponding to the identified critical components in C_1, C_2, or C_2', each of which contains 20 components, about 3.44% of the 582 total number of components. For example, we can multiply the elements of the columns of $\boldsymbol{B}_g$ corresponding to the ith element in C_1, C_2, or C_2' by $\alpha = i/40$, in which case the higher the ranking of a critical component, the more upgrade is implemented so that the probability that it may fail will reduce more significantly. We simulate $10M$ cascades under different $\boldsymbol{B}_g$'s. The overall offspring mean of the branching process

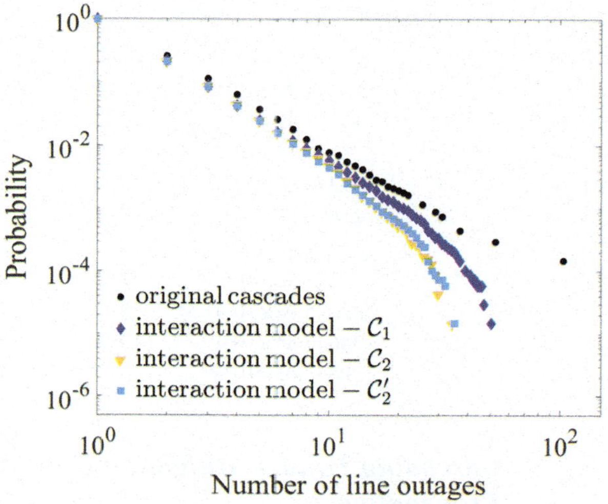

Fig. 3.30 CCDs of the number of line outages under different mitigation strategies

Table 3.4 Probability of different cascade sizes under different mitigation strategies

Mitigation strategy	Small cascade ($Y \leq 10$)	Medium cascade ($10 < Y \leq 50$)	Large cascade ($Y > 50$)
No mitigation	0.9930	0.0067	2.99×10^{-4}
C_1-based	0.9952	0.0048	1.50×10^{-5}
C_2-based	0.9961	0.0039	0
C_2'-based	0.9964	0.0036	0

defined in (3.42) for the cases without mitigation, with C_1-based mitigation, C_2-based mitigation, and C_2'-based mitigation, is, respectively, 0.273, 0.190, 0.184, and 0.179.

As shown in Fig. 3.30, all mitigation measures can reduce the risk of cascading, but the mitigation effect using C_2' is better than that using C_2 or C_1. In order to better quantify the mitigation effect, in Table 3.4 we list the probability for having different cascade sizes (the total number of line outages Y) with or without mitigation strategies. It is clearly seen that under all three mitigation strategies the probabilities for both medium size and large size cascades are reduced. Specifically, the C_1-based mitigation reduces the probability for large cascades to 1.50×10^{-5} from 2.99×10^{-4} for the case without any mitigation, and more impressively the C_2'- and C_2-based mitigation reduces that probability to zero. Furthermore, compared with C_2-based mitigation, C_2'-based mitigation reduces the probability for medium size cascades from 0.0039 to 0.0036.

Table 3.5 Comparison of critical components

Rank	C_e	C_d (I_d^g)	Rank	C_e	C_d (I_d^g)
1	83	24 (1.54×10^4)	11	201	**41** (6.86×10^3)
2	17	126 (1.47×10^4)	12	61	61 (6.48×10^3)
3	234	83 (1.15×10^4)	13	56	**179** (6.48×10^3)
4	24	92 (9.60×10^3)	14	59	**4** (5.99×10^3)
5	76	**116** (9.52×10^3)	15	126	76 (5.98×10^3)
6	2	17 (8.46×10^3)	16	26	42 (5.91×10^3)
7	85	2 (8.20×10^3)	17	92	101 (5.87×10^3)
8	8	234 (8.02×10^3)	18	73	59 (5.70×10^3)
9	101	**446** (7.86×10^3)	19	126	**75** (5.64×10^3)
10	187	**23** (7.02×10^3)	20	42	**13** (5.20×10^3)

3.9.2 Cascading Failure Mitigation Considering Spatial Propagation

Mitigation based on critical components is implemented and the spatial propagation features after mitigation are analyzed. The two sets of critical components, C_e based on the expected number of outages in Sect. 3.6 and C_d based on the total spatial propagation in Sect. 3.7, are listed in Table 3.5. It is seen that the critical components in C_d are different from those in C_e. Compared with C_e, eight new components (highlighted by bold font) are identified as critical components in C_d.

Cascades are generated under the mitigation based on the two critical component sets. Figure 3.31 shows the CCDs of the number of generations of each cascade with and without mitigation. Compared with the case without mitigation, two mitigation strategies can restrict failure propagation by decreasing the number of generations. In Fig. 3.32, the CCDs of $d_{\text{total}}^{(m)}$ with and without mitigation are shown. Note that only $d_{\text{total}}^{(m)} > 129$ (95th percentile of $d_{\text{total}}^{(m)}$) is displayed. Compared with C_e mitigation, the C_d mitigation can more significantly suppress cascading failure spatial propagation.

To illustrate the mitigation effect on $\bar{v}_{g\to g+1}$, this section generates $10M$ cascades and groups the generations following the same rules in Sect. 3.7. Figure 3.33 shows the number of outages in the grouped generations. The C_d mitigation strategy has a similar number of outages in the grouped generations as that under the C_e mitigation that only focuses on the expected number of outages. In Fig. 3.34, the $\bar{v}_{g\to g+1}$ for the grouped generations with and without mitigation is shown. Compared to the case without mitigation, both C_d and C_e mitigation can suppress the spatial propagation of cascading failures. The C_d mitigation has the smallest average spatial propagation velocity, thus validating its effectiveness.

To design the mitigation strategy for reducing the spatial propagation, this section identifies 20 critical components with the highest I_d values, and the set of these critical components is denoted by C_d. Mitigation based on C_d could be implemented

Fig. 3.31 CCDs of the number of generations of each cascade with and without mitigation

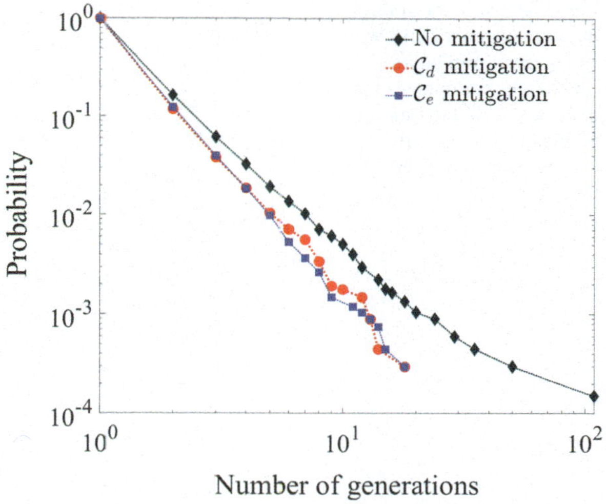

Fig. 3.32 CCDs of large $d_{\text{total}}^{(m)}$'s with and without mitigation

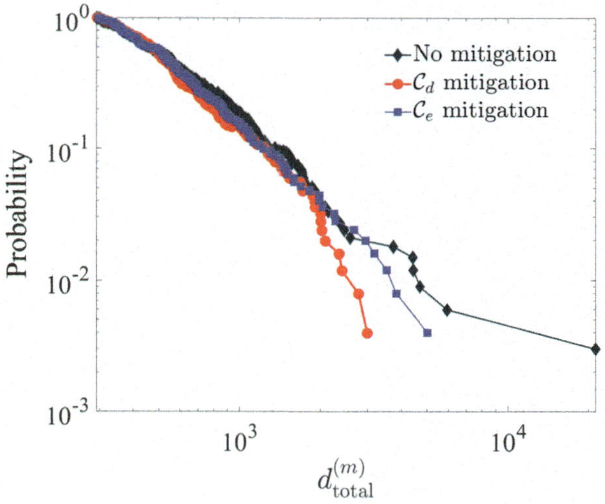

by reducing the failure probabilities of the critical components in $\boldsymbol{B}_g$. Specifically, for the jth component in C_d in descending order of I_a, the corresponding column in $\boldsymbol{B}_g$ is multiplied by $\alpha_j = j/40$. The greater I_d is, the heavier the suppression is on the corresponding component. Then the mitigated interaction matrices $\boldsymbol{B}_g^d$ are used by the generation-dependent interaction model in Fig. 3.26 to generate mitigated cascades. By recalculating the spatial propagation velocities based on the generated cascades, the mitigation effect will be validated.

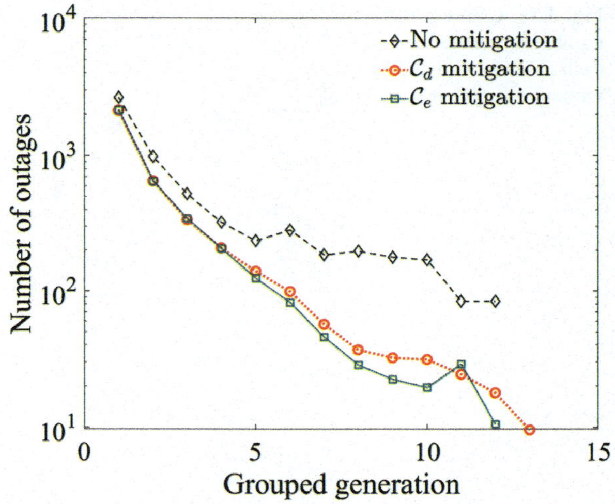

Fig. 3.33 Number of outages in grouped generations with and without mitigation (the number of outages for the cases with mitigation is divided by 10 since $10M$ cascades are generated for these cases)

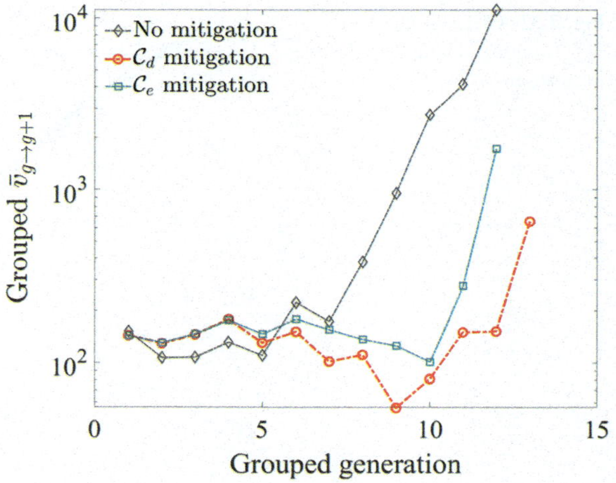

Fig. 3.34 $\bar{v}_{g\to g+1}$ for grouped generations with and without mitigation

3.9.3 *Cascading Failure Mitigation on Coupled Interaction Network*

Similar to the interaction network only for line outages, cascading failure mitigation can be performed based on the coupled interaction network [36].

- Critical links can be identified based on the coupled interaction network, considering the consequences in both line outages and the load shed. The severity of a link $l : S \to T$ is measured by the line outages and load shed propagated through l, where S and T are, respectively, the source outage and target outage and could be a line outage or a bus with load shedding. For each link $l : S \to T$, an acyclic interaction subgraph $\mathcal{G}_l(\mathcal{C}_l, \mathcal{L}_l)$ in which there is a path from vertex T to any other vertex can be obtained from the coupled interaction network. Different types of links ($\mathsf{L} \rightarrow \mathsf{L}$, $\mathsf{L} \rightarrow \mathsf{B}$, $\mathsf{B} \rightarrow \mathsf{L}$, $\mathsf{B} \rightarrow \mathsf{B}$) may play dramatically different roles in outage propagation.

- Mitigation strategies can further be implemented based on the identified critical links to reduce the risk of blackouts. One mitigation strategy can be implemented based on the identified $\mathsf{L} \rightarrow \mathsf{L}$ critical links by blocking Zone 3 relay [29]. For example, for critical link l : line $i \rightarrow$ line j, when line i fails, the relay of line j is blocked to reduce its tripping probability to 10% of its original probability so that the control center could perform remedy control and stop failure propagation.

Appendix: Discretization Unit for Each Load Bus

Let $\tilde{\lambda}_{iv}^{(k+1)}$ and $\tilde{\sigma}_{iv}^{2(k+1)}$ be the mean and the variance of the discretized load shed at bus v following the outage of line i when $\Delta_v^{(k)}$ is chosen as the discretization unit for bus v. If we change the discretization unit for bus v to $\Delta_v^{(k+1)}$, the mean changes to $\lambda_{iv}^{(k+1)} = \tilde{\lambda}_{iv}^{(k+1)}\Delta_v^{(k)}/\Delta_v^{(k+1)}$, and the variance changes to $\sigma_{iv}^{2(k+1)} = \tilde{\sigma}_{iv}^{2(k)}(\Delta_v^{(k)})^2/(\Delta_v^{(k+1)})^2$.

Note that the mean of the Poisson distribution equals its variance. A better choice of $\Delta_v^{(k+1)}$ should make the mean $\lambda_{iv}^{(k+1)}$ close to the variance $\sigma_{iv}^{2(k+1)}$ as much as possible. Moreover, the choice of $\Delta_v^{(k+1)}$ should consider all the lines whose outages have interactions with the load shed at bus v. Thus, to choose the optimal discretization unit $\Delta_v^{(k+1)}$ for bus v, the following optimization problem (3.46) can be solved:

$$\min\ f(\Delta_v^{(k+1)}) = \sum_{i \in \mathcal{S}_v^{\mathsf{LB}}} N_i^{\mathsf{L}} \left(\frac{\lambda_{iv}^{(k+1)}}{\sigma_{iv}^{2(k+1)}} + \frac{\sigma_{iv}^{2(k+1)}}{\lambda_{iv}^{(k+1)}} \right) \tag{3.46}$$

$$\text{s.t.}\quad \lambda_{iv}^{(k+1)} = \frac{\Delta_v^{(k)}}{\Delta_v^{(k+1)}} \tilde{\lambda}_{iv}^{(k+1)} \tag{3.47}$$

$$\sigma_{iv}^{2(k+1)} = \left(\frac{\Delta_v^{(k)}}{\Delta_v^{(k+1)}} \right)^2 \tilde{\sigma}_{iv}^{2(k+1)}. \tag{3.48}$$

It is easy to obtain the optimal solution as

$$\Delta_v^{(k+1)} = \Delta_v^{(k)} \sqrt{\frac{\sum_{i \in \mathcal{S}_v^{\mathsf{LB}}} N_i^{\mathsf{L}} \frac{\tilde{\sigma}_{iv}^{2(k+1)}}{\tilde{\lambda}_{iv}^{(k+1)}}}{\sum_{i \in \mathcal{S}_v^{\mathsf{LB}}} N_i^{\mathsf{L}} \frac{\tilde{\lambda}_{iv}^{(k+1)}}{\tilde{\sigma}_{iv}^{2(k+1)}}}}. \tag{3.49}$$

If we consider $\tilde{\lambda}_{iv}^{(k+1)} \approx \tilde{b}_{iv}^{\mathsf{LB}(k+1)}$ and $\tilde{\sigma}_{iv}^{2(k+1)} \approx S_{iv}^{2(k+1)}$, then $\Delta_v^{(k+1)}$ can be chosen as

$$\Delta_v^{(k+1)} = \Delta_v^{(k)} \sqrt{\frac{\sum_{i \in \mathcal{S}_v^{\mathsf{LB}}} N_i^{\mathsf{L}} \frac{S_{iv}^{2(k+1)}}{\tilde{b}_{iv}^{\mathsf{LB}(k+1)}}}{\sum_{i \in \mathcal{S}_v^{\mathsf{LB}}} N_i^{\mathsf{L}} \frac{\tilde{b}_{iv}^{\mathsf{LB}(k+1)}}{S_{iv}^{2(k+1)}}}}.$$

References

1. K. Sun, Y. Hou, W. Sun, J. Qi, *Power System Control under Cascading Failures: Understanding, Mitigation, and Restoration* (Wiley-IEEE Press, 2019)
2. J. Qi, *Smart Grid Resilience: Extreme Weather, Cyber-Physical Security, and System Interdependency* (Springer Nature, 2023)
3. P. Praks, V. Kopustinskas, M. Masera, Monte-Carlo-based reliability and vulnerability assessment of a natural gas transmission system due to random network component failures. Sustain. Resilient Infrastruct. **2**(3), 97–107 (2017)
4. M. Theoharidou, M. Kandias, D. Gritzalis, Securing transportation-critical infrastructures: Trends and perspectives, in *Global Security, Safety and Sustainability & e-Democracy* (Springer, 2011), pp. 171–178
5. S. Hong, H. Yang, T. Zhao, X. Ma, Epidemic spreading model of complex dynamical network with the heterogeneity of nodes. Int. J. Syst. Sci. **47**(11), 2745–2752 (2016)
6. S.V. Buldyrev, R. Parshani, G. Paul, H.E. Stanley, S. Havlin, Catastrophic cascade of failures in interdependent networks. Nature **464**(7291), 1025–1028 (2010)
7. B. Liscouski, W. Elliot, Final report on the august 14, 2003 blackout in the United States and Canada: Causes and recommendations. A Report to US Department of Energy, vol. 40(4), p. 86 (Apr. 2004)
8. Federal Energy Regulatory Commission, Arizona-Southern California outages on September 8, 2011: Causes and recommendations, *FERC and NERC Staff*, Apr. 2012
9. L.L. Lai, H.T. Zhang, C.S. Lai, F.Y. Xu, S. Mishra, Investigation on July 2012 Indian blackout, in *International Conference on Machine Learning and Cybernetics*, vol. 1, Jul. 2013, pp. 92–97
10. P. Henneaux, E. Ciapessoni, D. Cirio, E. Cotilla-Sanchez, R. Diao, I. Dobson, A. Gaikwad, S. Miller, M. Papic, A. Pitto, et al., Benchmarking quasi-steady state cascading outage analysis methodologies, in *2018 IEEE International Conference on Probabilistic Methods Applied to Power Systems (PMAPS)* (IEEE, Jun. 2018), pp. 1–6
11. C. Grigg, P. Wong, P. Albrecht, R. Allan, M. Bhavaraju, R. Billinton, Q. Chen, C. Fong, S. Haddad, S. Kuruganty, et al., The IEEE reliability test system-1996. a report prepared by the reliability test system task force of the application of probability methods subcommittee. IEEE Trans. Power Syst. **14**(3), 1010–1020 (1999)

12. D.S. Kirschen, D. Jayaweera, D.P. Nedic, R.N. Allan, A probabilistic indicator of system stress. IEEE Trans. Power Syst. **19**(3), 1650–1657 (2004)
13. A. Phadke, J.S. Thorp, Expose hidden failures to prevent cascading outages [in power systems]. IEEE Comput. Appl. Power **9**(3), 20–23 (1996)
14. J. Chen, J.S. Thorp, I. Dobson, Cascading dynamics and mitigation assessment in power system disturbances via a hidden failure model. Int. J. Elec. Power **27**(4), 318–326 (2005)
15. I. Dobson, B.A. Carreras, D.E. Newman, A loading-dependent model of probabilistic cascading failure. Probab. Eng. Inf. Sci. **19**(1), 15–32 (2005)
16. H. Ren, I. Dobson, B.A. Carreras, Long-term effect of the n-1 criterion on cascading line outages in an evolving power transmission grid. IEEE Trans. Power Syst. **23**(3), 1217–1225 (2008)
17. J. Qi, S. Mei, F. Liu, Blackout model considering slow process. IEEE Trans. Power Syst. **28**(3), 3274–3282 (2013)
18. B.A. Carreras, D.E. Newman, I. Dobson, N.S. Degala, Validating OPA with WECC data, in *2013 46th Hawaii International Conference on System Sciences (HICSS)* (IEEE, 2013), pp. 2197–2204
19. S. Mei, Y. Ni, G. Wang, S. Wu, A study of self-organized criticality of power system under cascading failures based on AC-OPF with voltage stability margin. IEEE Trans. Power Syst. **23**(4), 1719–1726 (2008)
20. P. Henneaux, P.-E. Labeau, J.-C. Maun, Blackout probabilistic risk assessment and thermal effects: Impacts of changes in generation. IEEE Trans. Power Syst. **28**(4), 4722–4731 (2013)
21. J. Song, E. Cotilla-Sanchez, G. Ghanavati, P.D. Hines, Dynamic modeling of cascading failure in power systems. IEEE Trans. Power Syst. **31**(3), 2085–2095 (2016)
22. I. Dobson, A. Flueck, S. Aquiles-Perez, S. Abhyankar, J. Qi, Towards incorporating protection and uncertainty into cascading failure simulation and analysis, in *IEEE Int. Conf. Probabilistic Methods Applied to Power Systems (PMAPS)*, Jun. 2018, pp. 1–5
23. J. Qi, S. Pfenninger, Controlling the self-organizing dynamics in a sandpile model on complex networks by failure tolerance. EPL (Europhys. Lett.) **111**(3), 38006 (2015)
24. S.R. Khazeiynasab, J. Qi, Resilience analysis and cascading failure modeling of power systems under extreme temperatures. J. Modern Power Syst. Clean Energy **9**(6), 1446–1457 (2021)
25. J. Kim, K.R. Wierzbicki, I. Dobson, R.C. Hardiman, Estimating propagation and distribution of load shed in simulations of cascading blackouts. IEEE Syst. J. **6**(3), 548–557 (2012)
26. J. Qi, I. Dobson, S. Mei, Towards estimating the statistics of simulated cascades of outages with branching processes. IEEE Trans. Power Syst. **28**(3), 3410–3419 (2013)
27. J. Qi, W. Ju, K. Sun, Estimating the propagation of interdependent cascading outages with multi-type branching processes. IEEE Trans. Power Syst. **32**(2), 1212–1223 (2017)
28. P.D. Hines, I. Dobson, E. Cotilla-Sanchez, M. Eppstein, "dual graph" and "random chemistry" methods for cascading failure analysis, in *2013 46th Hawaii International Conference on System Sciences* (IEEE, Jan. 2013), pp. 2141–2150
29. J. Qi, K. Sun, S. Mei, An interaction model for simulation and mitigation of cascading failures. IEEE Trans. Power Syst. **30**(2), 804–819 (2015)
30. W. Ju, J. Qi, K. Sun, Simulation and analysis of cascading failures on an NPCC power system test bed, in *2015 IEEE Power Energy Society General Meeting*, Jul. 2015, pp. 1–5
31. P.D. Hines, I. Dobson, P. Rezaei, Cascading power outages propagate locally in an influence graph that is not the actual grid topology. IEEE Trans. Power Syst. **32**(2), 958–967 (2017)
32. W. Ju, K. Sun, J. Qi, Multi-layer interaction graph for analysis and mitigation of cascading outages. IEEE J. Emerg. Sel. Top. Circuits Syst. **7**(2), 239–249 (2017)
33. J. Qi, J. Wang, K. Sun, Efficient estimation of component interactions for cascading failure analysis by EM algorithm. IEEE Trans. Power Syst. **33**(3), 3153–3161 (2018)
34. K. Zhou, I. Dobson, Z. Wang, A. Roitershtein, A.P. Ghosh, A Markovian influence graph formed from utility line outage data to mitigate large cascades. IEEE Trans. Power Syst. **35**(4), 3224–3235 (2020)
35. J. Qi, Utility outage data driven interaction networks for cascading failure analysis and mitigation. IEEE Trans. Power Syst. **36**(2), 1409–1418 (2021)

36. L. Wang, J. Qi, B. Hu, K. Xie, A coupled interaction model for simulation and mitigation of interdependent cascading outages. IEEE Trans. Power Syst. **36**(5), 4331–4342 (2021)
37. S. Huang, J. Qi, Analysis and mitigation of cascading failure spatial propagation in real utility outage data, in *2023 IEEE Power & Energy Society General Meeting (PESGM)* (IEEE, Jul. 2023), pp. 1–5
38. J. Bialek, E. Ciapessoni, D. Cirio, E. Cotilla-Sanchez, C. Dent, I. Dobson, P. Henneaux, P. Hines, J. Jardim, S. Miller, et al., Benchmarking and validation of cascading failure analysis tools. IEEE Trans. Power Syst. **31**(6), 4887–4900 (2016)
39. I. Dobson, Estimating the propagation and extent of cascading line outages from utility data with a branching process. IEEE Trans. Power Syst. **27**(4), 2146–2155 (2012)
40. B.A. Carreras, D.E. Newman, I. Dobson, North American blackout time series statistics and implications for blackout risk. IEEE Trans. Power Syst. **31**(6), 4406–4414 (2016)
41. I. Dobson, B.A. Carreras, D.E. Newman, J.M. Reynolds-Barredo, Obtaining statistics of cascading line outages spreading in an electric transmission network from standard utility data. IEEE Trans. Power Syst. **31**(6), 4831–4841 (2016)
42. Bonneville Power Administration Transmission Services Operations & Reliability website. [Online]. Available: http://transmission.bpa.gov/Business/Operations/Outages
43. North American Electric Reliability Corporation, Reliability terminology, Aug. 2013. [Online]. Available: http://www.nerc.com/AboutNERC/Documents/Terms%20AUG13.pdf
44. L. Wang, J. Qi, Optimal decomposition of utility outage sequence for cascading failure interaction estimation, in *2022 17th International Conference on Probabilistic Methods Applied to Power Systems (PMAPS)* (IEEE, 2022), pp. 1–6
45. A. Shandilya, H. Gupta, J. Sharma, Method for generation rescheduling and load shedding to alleviate line overloads using local optimisation. IEE Proc. C Gener. Transm. Distrib. **140**(5), 337–342 (1993)
46. D. Novosel, R.L. King, Using artificial neural networks for load shedding to alleviate overloaded lines. IEEE Trans. Power Deliv. **9**(1), 425–433 (1994)
47. A.P. Dempster, N.M. Laird, D.B. Rubin, Maximum likelihood from incomplete data via the EM algorithm. J. R. Stat. Soc. Ser. B Stat. Methodol. **39**(1), 1–22 (1977)
48. C.B. Do, S. Batzoglou, What is the expectation maximization algorithm? Nature Biotechnol. **26**(8), 897 (2008)
49. C.T. Kelley, *Iterative Methods for Optimization* (SIAM, 1999)
50. Power Systems Test Case Archive. Univ. Washington. [Online]. Available: http://www.ee.washington.edu/research/pstca/
51. I. Dobson, B. Carreras, V. Lynch, D. Newman, An initial model for complex dynamics in electric power system blackouts, in *Proc. 34th Hawaii Int. Conf. System Sciences*, 2001, pp. 710–718
52. S. Huang, J. Qi, Learning cascading failure interactions by deep convolutional generative adversarial network, in *2022 IEEE International Conference on Communications, Control, and Computing Technologies for Smart Grids (SmartGridComm)* (IEEE, Oct. 2022), pp. 21–26
53. J. Zhao, D. Li, H. Sanhedrai, R. Cohen, S. Havlin, Spatio-temporal propagation of cascading overload failures in spatially embedded networks. Nature Commun. **7**, 10094 (2016)
54. M. Vaiman, K. Bell, Y. Chen, B. Chowdhury, I. Dobson, P. Hines, M. Papic, S. Miller, P. Zhang, Risk assessment of cascading outages: Methodologies and challenges. IEEE Trans. Power Syst. **27**(2), 631 (2012)
55. V.J. Easton, J.H. McColl, Statistics glossary (2002)
56. I. Dobson, J. Kim, K.R. Wierzbicki, Testing branching process estimators of cascading failure with data from a simulation of transmission line outages. Risk Anal. **30**(4), 650–662 (2010)
57. S.-I. Lim, C.-C. Liu, S.-J. Lee, M.-S. Choi, S.-J. Rim, Blocking of zone 3 relays to prevent cascaded events. IEEE Trans. Power Syst. **23**(2), 747–754 (2008)

Chapter 4
Probabilistic Analytics of Cascading Failures: Modeling, Assessment, and Application

Qinfei Long, Jinpeng Guo, Yunhe Hou, and Feng Liu

Nomenclature

MCS	Monte Carlo simulation.
IS	Importance sampling.
SIS	Sequential importance sampling.
CFP	Component failure probability.
BR	Blackout risk.
CoFPF	Component failure probability function.
DTR	Dynamic thermal rating.
i.i.d	Independently identically distributed.
x_j^i	System state at the j-th stage of the i-th sampling of a cascading failure.
y_M^i	Load shedding.
$\mu(A)$	Estimation of true probability.
$\delta_{\{y_M^i \geq Y_0\}}$	The indicator function of set $\{\cdot\}$, indicating whether $y_M^i \geq Y_0$.
$Risk\,(Y_0)$	Risk estimation.
$X_t := (U_t, V_t)$	State variable of the cascading failure at time t, where U_t and V_t are discrete variables and continuous variables of the cascading failure at time t, respectively.
Z	A subsequence of the state X_t.
π_j	A transition-probability-like function.

Q. Long · Y. Hou
Department of Electrical and Electronic Engineering, University of Hong Kong, Hong Kong, China
HKU Shenzhen Institute of Research and Innovation, Shenzhen, China
e-mail: qflong@eee.hku.hk; yhhou@eee.hku.hk

J. Guo · F. Liu (✉)
Department of Electrical Engineering, Tsinghua University, Beijing, China
e-mail: guojinpeng2009@163.com; lfeng@tsinghua.edu.cn

K. Sun (ed.), *Cascading Failures in Power Grids*, Power Electronics and Power Systems, https://doi.org/10.1007/978-3-031-48000-3_4

ϕ_j	A deterministic surjection.
φ_k	Fault probability function of component k.
$w(Z)$	Important sampling weight between $f(Z)$ and $g(Z)$.
$D(A)$	Estimation variance.
η	SIS amplification factor of the component's fault probability.
$R_g(Y_0)$	Blackout risk with respect to $g(Z)$ and Y_0.
φ_m	CoFPF of component m.
M	Failure database, where M^k is k-th cascading failure chain.
S_{f_i}	Line set that functioned normal in FG_{i-1}^k but failed in FG_i^k.
FG_i^k	System line state of i-th failure generation in M^k.
$\overline{S_{f_i}}$	Line set functioned normal in FG_i^k.
$\overline{S_{f_i}^{\text{non}}}$	No-DTR line set that functioned normal in FG_{i-1}^k but failed in FG_i^k.
$S_{f_i}^{\text{non}}$	No-DTR line set functioned normal in FG_i^k.
S_{L}	Line set.
S_{A}^{surv}	Survived lines in updated set S_{A} when failure ends.
S_{A}^{fail}	Failed lines in updated set S_{A} when failure ends.
BPI	Braess paradox indicator (the smaller the better).
ε	Absolute error of $RiskW_{\emptyset}$.
$\overline{F}(S_{\text{A}})$	Realistic value of $F(S_{\text{A}})$.
$\overline{RiskW_{\text{A}}}$	Realistic value of $RiskW_{\text{A}}$.
$WindP$	Integrated wind power (the larger, the better).
$RiskW_{\text{A}}$	Risk of whole failure chain with DTR placed in S_{A} (the smaller the better).
$P_e/P_e^{min}/P_e^{max}$	Current/minimum/maximum power flow in line e.
$\text{Pr}_e^{min}/\text{Pr}_e^{max}$	Line e's minimum/maximum of failure probability.
Y_{ext}	Load loss threshold.
μ	Approximate factor to avoid vanishing gradient.
d_e	Generation that line e fails.
η	Adjustment multiplier in submodular function.
γ	Adjustment factor in submodular function.
d_{max}	Max generation among chains in M.
D	Number of unidentical chains in M.
α	DTR improved parameter for power transmission threshold, related to ambient weather.
$\varphi_e(FG_i^k)$	Line e's original failure probability in FG_i^k.
f^k	Original probability of a failure chain M^k.
g_{A}^k	Updated probability of a failure chain M^k with DTR placed in S_{A}.
$f(FG^k{}_i \mid FG^k{}_{i-1})$	Transition probability from failure generation $i-1$ to i in M^k.
$Risk_{\text{A}}^k$	Risk of failure chain M^k with DTR placed in S_{A}.

4.1 Introduction

While rare in power grids, cascading failures draw significant attention due to their potentially catastrophic consequences and far-reaching impact [1–3]. Given the inherently unpredictable nature of cascading failures, statistical and probabilistic analyses have become critical mathematical tools for their study based on historical data [4, 5]. These probabilistic models, such as the Markov chain model, CASCADE model, and branching process model [6], serve to mathematically abstract the complex physical processes underlying cascading failures and estimate the probability distribution of failure scale using statistical inference theory and sampling data. This approach also allows researchers to investigate cascading failures' macro characteristics and uncertain factors. In this section, we will introduce probabilistic analytics in cascading failure analyses.

4.1.1 Probabilistic Models for Cascading Failure

As a result of the stochastic nature of cascading failures, probability analyses are widely deployed as fundamental mathematic tools for examining cascading failures based on historical data [4, 7–9]. This probabilistic model usually abstracts the complex physical process of cascading failure mathematically. It estimates the probability distribution of the failure scale based on statistical inference theory with sampling data to analyze the macrocharacteristics and uncertain factors of cascading failures [10]. However, it can be quite challenging to obtain precise and sufficient information in reality, given the infrequency of blackouts and the paucity of recorded data available to researchers thus far. Various statistical models have been proposed to analyze cascading failures, including the Markov Chain Model, CASCADE model, branching process model, and so on [6]. A concise overview of these probabilistic models for cascading failure analysis is presented below.

4.1.1.1 Markov Chain Models

Markov chain is a stochastic process that describes a sequence of possible events with transition probability that depends only on the previous state [11]. In cascading failure application, it usually assumes that the current state of failure evolution depends only on the previous state, leading to various cascading failure evolution routes with different transition probabilities [12, 13]. This model's outcome can be used to assess the overall probabilities of all states that depict the cascading failure. Specifically, a sensitivity model based upon the Markov chain in cascading failure has been established [12], and the hidden Markov model is also employed to infer a probabilistic estimation of the transmission line times series status in the failure process. Researches [8, 14–16] utilized the cascading failure data to construct the Markovian influence graph, which captures the hidden influence of interactions on cascading failure risk.

4.1.1.2 CASCADE Model

The CASCADE model is an analytically tractable model used to capture the defining characteristics of large blackouts by utilizing the load distribution following failure event [17]. The subsequent generation of failure evolution is determined by detecting overloading lines in the cascading failure process, and the cascading failure ceases when no line loads exceed the threshold value. In [17], the CASCADE model describes the blackout size distribution and the impact of loading during the failure process, demonstrating a power law distribution of failed component numbers. The same group also applied the CASCADE model to analyze the self-organized criticality and find the load-dependent critical point [18]. Reference [19] derived the CASCADE model parameters to predict the probability distribution of the number of motors and assess the risk of large numbers of stalled motors. Researchers in [20] employed the CASCADE model to formulate the system reliability indices, facilitating the decision-making on the stopping actions of cascading failures.

4.1.1.3 Branching Process Model

The branching process constitutes a stochastic process encompassing collections of independent and identically distributed random variables, systematically indexed by the natural numbers [21]. During the branching process, the components in the current generation will generate a random number of components in the succeeding generation, which are based on a specific probability distribution. Its application can be observed in the surnames spread in genealogy [22] and the neutrons propagation in a nuclear reactor [23]. In research works [1, 24, 25], the branching process model was earlier introduced for cascading failure analysis. In detail, with a fixed number of initial failures, the failure number in the next generation will be produced by certain probability distributions, such as Poisson distribution [24] or Borel distribution [1], and the evolution of branching process will continue until it reaches the probability of ultimate extinction. The results in [1, 24, 25] illustrate the relationship between the power law distribution of failure numbers and the branching process parameter, indicating that the branching process model can reflect the cascading failure scale in reality.

4.1.2 Sampling Techniques in Probabilistic Models

It can be concluded that no existing probabilistic models can capture all the mechanisms during the cascading failure, while each model has its specific concentration and application. To sum up, some probabilistic models only consider the most possible uncertainties but do not simulate the dynamics of power systems [2]. Some models aim to capture the macroscopic features of the overall system in the sense of statistics while omitting the details of the cascading outage process. To achieve a closer sight into cascading outages, researchers, however, consider picking up such details back, including the uncertain occurrence of initial disturbances, the action of protection, and the dispatch of the control center. This consequentially results in different blackout models, such as hidden failure model [26, 27], ORNL-PSerc-Alaska (OPA) model [28], and AC-OPA model [29]. As this kind of approach is capable of analyzing the cascading outage process in a more detailed way, it is expected to exploit mechanisms behind cascading outages by carrying massive simulations on these models [10]. These simulation processes can be categorized as sampling techniques, a method of selecting individual members or a subset of the population to draw statistical inferences and estimate the characteristics of the entire population [30]. Some of them are listed below.

4.1.2.1 Monte Carlo Simulation

Regarding every simulation as an independent identically distributed (i.i.d.) sample, the simulation-based cascading failure analysis is essentially a statistic analysis based on a sample set produced by Monte Carlo simulation (MCS) [31]. However, the intrinsic deficiency of the MCS seriously limits its practicability and deployments. The main obstacle stems from the notorious "curse of computational dimensionality" [32]. It is recognized that a realistic large-scale power system is always composed of numerous components, such as transmission lines, transformers, and generators. The possible propagation paths of cascading outages diverge dramatically [33]. Hence, a specific cascading outage with severe consequences is indeed a rare event. In this context, the MCS analysis turns out to be computationally intractable as a vast number of simulations are required to achieve a reliable estimation of the probability distribution of cascading outages. Empirical results also confirm that the estimation variance can remain unacceptably large even if thousands of simulations have been conducted for a system with only tens of buses [34].

4.1.2.2 Splitting Method

In the literature, the splitting method is recognized as one of the most effective methods to reduce the variance of cascading blackout simulations [35–37]. The basic idea of the splitting method is to divide the simulation process into several sub-levels according to the importance function and create separate copies at the beginning of each level so that more simulations can be conducted near the rare events and the estimation variance can be reduced therein. Research [35–37] apply the splitting method to estimate the probability distribution of blackout size efficiently. Specifically, [35] provide an optimization-based method to decide the number of splitting levels. Shortle [36] examines the impact of various factors within the splitting method on simulation efficiency, and [37] study the optimal choice of sample size. While the splitting method can significantly improve computational efficiency and reduce estimation variance, it inherently depends on certain posterior information, such as the probability of blackouts and the simulation cost at each level. Such information is usually estimated by conducting pre-simulations. In addition, it needs substantial storage resources to store separate copies of simulations at each level and hence requires additional time to write into or read from the storage. As an efficient simulation method, the splitting method has also been deployed for other purposes, such as reliability evaluation [38].

4.1.2.3 Importance Sampling

Importance sampling (IS) is an effective method to improve the efficiency of the MCS, which approximates the targeted distribution by a weighted average of random draws from another distribution [39, 40]. This technique has already been successfully deployed in various fields, including power systems, such as reliability evaluation of composite power systems [41] and risk management in the electricity market [42]. It has also been intuitively used in cascading outage simulations in a heuristic manner [43, 44]. Nevertheless, the absence of a solid mathematical formulation of cascading outages makes it challenging to conduct the analytic investigation rigorously. And it is unknown what are the scope and the conditions of applying the IS strategy in cascading outage simulations and how to set the parameters of the IS in simulations. Similar to MCS, IS also encounters the problem of estimation variance reaching intolerably large values. Designing an appropriate biasing strategy in the IS method is beneficial to solve the variance problem. Nevertheless, how to design an effective biasing scheme for practical applications, such as simulating long-chain cascading failures, is yet another challenge, which will be discussed later.

4.1.2.4 Sequential Importance Sampling

Sequential importance sampling (SIS) is an extension of the IS method, which decomposes IS into a sequence of consecutive sampling steps when dealing with a multi-period random process [39, 45]. Reference [46] leverages this idea to analyze the security of the electricity markets by sampling the statuses of the electricity market and the related statuses of market participants sequentially. Research [47] proposes a cross-entropy-based SIS method for evaluating the short-term reliability of the composite system. The cross-entropy-based approach can help to determine the parameters during the sampling on the transition paths of the system status and improve the analysis efficiency notably. Moreover, [48] show that SIS effectively reduces scenarios in co-planning a combined electricity and gas market.

4.1.3 Applications of Cascading Failure Probabilistic Model

A critical goal of investigating the probabilistic model of cascading failure is to assist operators in designing strategies to prevent or decrease the likelihood of cascading failure, ensuring the safe operation of the power system. The probabilistic model enables implementing and comparing various cascading failure aversion strategies in a unified framework. The following are some such applications.

4.1.3.1 Critical Components Selection

Critical components or initial fault identification represents a unique classification of research in the context of cascading failure analysis. This type of research focuses on efficiently identifying those components or fault sets that contribute the most to cascading failures [49, 50]. As a result, the analysis results can provide a reference for selecting critical components. In [51], researchers utilized a simulation model to establish the relationship between the failure chain path and system status, while reference [52] focused on approximating the correlation between the initial state and final outage risk using a Markov tree model and a sensitivity analysis tool. Additionally, research [53] proposed a stochastic "Random Chemistry" algorithm to identify extensive collections of multiple $N - K$ contingencies that initiate severe cascading failures in a simulated power system. Study [14] put forth a PageRank-based fast screening method that was derived to determine vulnerable lines, while [8] establish an influence graph via utility outage data to identify the lines most involved in large cascades.

4.1.3.2 Risk Control

The risk control of failure propagation is among the most critical applications in practical cascading failure research. The implemented risk control strategies include $N - K$ safety constraints, low-frequency and low-voltage load shedding, active splitting actions, and so forth [6]. These strategies have played a vital role in power system safety operations. However, their effectiveness remains unsatisfactory in various scenarios involving complex cascading failures with high uncertainties. In literature [54], it thoroughly analyzed and compared the long-term effects of the $N - 1$ criterion on the probability distribution of outage size and the grid utilization, which can remarkably affect the risk estimation. Research [9] designed a statistics-based concept, waiting time between successive failure generations, to establish a risk control strategy. Moreover, study [55] introduced the CPINDEX, a security-oriented stochastic risk management technique that computes cyber-physical security metrics to measure the underlying cyber-physical environment's security level. The objective of [56] is to compare two established cascade failure models, i.e., the Manchester model and the OPA model, to determine their consistency with risk-based decision-making in the presence of cascading failure.

4.1.4 Summary

This part comprehensively overviews probability analytics research and practice in cascading failure. It elucidates some typical probabilistic models for cascading failures and introduces related sampling techniques applied. Moreover, it highlights potential applications of the cascading failure probabilistic model. Nevertheless, despite significant interest in cascading failure probabilistic models due to the intricate nature of cascading failures, there are still some challenges in both theory and practical application. These challenges include (1) an incomplete basic model and theory, (2) a lack of practical and efficient outage probability or risk assessment method, and (3) difficulties in risk control of cascading failure.

Therefore, this chapter is motivated by the desire to establish a rigorous mathematical framework for probabilistic analysis of cascading failures and offer tools for designing effective mitigation strategies that ensure the safe operation of power systems considering cascading failures.

4.2 Probabilistic Modeling of Cascading Failures

This subsection addresses how to improve computational efficiency and estimation reliability in cascading failure analysis. Using the Markov property of cascading failures, a cascading failure can be formulated as a Markov process with specific state space and transition probabilities, providing a suitable mathematical framework for probabilistic and statistical analysis of cascading failures. Then, a sequential importance sampling (SIS)-based strategy is derived for cascading failure simulations and blackout risk analysis with a solid theoretical foundation. Numerical experiments manifest that the proposed SIS strategy can significantly enhance the efficiency of simulations and reduce the estimation variance of blackout probability/risk compared with the traditional Monte Carlo simulation strategy.

4.2.1 Monte Carlo Simulations for Cascading Failure Analysis

In cascading failure analysis, load shedding is usually adopted to evaluate the severity of cascading outages. To characterize the load shedding distribution due to cascading failures, various blackout models are built by emulating the cascading failure process in a "*descriptive*" way. Massive simulations on such models can provide a number of i.i.d. samples for statistical analyses. This approach is essentially based on the Monte Carlo simulation (MCS) if one regards each simulation as a sample. Though many blackout models exist, the simulation principles are similar. In simple terms, at the j-th stage of the i-th sampling of a cascading failure, the blackout model for simulation determines the outage probability of each component in the system at the stage, depending on the system states x_j^i and other related factors, such as weather and maintenance conditions. Then the outage components at stage j are sampled, and the system state x_j^i transits to x_{j+1}^i. Repeating the above steps until no new outages occur, one simulation run is completed. It generates a sample of load shedding, Y, denoted by y_M^i. Here, the subscript M stands for samples given by the MCS. The load shedding, Y, is a random variable that will be strictly defined later.

Define the sample set as $Y_M := \{y_M^i, i = 1, \cdots, N_M\}$ obtained after N_M simulations. Then, the probability distribution of load shedding can be estimated by statistics based on Y_M. We are interested in the probability of a given event A, which means the load shedding is greater than a certain level Y_0. The unbiased estimation of true probability $\mu(A)$ is given by

$$\tilde{\mu}(A) = \frac{1}{N_M} \sum_{i=1}^{N_M} \delta_{\{y_M^i \geq Y_0\}}, \tag{4.1}$$

where $\delta_{\{\cdot\}}$ is the indicator function of set $\{\cdot\}$, which means $\delta_{\{y_M^i \geq Y_0\}} = 1$ if $y_M^i \geq Y_0$; otherwise, $\delta_{\{y_M^i \geq Y_0\}} = 0$. It is easy to see $\delta_{\{y_M^i \geq Y_0\}} = \delta^2_{\{y_M^i \geq Y_0\}}$.

The estimation variance on N_M samples is given by

$$\sigma^2(A) = D(A) = \frac{1}{N_M}(\mu(A)(1 - \mu(A))). \tag{4.2}$$

In addition to the probability distribution of load shedding, the blackout risk of cascading failures is also of interest. Theoretically, the blackout risk of a power system can be defined as the expectation of load shedding greater than the given level, Y_0. That is

$$Risk\,(Y_0) = \mathbb{E}\left(Y \cdot \delta_{\{Y \geq Y_0\}}\right). \tag{4.3}$$

Similar to probability estimation, it can be estimated by

$$\tilde{Risk}\,(Y_0) = \frac{1}{N_M} \sum_{i=1}^{N_M} y_M^i \cdot \delta_{\{y_M^i \geq Y_0\}}. \tag{4.4}$$

The definition of the blackout risk in (4.3) represents the risk of cascading failures with severe consequences. It is closely related to the well-known risk measures, value at risk (*VaR*) and conditional value at risk (*CVaR*). Actually, the risk measure, *Risk*, defined in (4.3) is $CVaR_\alpha$ times $(1 - \alpha)$, provided VaR_α is known as the risk associated to the given load shedding level Y_0 with a confidence level of α. Particularly when $Y_0 = 0$, (4.3) is simply the expectation of load shedding, frequently used to evaluate risk.

With the sample set obtained by repeatedly carrying out simulations on the cascading failure model, the probability of load shedding and the blackout risk can be estimated by using (4.1) and (4.4), respectively. However, it should be noted that, to achieve acceptably small estimation variances, a tremendous number of samples are usually required, even if the system merely has tens or hundreds of buses.

4.2.2 *Markov-Sequence-Based Cascading Failure Analysis*

A cascading failure is always triggered by one or several initial disturbances or component-wise failures. Consequently, the protection devices and/or the control center take actions to trip the failed components, and then the system state changes sequentially according to these actions. In the model, aggregate state variables are used as state variables, such as the number of failed lines, the maximum capacities of all fallen lines, etc. Meanwhile, state transition rates are formulated in terms of specific parameters. In this part, the cascading failure is formulated more rigorously. To this end, discrete and continuous variables are first considered in modeling cascading failures. With pragmatic assumptions, it will be shown how the physic process can be abstracted and strictly formulated as a Markov sequence.

4.2.2.1 Mathematical Formulation

Noting that both discrete and continuous states are involved during a cascading process, the propagation of cascading failures is essentially a dynamic process over the time horizon. Therefore, a cascading failure can be abstractly represented by $\{X_t, t \geq 0\}$, where t is the time variable and X_t is the state variable of the cascading failure at time t. For simplicity, assume the initial state, X_0, is determinate.

To distinguish the continuous and discrete variables, we further define $X_t := (U_t, V_t)$, where U_t and V_t as discrete variables and continuous variables of the cascading failure at time t, respectively. U_t represents (controlled or uncontrolled) noncontinuous changes, such as line tripping, shunt capacitor switching, and On-Load Tap Changer (OLTC) regulation. In contrast, V_t represents continuous quantities, such as power flows, active/reactive power of generators, trajectories after a perturbation, etc. In this part, we only consider the randomness in U_t, which may be caused by many practical factors (e.g., hidden failures) and usually promotes the propagation of cascading failures.

Since the random changes of discrete variables happen sequentially during a cascading process, there exists a (discrete) subsequence of times t, denoted by $\{t_1, \ldots, t_j, \ldots, t_n, n \in \mathbb{Z}^+\}$ at which the discrete variable U_t changes randomly. In this context, the time horizon can be divided into a series of time intervals. During

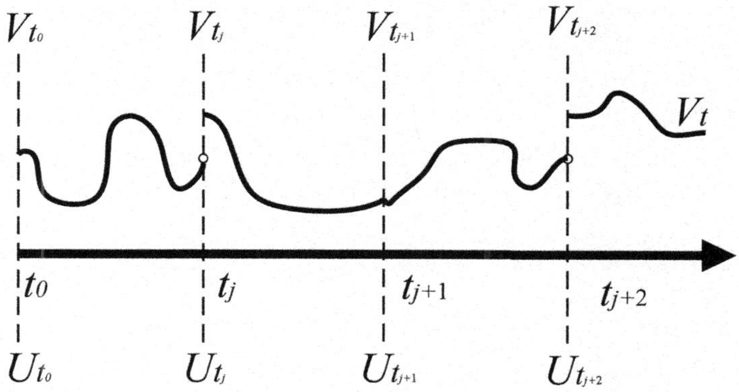

Fig. 4.1 The propagation of cascading failures

each interval $[t_j, t_{j+1})$, U_t stays unchanged, while the continuous variable V_t either remains constant or varies continuously, up to specific system characteristics. The propagation of a cascading failure can be depicted in Fig. 4.1.

As shown in Fig. 4.1, V_t changes continuously (or remains unchanged) from time t_j until time t_{j+1} when the discrete variable U_{t_j} jumps to $U_{t_{j+1}}$ (due to relay actions or human manipulations, for instance, a line being tripped by overload protection). And the new discrete variable $U_{t_{j+1}}$ will further determine a new initial state, $V_{t_{j+1}}$, of the continuous variable V_t for the next interval $[t_{j+1}, t_{j+2})$. This process repeats until the cascading failure terminates.

Note that in cascading failure analysis, one mainly cares about the paths of cascading events (random changes of U_t) and the consequential load shedding caused by these paths. However, different mechanisms between two successive events lead to different events, wherein the inclusion of continuous variables complicates the associated analysis. For example, a line tripping by overload protection depends on the continuous evolution of line flow over time. But when assuming the relays operate independently, the state transition probability becomes a function of the line transition rate [57]. Following this idea, the cascading failure can be represented abstractly as a subsequence of the state X_t with the initial time $t_0 = 0$. This representation takes into account the event path that primarily concentrates on the variables in each generation, which is

$$\begin{aligned} Z &:= \left\{X_{t_0}, \ldots, X_{t_j}, \ldots, X_{t_n}\right\} \\ &= \left\{\left(U_{t_0}, V_{t_0}\right), \ldots, \left(U_{t_j}, V_{t_j}\right), \ldots, \left(U_{t_n}, V_{t_n}\right)\right\}. \end{aligned}$$

Denote the state spaces of U_{t_j} and V_{t_j} as $\mathcal{U}_j$ and $\mathcal{V}_j$, respectively, representing the set of all possible states of U_{t_j} and V_{t_j} at stage j of the cascading failure. Denote $|\cdot|$ as the cardinality of a set. Then, it holds that $|\mathcal{V}_0|$ is finite, provided a finite number of initial failures are considered. Moreover, $|\mathcal{U}_j|$ for arbitrary finite stage j

must be finite as both the number of the components and the discrete states of each component are finite. However, $|\mathcal{V}_j|$ may not be the case because V_{t_j} is determined by continuous states. We will show that it is finite with the following two pragmatic assumptions.

A1 The probability function of $U_{t_{j+1}}$, denoted by π_j, is only determined by $X_{t_j} = (U_{t_j}, V_{t_j})$ and independent of the historical $X_{t_0}, \ldots, X_{t_{j-1}}$. More precisely,

$$\Pr\left(U_{t_{j+1}} = u_{t_{j+1}} \mid U_{t_j} = u_{t_j}, V_{t_j} = v_{t_j}\right) = \pi_j\left(u_{t_{j+1}}, u_{t_j}, v_{t_j}\right). \tag{4.5}$$

Here, π_j severs as a transition-probability-like function.

A2 $V_{t_{j+1}}$ is only determined by V_{t_j} and $U_{t_{j+1}}$, denoted by

$$V_{t_{j+1}} = \phi_j\left(U_{t_{j+1}}, V_{t_j}\right). \tag{4.6}$$

Here, the mapping $\phi_j : \mathcal{U}_{j+1} \times \mathcal{V}_j \longrightarrow \mathcal{V}_{j+1}$ is a *deterministic* surjection.

The two assumptions above assume that the subsequence Z has a Markov property, which is pragmatic and has been widely used in various blackout models [26, 28, 29]. They can be understood according to physical reality. For Assumption A1, given a discrete state U_{t_j} at time t_j, if the continuous state V_{t_j} (e.g., the power flow) exceeds some preset thresholds, the protection or controller will be triggered to change the discrete state U_{t_j} to $U_{t_{j+1}}$ (up to V_{t_j}). Since no randomness except the change of U_{t_j} is involved in this process, t_{j+1} is uniquely determined. Moreover, to incorporate the random change of U_{t_j}, π_j is set as a probability function independent of t_{j+1}. Afterward, the continuous regulation will accordingly act to drive the continuous state from V_{t_j} to $V_{t_{j+1}}$. This process is deterministically decided by ϕ_j, as stated in Assumption A2. Note that $|\mathcal{U}_j|$ and $|\mathcal{V}_j|$ represent the cardinality of the sets at stage j, whereas U_{t_j} and V_{t_j} signify the values of variable.

With the assumptions mentioned above, one essential property can be derived as follows.

Property 1 $|\mathcal{V}_{j+1}|$ is finite for any finite $j \in \mathbb{N}$.

To show Property 1, which implies the reduction of the state space of continuous variables, we consider the image domain of the mapping ϕ_j. As ϕ_j is a deterministic surjection as assumed in assumption A2, $V_{t_{j+1}}$ is uniquely determined by $\phi_j\left(U_{t_{j+1}}, V_{t_j}\right)$. Therefore, there is

$$\begin{aligned}
\left|\mathcal{V}_{j+1}\right| &\leq \left|\mathcal{V}_j\right| \times \left|\mathcal{U}_{j+1}\right| \\
&\leq \left|\mathcal{V}_{j-1}\right| \times \left|\mathcal{U}_j\right| \times \left|\mathcal{U}_{j+1}\right| \\
&\leq \cdots \\
&\leq |\mathcal{V}_0| \times \prod_{s=0}^{j+1} |\mathcal{U}_s|,
\end{aligned} \tag{4.7}$$

which indicates $|\mathcal{V}_{j+1}|$ is finite by noting that $|\mathcal{V}_0|$ and $|\mathcal{U}_s|(s = 0, 1, \cdots, j+1)$ are finite as discussed previously.

With Property 1, it is straightforward that the possible paths of cascading failures starting with finite initial failures in a power system are finite under the assumptions A1 and A2, although the number of cascading failure paths can be huge. One may notice that dealing directly with such a huge state space is impractical for large-scale real systems. We emphasize that this treatment is only for the theoretic rigorousness of probabilistic analysis of cascading failures, as explained later in Remark 5. The practical treatment will be derived in the next section.

Denote the probability of a specific cascading failure as $f(Z) = f\left(X_{t_0}, \cdots X_{t_n}\right)$. Here, $f(Z)$ serves as a distribution series, which can be further simplified using the Markov property in subsequence Z.

With the well-defined sequence Z, finite state spaces $\mathcal{U}_j$ and $\mathcal{V}_j$, as well as distribution series $f(Z)$, a cascading failure can be abstractly formulated as a Markov model. According to the assumptions, the changes in state variables are not influenced by the exact value of event times. Therefore, the time symbols are further simplified. Specifically, denoting $N := \{0, 1, \cdots, n, (n \in \mathbb{N})\}$ and omitting time symbols, an n-stage cascading failure is defined below.

Definition 1 An n-stage cascading failure is a Markov sequence $Z := \{X_0, X_1, \ldots, X_j, \ldots, X_n, X_j \in \mathcal{V}_j \times \mathcal{U}_j, \forall j \in N\}$ with respect to the (finite) random state space $\mathcal{X} := \bigcup_{j=1}^{n} \mathcal{V}_j \times \mathcal{U}_j$ and a given distribution series $f(Z) = f(X_n, \cdots X_1, X_0)$.

In the definition above, j is the stage label of the cascading failure, while n is the total number of stages or the length of the cascading failure. As n is always finite, we use a finite set $\mathcal{Z}$ to denote all possible (paths of) cascading failures in a power system. Meanwhile, the associated load shedding for a given n-stage cascading failure is merely a stochastic variable being a function of the stochastic sequence, denoted by $Y = h(X_0, \cdots, X_n) = h(Z)$.

4.2.2.2 Sequential Implementation of the Markov Model

It is worth noting that $f(Z)$ is practically difficult to obtain, even if the fault probability functions of individual components are known. However, invoking the conditional probability formula and the Markov property, $f(Z)$ can be rewritten as

$$\begin{aligned} f(Z) =& f(X_n, \cdots, X_1, X_0) \\ =& f_n(X_n \mid X_{n-1} \cdots X_0) \cdot f_{n-1}(X_{n-1} \mid X_{n-2} \cdots X_0) \\ & \cdots f_1(X_1 \mid X_0) \cdot f_0(X_0) \\ =& f_n(X_n \mid X_{n-1}) \cdot f_{n-1}(X_{n-1} \mid X_{n-2}) \cdots f_0(X_0), \end{aligned} \tag{4.8}$$

where $f_{j+1}\left(X_{j+1} \mid X_j\right)$ is the related conditional probability.

Assume in the sampling process, x_i^j is the sample of the state at stage j of the i-th sampling, while the length of the cascading failure in the i-th sampling is n^i. Then there is

$$\begin{aligned}&\mathbf{Pr}\left(X_{n^i}=x_n^i \mid X_{n^i-1}=x_{n^i-1}^i,\cdots,X_1=x_1^i,X_0=x_0^i\right)\\=&\mathbf{Pr}\left(X_{n^i}=x_n^i \mid X_{n^i-1}=x_{n^i-1}^i\right).\end{aligned} \tag{4.9}$$

Equations (4.8) and (4.9) indicate that a cascading failure can be simulated following the sequential conditional probability, other than directly using $f(Z)$.

Next, we explain how to implement it. For simplicity, we only consider using discrete variables to represent the ON/OFF statuses of components in this analysis. However, it can be easily generalized to incorporate other types of discrete states, such as tap ratios of OLTC.

Assume the fault probability of component k at stage j of the i-th sampling is represented by

$$p_{j,k}^i=\varphi_k\left(x_j^i\right), \tag{4.10}$$

where φ_k stands for the fault probability function of component k. According to Eq. (4.10), when x_j^i is known, the fault components can be determined concerning Eq. (4.10) in a random manner. Such fault components constitute the set of fault components at stage j of the i-th sampling, denoted by F_j^i. Accordingly, the other components constitute the set of normal components, $\bar{F}_j^i$, and they will still operate normally at the next stage, i.e., stage $j+1$.

Then the transition probability, $\hat{p}_{j,j+1}^i$, from state x_j^i to state x_{j+1}^i can be described as

$$\hat{p}_{j,j+1}^i=f_{j+1}\left(x_{j+1}^i \mid x_j^i\right)=\prod_{k\in F_j^i}p_{j,k}^i\prod_{k\in\bar{F}_j^i}\left(1-p_{j,k}^i\right). \tag{4.11}$$

Based on (4.11), the probability of the i-th sample of the (paths of) cascading failure, denoted by p_c^i, is given by

$$p_c^i=\prod_{j=0}^{n^i-1}\hat{p}_{j,j+1}^i. \tag{4.12}$$

Equation (4.12) indicates that the probability of a cascading failure can be tiny as it is the product of a series of small probabilities. It is the leading cause why blackout events can hardly be captured using the traditional MCS. Consequently, insufficient samples of rare events may give unreliable estimation results of the blackout risk with biased expectations and/or unexpected significant variance. This problem, theoretically, cannot be alleviated effectively in large-scale systems by merely increasing the number of simulations, as the variability and rarity of evolutionary trajectories across the state space $\mathcal{Z}$ lead to substantial variation in the size and rarity of results. It motivates us to develop a high-efficiency simulation approach, as presented in Sect. 4.3.

Remark 1 This sequential treatment has been heuristically used in some cascading failure simulations, such as the OPA model. This work provides a convincing interpretation of such treatment by abstractly defining cascading failures as the Markov model under pragmatic assumptions. Based on the model, we will develop a high-efficiency simulation strategy and blackout risk analysis with a solid theoretical foundation in the following section.

Remark 2 In this subsection, we only focus on the uncertainty of the discrete variables during the propagation of cascading failures with a fixed initial state, as many others did [26, 28, 37]. However, it is straightforward to consider the randomness of initial conditions further as long as the estimation with a fixed initial state is obtained, as the probability distribution of initial states should have been known beforehand (by distribution assumption or statistics from historical data). When processing a trial of cascading simulation, only one sample of the initial state is generated every time following the probability distribution to trigger the failure propagation. Hence, one can adopt Monte Carlo sampling to generate a number of initial states, trigger a batch of SIS simulations, and figure out the cascading blackout risk accordingly.

4.2.3 Example: The OPA Model

To illustrate how to deploy the proposed formulation in practice, a simplified OPA model (omitting the slow dynamic) is taken as an example. For the convenience of reading, the basic steps of the simplified OPA model are summarized as follows:

- **Step 1: Initialization**. Determine the system's initial state (stage 0), including the admittance matrix, power injections at each bus, power flow, etc.
- **Step 2: Protection emulation**. At stage j of the i-th sampling, components are tripped with probabilities according to the power flow and the fault probability functions of components, which functionally emulate the action of protection. Afterward, the topology of the system may change.
- **Step 3: Dispatch emulation**. Calculate (optimal) power flow based on the current system topology, which functionally emulates dispatch actions of the control center. Then the system state changes to the one at stage $j + 1$.
- **Step 4: Termination judgment**. If the system state at stage j is the same as at stage $j + 1$, the total load shedding is recorded, and the simulation terminates. Otherwise, go back to Step 2.

Concerning the simplified OPA simulation process, the procedure of abstracting a cascading failure as a Markov model can be briefed as follows:

1. In Step 1, the state variables are chosen as the ON/OFF statuses of components and corresponding power flow. The former are discrete variables, while the latter are continuous ones. Then the initial state of the i-th sampling can be represented as $x_0^i = (u_0^i, v_0^i)$.
2. At the beginning of Step 2 (in terms of stage j of the i-th sampling), the state variables are known as $x_j^i = \left(u_j^i, v_j^i\right)$. The fault probability of component k at stage j of sample i in the proposed model is $\varphi_k\left(x_j^i\right)$, as shown in (4.10). After sampling at stage j, the corresponding sets F_j^i and $\bar{F}_j^i$ are determined, and the discrete variables change to u_{j+1}^i, which indicates some components have new ON/OFF statuses. Then the transition probability from stage j to $j+1$ can be calculated according to (11). It is worth noting that this process is abstracted as Assumption A1 holds in the proposed model.
3. In Step 3, (optimal) power flow is calculated with v_j^i and u_{j+1}^i. Afterward, the system state changes to $x_{j+1}^i = (u_{j+1}^i, v_{j+1}^i)$, the state at stage $j+1$ of the i-th sampling. This process is abstracted as Assumption A2 holds.
4. Finally, comparing x_{j+1}^i with x_j^i in Step 4 determines the termination of one sampling. If they are different, go to Step 2; otherwise, the i-th sampling is completed, and the Markov sequence $z^i = \left\{x_0^i, \cdots, x_j^i\right\}$ is obtained. Comparing x_0^i with x_j^i, the load shedding y^i can also be obtained.

When other simulation models are deployed, some adaptations may turn out to be necessary, albeit straightforward. For example, if an AC-based simulation model is adopted, AC (optimal) power flow should be used instead of the DC one. Hence, the continuous variables should be extended to quantities of AC power flow. Similarly, when characteristics of OLTC are considered, the discrete state variables should be augmented to include the tap ratios of OLTC. However, the modeling procedure remains similar.

4.3 Cascading Failures Probabilistic Analysis

4.3.1 *Importance Sampling for Cascading Failure*

Under certain conditions, the importance sampling (IS) technique is recognized as an effective tool for improving the sampling efficiency and depressing the estimation variance. One such condition entails implementing a well-designed biasing scheme within the IS that concentrates on sampling the portion with lower variance. In detail, the basic idea of IS is to sample the stochastic process under a proposal distribution series $g(X_n, X_{n-1}, \ldots, X_0)$ ($g(Z)$ for short) instead of the true one,

$f(Z)$. Especially, the probability of arbitrary possible cascading failure under the proposal distribution series needs to be positive, i.e., $g(Z) \in [\varepsilon, 1], \forall Z \in \mathcal{Z}$, where ε is a given small positive number. Then, after conducting N_s i.i.d. simulations, we can obtain a sample set of cascading failures, $Z_s := \{z_s^i, i = 1, 2, \cdots, N_s\}$, where $z_s^i = \{x_0^i, x_1^i, \cdots, x_{n^i}^i\}$ is the i-th sample of cascading failures; n^i is the length of the sampled cascading failure in the i-th simulation; x_j^i is the sampled state at stage j of the i-th simulation. Afterward, we can obtain the sample set of the load shedding, $Y_s := \{y_s^i, i = 1, \cdots, N_s\}$, where $y_s^i = h(z_s^i)$. For simplicity, we abuse the notation δ_{Y_0} throughout to stand for $\delta_{\{h(Z) \geq Y_0\}}$. As $|\mathcal{Z}|$ is finite, the true probability of event A defined previously is given by

$$\mu(A) = \sum_{Z \in \mathcal{Z}} \delta_{Y_0} f(Z). \tag{4.13}$$

As $\mu(A)$ is very difficult to obtain in large-scale systems in practice, particularly when hidden failures are taken into account, we alternatively estimate it through Eq. (4.1). The estimation variance, $D(A)$, is given by (4.2).

We are interested in the expectation and variance based on the IS under the proposal distribution series $g(Z)$. To this end, we let $w(Z) = f(Z)/g(Z)$, yielding

$$\mu(A) = \sum_{Z \in \mathcal{Z}} \delta_{Y_0} w(Z) g(Z). \tag{4.14}$$

As the IS with $g(Z)$ is deployed, the unbiased estimation of $\mu(A)$ based on N_s samples turns to be

$$\tilde{\mu}_{IS}(A) = \frac{1}{N_s} \left(\sum_{i=1}^{N_s} w\left(z_s^i\right) \cdot \delta_{\{h(z_s^i) \geq Y_0\}} \right), \tag{4.15}$$

where $w\left(z_s^i\right) > 0$ is the sampling weight subject to

$$w\left(z_s^i\right) = \frac{f\left(z_s^i\right)}{g\left(z_s^i\right)}. \tag{4.16}$$

Moreover, the estimation variance of the blackout probability is

$$\begin{aligned} D_{IS}(A) &= \frac{\mathbb{E}_g \left\{ \delta_{Y_0} w(Z) - \mathbb{E}_g \left[\delta_{Y_0} w(Z) \right] \right\}^2}{N_s} \\ &= \frac{\mathbb{E}_g \left\{ \left[\delta_{Y_0} w(Z) \right]^2 \right\} - \left\{ \mathbb{E}_g \left[\delta_{Y_0} w(Z) \right] \right\}^2}{N_s} \end{aligned} \tag{4.17}$$

$$= \frac{\sum_{Z\in\mathcal{Z}} \delta_{Y_0}^2 w^2(Z)g(Z) - \left[\sum_{Z\in\mathcal{Z}} \delta_{Y_0} w(Z)g(Z)\right]^2}{N_s},$$

where $\mathbb{E}_g$ indicates the related expectation is with respect to g instead of f.

Let

$$w_0 = \frac{\sum_{Z\in\mathcal{Z}} \delta_{Y_0}^2 w^2(Z)g(Z)}{\sum_{Z\in\mathcal{Z}} \delta_{Y_0}^2 w(Z)g(Z)} = \frac{\sum_{Z\in\mathcal{Z}^*} w^2(Z)g(Z)}{\sum_{Z\in\mathcal{Z}^*} w(Z)g(Z)}, \tag{4.18}$$

which is referred to as the equivalent sampling weight.

Let $\mathcal{Z}^* := \{Z \in \mathcal{Z} \mid h(Z) \geq Y_0\}$. Then substituting (4.14) and (4.18) into (4.17) yields

$$\begin{aligned} D_{IS}(A) &= \frac{1}{N_s}\left(w_0 \sum_{z\in\mathcal{Z}} \delta_{Y_0} w(z)g(z) - \mu^2(A)\right) \\ &= \frac{1}{N_s}\left(w_0\mu(A) - \mu^2(A)\right). \end{aligned} \tag{4.19}$$

Next, some key propositions are presented.

Proposition 1 *Given $g(Z)$, $w(Z)$, and $\mathcal{Z}^*$, w_0 given by (4.18) satisfies*

$$w_0 \in \left[\min_{Z\in\mathcal{Z}^*} w(Z),\ \max_{Z\in\mathcal{Z}^*} w(Z)\right].$$

Proof Since $w(Z)$ and $g(Z)$ are non-negative, there is

$$w_0 \leq \frac{\sum_{Z\in\mathcal{Z}^*} \delta_{Y_0}^2 w(Z)g(Z) \max_{Z\in\mathcal{Z}^*} w(Z)}{\sum_{Z\in\mathcal{Z}^*} \delta_{Y_0}^2 w(Z)g(Z)} = \max_{Z\in\mathcal{Z}^*} w(Z).$$

Similarly, it has

$$w_0 \geq \frac{\sum_{Z\in\mathcal{Z}^*} \delta_{Y_0}^2 w(Z)g(Z) \min_{Z\in\mathcal{Z}^*} w(Z)}{\sum_{Z\in\mathcal{Z}^*} \delta_{Y_0}^2 w(Z)g(Z)} = \min_{Z\in\mathcal{Z}^*} w(Z).$$

□

Proposition 2 *Let $D_{IS}(A)$ and $D(A)$ be the variances of the probability estimation of event A defined previously by using the IS and the MCS, respectively. If $N_s = N_M$, then $D_{IS}(A) < D(A)$ holds if and only if the proposal distribution series $g(Z)$ satisfies $w_0 < 1$, or equivalently,*

$$w_0\mu(A) < \mu(A). \tag{4.20}$$

Moreover, the relative reduction of variance is

$$\frac{D(A) - D_{IS}(A)}{D(A)} = \frac{1 - w_0}{1 - \mu(A)}.$$

Proposition 3 *Let $D_{IS}(A)$ and $D(A)$ be the variances of the probability estimation of event A defined previously by using the IS and the MCS, respectively. If $D_{IS}(A) = D(A)$, then $N_{IS} < N_M$ holds if and only if the proposal distribution series $g(Z)$ satisfies $w_0 < 1$, or equivalently,*

$$w_0\mu(A) < \mu(A). \tag{4.21}$$

Moreover, the relative reduction of sample size is

$$\frac{N_M - N_{IS}}{N_M} = \frac{1 - w_0}{1 - \mu(A)}.$$

Propositions 2 and 3 can be proved by directly using (4.2) and (4.19), which are omitted here.

Remark 3 In practice, it is usually difficult to check the conditions (4.20) or (4.21). A more convenient yet conservative way is to use the following sufficient condition:

$$g(Z) > f(Z), \forall Z \in \mathcal{Z}^* = \{Z \in \mathcal{Z} | \, h(Z) > Y_0\}, \tag{4.22}$$

which is an immediate corollary of Propositions 1–3. As Eq. (4.22) can be rewritten as

$$w(Z) = \frac{f(Z)}{g(Z)} < 1, \forall Z \in \{Z \in \mathcal{Z} | \, h(Z) > Y_0\}, \tag{4.23}$$

which indicates

$$\max_{Z \in \mathcal{Z}^*} w(Z) < 1. \tag{4.24}$$

Then, according to Proposition 1, there is

$$w_0 \le \max_{Z \in Z^*} w(Z) < 1,$$

which indicates conditions (4.20) and (4.21) are satisfied.

Similar conclusions can be drawn for the blackout risk assessment. Given $g(Z)$ for the IS, then the blackout risk is

$$Risk_{IS}(Y_0) = \mathbb{E}_g(h(Z) \cdot w(Z) \cdot \delta_{Y_0}). \tag{4.25}$$

The estimation of blackout risk based on N_s samples is

$$\tilde{Risk}_{IS}(Y_0) = \frac{1}{N_s}\left(\sum_{i=1}^{N_s} h(z_s^i)w(z_s^i)\delta_{\{h(z_s^i)\geq Y_0\}}\right). \tag{4.26}$$

Then the estimation variances of (4.4) and (4.26) are given by

$$D(R) = \frac{\sum\limits_{Z\in\mathcal{Z}} h^2(Z)\delta_{Y_0} f(Z) - Risk\,(Y_0)^2}{N_M} \tag{4.27}$$

and

$$D_{IS}(R) = \frac{\sum\limits_{Z\in\mathcal{Z}} w^2(Z)h^2(Z)\delta_{Y_0} g(Z) - Risk(Y_0)^2}{N_s}, \tag{4.28}$$

respectively. According to (4.27) and (4.28), the condition of the variance reduction can be obtained accordingly.

The theoretical analysis indicates that the IS can reduce both the sample size and the estimation variance, providing an appropriately selected proposal distribution series $g(Z)$. Considering the unbiasedness of the estimation using the IS and the MCS, the lower variance indicates that the IS has a better estimation performance than the MCS.

4.3.2 Sequential Importance Sampling-Based Probabilistic Analysis

Similar to (4.8), $g(Z)$ can be expressed as

$$\begin{aligned} g(Z) &= g(X_n, \cdots X_1, X_0) \\ &= g_n(X_n|X_{n-1}) \cdot g_{n-1}(X_{n-1}|X_{n-2}) \cdots g_0(X_0). \end{aligned} \tag{4.29}$$

It means the proposal distribution series $g(Z)$ can be chosen sequentially at individual stages in a cascading failure. Thus, the problem of selecting $g(Z)$ turns out to be one of choosing the series $g_{j+1}(X_{j+1}|X_j)$ sequentially. To acquire more information about the cascading failure with severe load shedding, $g_{j+1}(X_{j+1}|X_j)$ should be carefully chosen to amplify the probability of cascading failures in future stages versus the original $f_{j+1}(X_{j+1}|X_j)$. Heuristically, we modify the fault probability of components given in (4.10) to

$$q_{j,k}^i = \min(\eta p_{j,k}^i, \max(\varphi_k)), \tag{4.30}$$

where $q^i_{j,k}$ is the modified component's fault probability; η is the SIS parameter that stands for the amplification factor of the component's fault probability; $\max(\varphi_k)$ is the maximal value of the fault probability function of component k. Correspondingly, the modified transition probability becomes

$$\hat{q}^i_{j,j+1} = \prod_{k \in F^i_j} q^i_{j,k} \prod_{k \in \overline{F}^i_j} \left(1 - q^i_{j,k}\right). \tag{4.31}$$

For the i-th sample, the original load shedding probability p^i_c is given by (4.12), while the modified probability q^i_c is given by

$$q^i_c = \prod_{j=0}^{n^i-1} \hat{q}^i_{j,j+1}. \tag{4.32}$$

The corresponding sampling weight is

$$w(z^i_s) = \frac{p^i_c}{q^i_c} = \prod_{j=0}^{n^i-1} \frac{\hat{p}^i_{j,j+1}}{\hat{q}^i_{j,j+1}}. \tag{4.33}$$

Simulating cascading failures with sampling weights given by (4.33), the load shedding probability and the blackout risk can be estimated by using (4.15) and (4.26), respectively. According to the previous analyses, both the number of simulations and the estimation variances can be reduced, provided appropriately selected $g(Z)$.

Remark 4 To achieve high efficiency of SIS-based simulations, η should be chosen carefully so that (4.20) or (4.22) is satisfied. Unfortunately, it is not really a trivial task because, according to the necessary and sufficient condition (4.20), w_0 cannot be known a priori. However, noticing $p^i_{j,k}$ in (4.10) and $q^i_{j,k}$ in (4.30) are usually very small, we have $(1 - p^i_{j,k}) \approx (1 - q^i_{j,k}) \approx 1$. It implies that if η is selected such that $\eta > 1$, the inequality

$$w(z^i_s) = \prod_{j=0}^{n^i} \frac{\hat{p}^i_{j,j+1}}{\hat{q}^i_{j,j+1}} \approx \prod_{j=0}^{n^i} \frac{\prod_{k \in F^i_j} p^i_{j,k}}{\prod_{k \in F^i_j} q^i_{j,k}} < 1$$

holds for most samples, especially when j is not small, meaning the blackouts of interest are rare. Then the sufficient condition (4.22) can hold approximately. Numerical experiments empirically support this conjecture.

The algorithm of the SIS-based strategy is given as follows:

- **Step 1: Data preparation**. Initialize the system data and parameters. Specifically, choose $\eta > 1$.
- **Step 2: Sampling states**. For the i-th sampling, according to the system state, x_j^i at stage j, and the fault probability of components based on (4.10) and (4.30), simulate the component outages and acquire the new state x_{j+1}^i at the next stage. Afterward, calculate the state transition probability and the sampling weight using (4.11) and (4.33), respectively.
- **Step 3: Termination judgment**. If x_j^i is the same as x_{j+1}^i, the i-th sample of cascading failure simulation is completed at stage j and the i-th sample $z_s^i = \left\{x_1^i \cdots x_j^i\right\}$ is obtained. If all N_s simulations are completed, the sampling process is ended; otherwise, let $i = i + 1$ and go back to Step (4.2).
- **Step 4: Data analysis**. According to (4.15) and (4.26), estimate the probability of load shedding and blackout risk.

Remark 5 While the SIS method can achieve high sampling efficiency, additional improvements warrant consideration:

1. Reference [58] (Section 3.3, "Importance Sampling," pp. 95–96) points out that IS/SIS can perform very poorly if the integral

$$\int_{\mathcal{Z}} \frac{f^2(Z)}{g(Z)} dZ = \int_{\mathcal{Z}} w^2(Z) g(Z) dZ$$

interpreted as the second-order moment of $w(Z)$ with respect to $g(Z)$ is infinite. Thanks to Assumptions A1 and A2, here $|\mathcal{Z}|$ is finite according to Property 1. Noting that $f(Z) \le 1$ and $g(Z) \in [\varepsilon, 1], \forall Z \in \mathcal{Z}$, there is

$$\int_{\mathcal{Z}} \frac{f^2(Z)}{g(Z)} dZ = \sum_{Z \in \mathcal{Z}} \frac{f^2(Z)}{g(Z)} \le \sum_{Z \in \mathcal{Z}} \frac{1}{\varepsilon} f(Z) = \frac{1}{\varepsilon}, \tag{4.34}$$

indicating that the second-order moment of $w(Z)$ is bounded. Although it can be very large, the boundedness guarantees that the IS/SIS techniques developed previously are theoretically valid, which lays a solid theoretical foundation for the probabilistic analysis based on these techniques.
2. When the second-order moment (or the variance) of $w(Z)$ is large, it implies that some samples have large weights, while others are very small. In this case, the IS/SIS method may suffer from sample degeneration. According to Eqs. (4.15) and (4.26), those samples with small weights have little influence on final estimations. When the number of such samples is large, the sampling efficiency will be remarkably degraded. Despite (4.34) indicating the second-

order moment of $w(Z)$ is bounded theoretically, it is difficult to know whether or not the second-order moment is small enough. A practical and effective way to circumvent the problem is to adopt a resampling scheme, which relies heavily on the Markov structure. As mentioned in Sect. 4.2.2, the Markov property is well preserved in the proposed model. Therefore, resampling is generally appropriate when further improvement of IS/SIS is needed.

3. When the randomness of continuous variables is taken into account, e.g., in highly renewable-integrated systems, $\mathcal{Z}$ is no longer a finite set, and the variance of $w(Z)$ may become infinite. It could remarkably undermine the performance of the IS/SIS. In this case, discretization of continuous variables or other techniques is needed. It, however, is beyond the topic of this chapter.

4.3.3 Example

For explanation, numerical cases on the IEEE 39-bus system are carried out using the simplified OPA model without slow dynamic. This tested system has 39 buses and 46 lines with a total load of 6254 MW.

4.3.3.1 Efficiency of Probability Distribution Estimation

In this case, the probability of load shedding in the IEEE 39-bus system is estimated using the MCS and the SIS ($\eta = 1.3$) method, respectively. The sample size of the MCS is 4000, while that of the SIS is only 800 since the former requires many more samples to reach a small variance of estimation. As mentioned previously, the MCS and the SIS strategies provide an unbiased estimate of the load shedding probability. According to the estimation results given in Fig. 4.2, the two strategies output almost the same estimations on the probability distribution as the load shedding less than 3500 MW. This result justifies that the SIS-based simulation can achieve a given estimation accuracy with fewer simulations, and thus, it is of higher efficiency than the MCS method.

As for the failures whose load shedding is larger than 3500 MW (the corresponding probability is less than 0.83%), the SIS strategy can discover them with a higher likelihood than the MCS method. Regarding the load shedding greater than 4000 MW (the corresponding probability is less than 0.13%), the MCS fails to find any event in 4000 simulations and cannot estimate such rare events. On the contrary, the SIS strategy successfully finds numerous rare events with load shedding up to 4000 MW in only 800 simulations. It suggests that the SIS strategy can considerably facilitate capturing rare events of cascading failures even with much fewer simulations. It also implies that the blackout risk analysis based on the MCS might not be reliable enough since the captured rare events are usually far from sufficient.

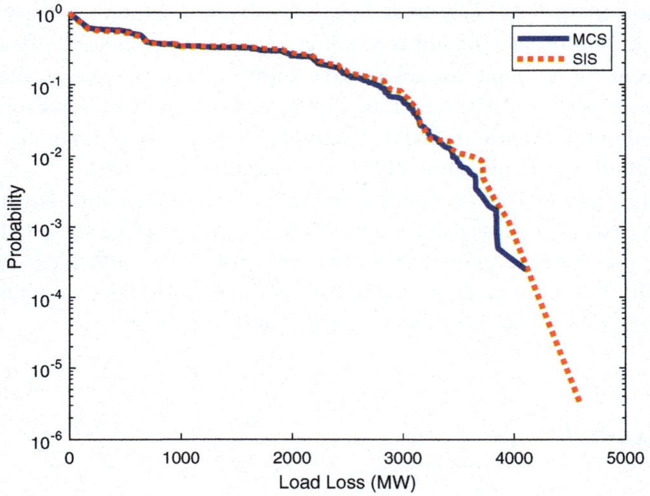

Fig. 4.2 Probability estimation of the load shedding with MCS and SIS

4.3.3.2 Variance of Probability Distribution Estimation

In this case, we compare the variance of probability estimation with the two methods (see Fig. 4.3) in the IEEE 39-bus system. Since the true variance of probability estimation cannot be obtained directly, the sample variance is used as a surrogate. Take the MCS as an example. Denote $\widetilde{\mu}^m$ as the estimation of m-th sample sets; then the sample variance is

$$\widetilde{D}(A) = \frac{1}{m_{\max} - 1} \sum_{m=1}^{m_{\max}} \left[\mu^m(A) - \left(\frac{1}{m_{\max}} \sum_{m=1}^{m_{\max}} \mu^m(A) \right) \right]^2,$$

where $m_{\max}$ is the number of i.i.d sample sets, which is set as 1000 here.

In simulation, the MCS and the SIS have the same sample sizes of 1000. The SIS parameter remains at $\eta = 1.3$. As shown in Fig. 4.3, the estimation variances of the MCS and the SIS both have a downward trend as load loss increases, although the estimation variance of the SIS is lower than that of the MCS. Meanwhile, we calculate the ratio that Eq. (4.22) is satisfied as well as $\widetilde{w}_0\widetilde{\mu}(A)$ based on the samples. The results are provided in Table 4.1 and Fig. 4.4. Table 4.1 indicates that Eq. (4.22) is satisfied in most samples, particularly when Y_0 is large. Actually, the ratio of $g(Z) > f(Z)$ is always fulfilled when $Y_0 > 1000\,MW$ in this case, which verifies Remark 4. It is interesting to observe in Fig. 4.4 that $\widetilde{w}_0\widetilde{\mu}(A) < \widetilde{\mu}(A)$ is still satisfied even when condition (4.22) does not hold (e.g., when $Y_0 \leq 500\,MW$). This is because Eq. (4.22) is merely a sufficient condition. Since the MCS cannot capture the failures of $Y_0 > 4000\,MW$, the curve of probability estimation variance of the

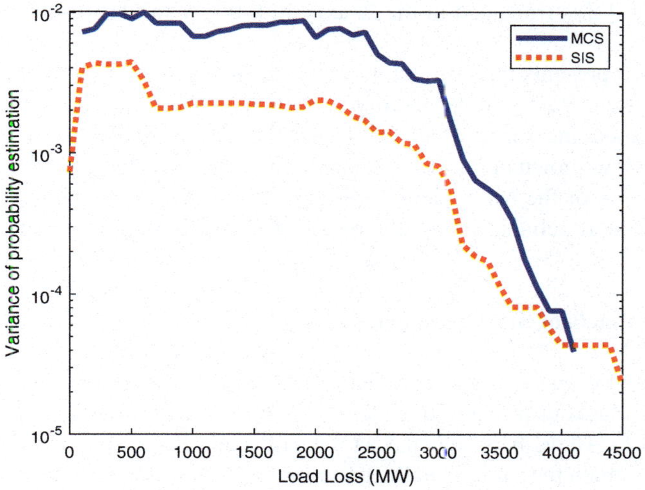

Fig. 4.3 Variance of probability estimation with MCS and SIS

Table 4.1 Ratio of $g(Z) > f(Z), Z \in \{Z \in \mathcal{Z}|H(Z) > Y_0\}$ (%)

Y_0	500	1000	1500	2000	2500	3000	3500	4000	4500	5000
Ratio	99.83	100	100	100	100	100	100	100	100	100

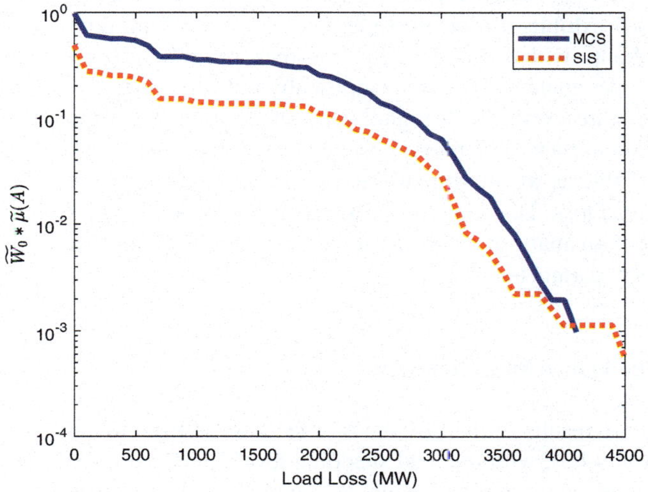

Fig. 4.4 $\widetilde{w}_0\widetilde{\mu}(A)$ for SIS and MCS ($\widetilde{w}_0 \equiv 1$ for MCS)

MCS in Fig. 4.3 is truncated in the X-axis of 4000. A similar result is observed in Fig. 4.4.

Figure 4.5 presents the estimation variances decrease along with the increase of the sample size. Here, the probability is estimated according to cascading failures with load shedding larger than (a) 1000 MW, (b) 2000 MW, and (c)3000 MW, respectively. As shown in Fig. 4.5, the variance of the SIS estimation decreases much faster than that of the MCS, demonstrating that the SIS-based simulation strategy can achieve more reliable estimation results with much fewer simulations.

4.3.3.3 Impacts of SIS Parameters η

In this part, we analyze the influence of the SIS parameter η on the estimation variance of blackout probabilities (see Fig. 4.6, in which a partially enlarged view is embedded). Here, η is selected as 1.1, 1.3, and 1.5, respectively, while other conditions remain the same as in the previous cases. It is found that η can impact the probability estimation of cascading failures twofold: On the one hand, as η grows greater, more detailed information about the rare events can be captured. From Fig. 4.6, it is observed that the SIS with $\eta = 1.5$ collects blackout samples with load shedding even over 4200 MW (the corresponding probability is nearly 10^{-4}), while the SIS with a smaller η, say 1.1 or 1.3, does not capture such rare events. In most load loss values, however, the variances of the SIS probability estimation with any η value are less than that of the MCS, which verifies our theoretical results well.

On the other hand, whereas more rare event samples are captured, the estimation variances of normal events with lower load shedding increase. In this case, the SIS with $\eta = 1.5$ exhibits a more considerable variance of the probability estimation of the blackouts with load shedding less than roughly 550 MW versus the SIS with smaller parameters. This case empirically indicates a larger SIS parameter can help capture rare events with higher load shedding at the expense of increasing the estimation variances of normal events. Nevertheless, this expense makes sense and is acceptable as we are primarily concerned with the potential blackouts with a quite large load loss. This feature also allows us to purposely adjust the resolution of cascading failure analysis according to desired levels of load shedding by carefully tuning the SIS parameter.

4.3.3.4 Blackout Risk Estimation

In this part, the results of the SIS and the MCS are utilized to analyze the blackout risk defined in (4.3), with the load shedding level Y_0 set as 0 MW. The mean value and the variance are computed using 25 sample sets. In each sample set, the same number of simulations are carried out independently with the SIS and the MCS. The curves of mean values and variances, along with the sample size, are shown in Fig. 4.7. It displays that the mean value of blackout risk of the SIS is close to that of the MCS, with an average error of 5.2%. However, the variance values of blackout

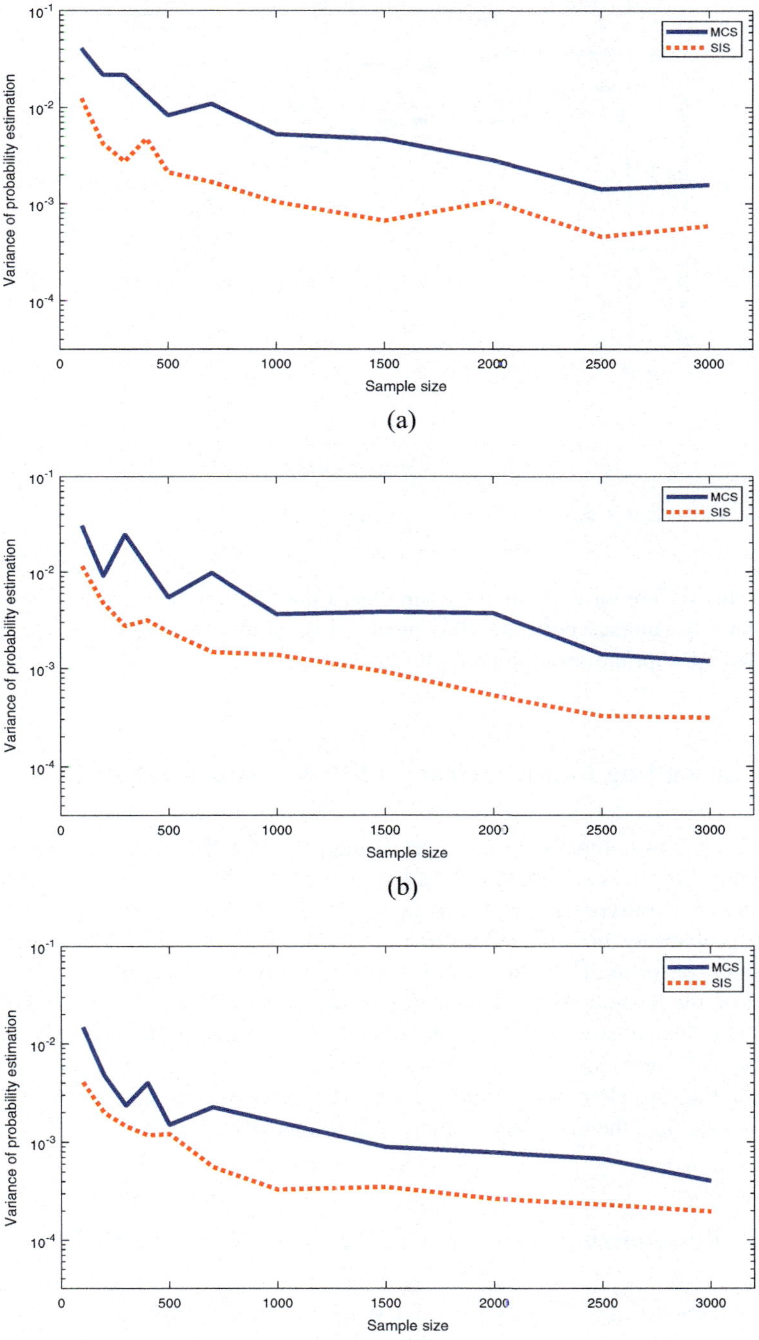

(a)

(b)

(c)

Fig. 4.5 Convergence of the variance of the probability estimation. (**a**) Loading shedding larger than 1000 MW. (**b**) Loading shedding larger than 2000 MW. (**c**) Loading shedding larger than 3000 MW

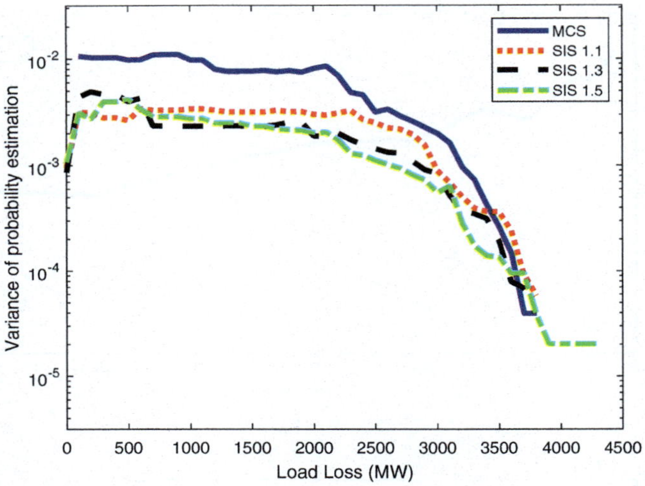

Fig. 4.6 Variance of probability estimation with different SIS parameters

risk of the SIS are virtually all less than that of the MCS, indicating that the SIS can considerably enhance both the efficiency and reliability of blackout risk analyses, provided appropriate sampling weights are selected.

4.4 Cascading Failures Probability and Blackout Risk

To quantify how component failure probability (CFP) influences blackout risk, this subsection proposes a sample-induced semi-analytic approach to characterize the relationship between CFP and blackout risk. To this end, a generic component failure probability function (CoFPF) is introduced to describe CFP with varying parameters or forms. Then, the exact relationship between blackout risk and CoFPFs is built on the abstract Markov sequence model of cascading failures. Leveraging a set of samples generated by blackout simulations, a sample-induced semi-analytic mapping is further established between the unbiased estimation of blackout risk and CoFPFs. Finally, when the CoFPFs change, an efficient algorithm is derived that can directly calculate the unbiased estimate of blackout risk.

4.4.1 Relationship Between CoFPFs and Blackout Risk

The propagation of a cascading failure is a complicated dynamic process during which many practical factors are involved, such as hidden failures of components,

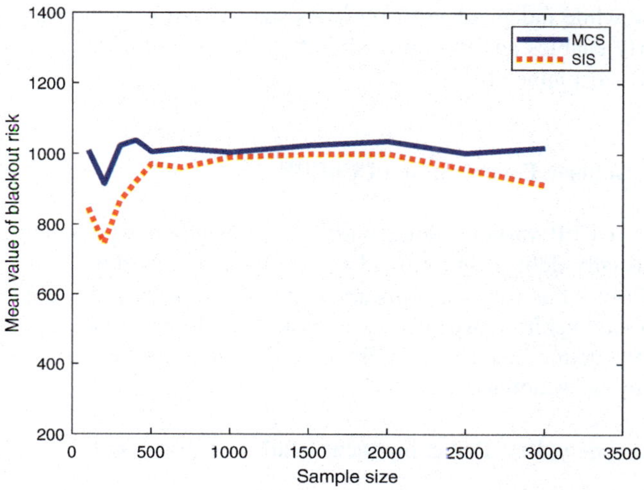

(a)

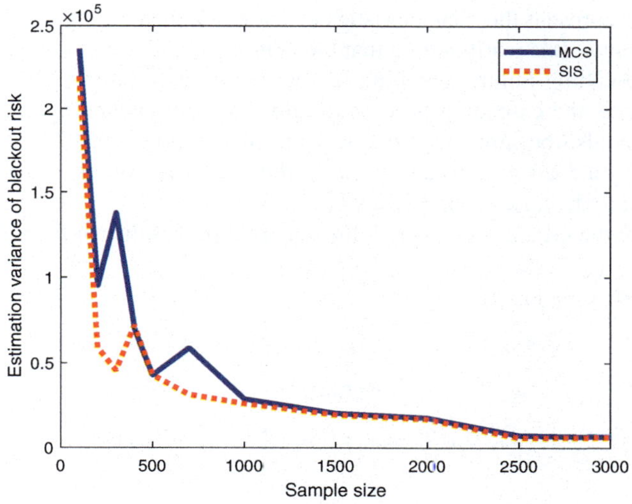

(b)

Fig. 4.7 Performance of blackout risk estimation with different sample sizes. (**a**) Mean value of blackout risk. (**b**) Estimation variance of blackout risk

actions of the dispatch/control center, etc. This part focuses on the influence of random component failures (or, more precisely, CFP) on the blackout risk, where a cascading failure can be simplified into a sequence of component failures with corresponding system states and usually emulated by steady-state models [26, 28, 59]. In this case, individual component failures are only related to the current system state while independent of previous states, known as the Markov property. This property enables an abstract model of cascading failures with a generic form of CoFPFs, as explained below.

4.4.1.1 A Generic Formulation of CoFPFs

To describe a CFP varying along with the propagation of cascading failures, a CoFPF is usually defined in terms of the component's working conditions. In the literature, CoFPF has various forms about specific scenarios [59, 60]. To generally depict the relationship between blackout risk and CFP with varying parameters or forms, we first define an abstract CoFPF here. Specifically, the CoFPF of component k, denoted by φ_k, is defined as

$$\varphi_k\left(s_k, \boldsymbol{\eta}_k\right) := \mathbf{Pr}\left(\text{ component } k \text{ fails at } s_k \text{ given } \boldsymbol{\eta}_k\right). \tag{4.35}$$

In (4.35), s_k represents the current working condition of component k, which can be load ratio, voltage magnitude, etc. $\boldsymbol{\eta}_k$ is the parameter vector. Both $\boldsymbol{\eta}_k$ and the form of φ_k represent the characteristics of the component k, e.g., the type and age of component k. It is worth noting that the working condition of component k varies during a cascading failure, resulting in changes in the related CFPs. On the other hand, whereas the cascading process usually does not change $\boldsymbol{\eta}_k$ and the form of φ_k, they can also be influenced due to controlled or uncontrolled factors, such as maintenance and extreme weather, etc. In this sense, Eq. (4.35) provides a generic formulation to depict such properties of CoFPFs.

Based on the previous chapter, define the joint probability of the failure path as

$$\begin{aligned} g(z) =& g\left(x_n, \cdots, x_1, x_0\right) \\ =& g_n\left(x_n \mid x_{n-1} \cdots x_0\right) \cdot g_{n-1}\left(x_{n-1} \mid x_{n-2} \cdots x_0\right) \\ & \cdots g_1\left(x_1 \mid x_0\right) \cdot g_0\left(x_0\right) \\ =& g_n\left(x_n \mid x_{n-1}\right) \cdot g_{n-1}\left(x_{n-1} \mid x_{n-2}\right) \cdots g_0\left(x_0\right), \end{aligned} \tag{4.36}$$

where

$$g_{j+1}\left(x_{j+1} \mid x_j\right) = \Pr\left(X_{j+1} = x_{j+1} \mid X_j = x_j\right)$$

$$g_0\left(x_0\right) = \Pr\left(X_0 = x_0\right) = 1.$$

Moreover, due to the intrinsic randomness of cascading failures, the load shedding, denoted by Y, is also a random variable up to the path-dependent propagation of cascading failures. Then, the blackout risk concerning $g(Z)$ is defined as the expectation of the load shedding greater than a given level, Y_0. That is

$$R_g\left(Y_0\right) = \mathbb{E}\left(Y \cdot \delta_{\{Y \geq Y_0\}}\right), \tag{4.37}$$

where $R_g\left(Y_0\right)$ stands for the blackout risk with respect to $g(Z)$ and Y_0; $\delta_{\{Y \geq Y_0\}}$ is the indicator function of $\{Y \geq Y_0\}$, given by

$$\delta_{\{Y \geq Y_0\}} := \begin{cases} 1 & \text{if } Y \geq Y_0 \\ 0 & \text{otherwise} \end{cases}.$$

4.4.1.2 Relationship Construction

The probability of cascading failures is first derived based on the generic form of CoFPFs. Then, the relationship between blackout risk and CoFPFs can be characterized.

At stage j, the working condition of component k can be represented as a function of the system state x_j, denoted by $\phi_k(x_j)$. That is, $s_k := \phi_k(x_j)$. Hence, the CFP of component k at stage j is $\varphi_k(\phi_k(x_j), \eta_k)$. Without causing confusion, thereafter we abuse the notation $\varphi_k(x_j)$ to stand for $\varphi_k(\phi_k(x_j), \eta_k)$ for simplicity.

Considering stages j and $(j+1)$, we have

$$g_{j+1}\left(x_{j+1} \mid x_j\right) = \prod_{k \in F(x_j)} \varphi_k\left(x_j\right) \cdot \prod_{k \in \bar{F}(x_j)} \left(1 - \varphi_k\left(x_j\right)\right). \tag{4.38}$$

In Eq. (4.38), $F(x_j)$ is the component set consisting of the components that are defective at x_{j+1} but normally work at x_j, while $\bar{F}\left(x_j\right)$ consists of components that work normally at x_{j+1}. With (4.38), Eq. (4.36) can be rewritten as

$$\begin{aligned} g(z) &= \prod_{j=0}^{n-1} g_{j+1}\left(x_{j+1} \mid x_j\right) \\ &= \prod_{j=0}^{n-1} \left[\prod_{k \in F(x_j)} \varphi_k\left(x_j\right) \cdot \prod_{k \in \bar{F}(x_j)} \left(1 - \varphi_k\left(x_j\right)\right) \right]. \end{aligned} \tag{4.39}$$

Furthermore, invoking (3), it yields

$$R_g\left(y_0\right) = \mathbb{E}\left(h(Z) \cdot \delta_{\{h(Z) \geq Y_0\}}\right) = \sum_{z \in \mathcal{Z}} g(z) h(z) \delta_{\{h(z) \geq Y_0\}}. \tag{4.40}$$

Theoretically, the relationship between blackout risk and CoFPFs can be established immediately by substituting (4.39) into (4.40). However, this relationship cannot be directly applied in practice. For this reason, according to (4.39), the component failures occurring at different stages on a path of cascading failures are correlated with one another. As a consequence, this long-range coupling, unfortunately, produces a complicated and nonlinear correlation between blackout risk and CoFPFs. In addition, since the number of components in a power system is usually quite large, the cardinality of $\mathcal{Z}$ can be extremely huge. Hence, it is practically impossible to accurately calculate blackout risk concerning the given CoFPFs by directly using (4.39) and (4.40). To circumvent this problem, a sample-based semi-analytic method is proposed to characterize the relationship by using an unbiased estimation of blackout risk as a surrogate.

4.4.2 Sample-Induced Semi-analytic Characterization

4.4.2.1 Unbiased Estimation of Blackout Risk

To estimate blackout risk, conducting MCS is the easiest and the most extensively used way. The first step is to generate i.i.d. samples of cascading failures. (Since the simulation of cascading failures is often independently carried out under the same condition, it is common to assume the samples are i.i.d. Note that a sample represents an entire cascading path here.) and corresponding load shedding concerning the joint probability series, $g(Z)$. Unfortunately, $g(Z)$ is indeed unknown in practice. In such a situation, one can heuristically sample the failed components at each stage of possible cascading failure paths regarding the corresponding system states and CoFPFs. Afterward, system states at the next stage are determined with the updated system topology. This process repeats until there is no new failure happening anymore. Then, a path-dependent sample is generated. This method essentially performs sampling sequentially using the *conditional component probabilities* instead of the *joint probabilities*. Equation (4.39) provides this method with a mathematical interpretation, which is an application of the Markov property of cascading failures.

Suppose N i.i.d. samples of cascading failure paths are obtained concerning $g(Z)$. Let $Z_g := \{z^i, i = 1, \cdots, N\}$ record the set of these samples. Then, the i-th cascading failure path contained in the set is expressed by $z^i := \{x_0^i, \cdots, x_{n^i}^i\}$, where n^i is the number of total stages of the i-th sample. For each $z^i \in Z_g$, the associated load shedding is given by $y^i = h(z^i)$. All y^i make up the set of load shedding with respect to $g(Z)$, denoted by $Y_g := \{y^i, i = 1, \cdots, N\}$. Then, the unbiased estimation of blackout risk is formulated as

$$\hat{R}_g(Y_0) = \frac{1}{N}\sum_{i=1}^{N} y^i \delta_{\{y^i \geq Y_0\}} = \frac{1}{N}\sum_{i=1}^{N} h\left(z^i\right) \delta_{\{h(z^i) \geq Y_0\}}. \tag{4.41}$$

Note that Eq. (4.41) applies to $g(Z)$ or the corresponding CoFPFs. It implies that the underlying relationship between blackout risk and CoFPFs relies on samples. Hence, whenever parameters or forms of the CoFPFs change, all samples need to be regenerated to estimate the blackout risk, which is highly time-consuming, even practically impossible. Next, a semi-analytic method is derived by building a mapping between CoFPFs and the unbiased estimation of blackout risk.

4.4.2.2 Sample-Induced Semi-analytic Characterization

Suppose the samples are generated concerning $g(Z)$. Then the sample set is Z_g, and the set of load shedding is Y_g. When changing $g(Z)$ into $f(Z)$ (both are defined on $\mathcal{Z}$), usually all samples of cascading failure paths need to be regenerated. However, inspired by the sample treatment in Importance Sampling, it is possible to avoid sample regeneration by revealing the underlying relationship between $g(Z)$ and $f(Z)$. Specifically, for a given path z^i, we define

$$w\left(z^i\right) := \frac{f\left(z^i\right)}{g\left(z^i\right)}, \quad (z^i \in \mathcal{Z}), \tag{4.42}$$

and then each sample in terms of $f(z)$ can be represented as the sample of $g(z)$ weighted by $w(z)$. Consequently, the unbiased estimation of blackout risk in terms of $f(Z)$ can be directly obtained from the sample generated concerning $g(Z)$, as explained below.

From Eqs. (4.41) and (4.42), there is

$$\hat{R}_f\left(Y_0\right) = \frac{1}{N}\sum_{i=1}^{N} w\left(z^i\right) h\left(z^i\right) \delta_{\{h(z^i)\geq Y_0\}}. \tag{4.43}$$

Obviously, when $w(z) \equiv 1, z \in \mathcal{Z}$, (4.43) is equivalent to (4.41). Moreover, Eq. (4.43) is an unbiased estimation by noting that

$$\begin{aligned} \mathbb{E}\left(\hat{R}_f\left(Y_0\right)\right) &= \mathbb{E}\left(\frac{f(Z)}{g(Z)}h(Z)\delta_{\{h(Z)\geq Y_0\}}\right) \\ &= \sum_{z\in\mathcal{Z}} g(z) \times \frac{f(z)}{g(z)}h(z)\delta_{\{h(z)\geq Y_0\}} \\ &= R_f\left(Y_0\right). \end{aligned} \tag{4.44}$$

Equation (4.44) holds since both $g(Z)$ and $f(Z)$ are defined on the same set $\mathcal{Z}$, which indicates that any possible cascading failures can be sampled concerning $g(Z)$ or $f(Z)$, provided the sample size is large enough.

The unbiasedness of Eq. (4.43) guarantees the effectiveness of (4.43). Moreover, in some special cases where the difference between $g(Z)$ and $f(Z)$ is huge, the size of Z_g can be enlarged to reduce the estimation error. Noting that only the information of samples in Z_g is required in Eq. (4.43), the blackout risk concerning $f(Z)$, i.e., R_f, can be estimated directly, with no need to regenerate cascading failure samples. That indicates (more) samples only need to be generated concerning $g(Z)$ instead of the varying $f(Z)$. This feature can further lead to an efficient algorithm to analyze blackout risk under varying CoFPFs, which will be discussed in the following subsection.

4.4.3 Estimating Blackout Risks with Varying CoFPFs

4.4.3.1 Changing a Single CoFPF

Initially, it is a simple case where only a single CoFPF change is considered. Suppose the CoFPF of component m changes from φ_m to $\bar{\varphi}_m$ (for simplicity, we use the notion $\bar{\varphi}_m$ to denote the new CoFPF of component m, which may have a new function form or parameters $\boldsymbol{\eta}_m$), and the corresponding joint probability series changes from $g(Z)$ to $f(Z)$. Considering a sample of cascading failure path generated concerning $g(Z)$, i.e., $z^i \in Z_g$, it has

$$\begin{aligned} f\left(z^i\right) &= \prod_{j=0}^{n^i-1} f_{j+1}\left(x^i_{j+1} \mid x^i_j\right) \\ &= \prod_{j=0}^{n^i-1}\left[\prod_{k\in F^m\left(x^i_j\right)} \varphi_k\left(x^i_j\right)\cdot \prod_{k\in \bar{F}^m\left(x^i_j\right)}\left(1-\varphi_k\left(x^i_j\right)\right)\right] \\ &\quad \cdots \Gamma\left(\bar{\varphi}_m, z^i\right), \end{aligned} \tag{4.45}$$

where

$$\Gamma\left(\bar{\varphi}_m, z^i\right) = \begin{cases} \prod_{j=0}^{n^i_m-1}\left(1-\bar{\varphi}_m\left(x^i_j\right)\right) & : \quad \text{if} \quad n^i_m = n^i \\ \bar{\varphi}_m\left(x^i_{n^i_m}\right)\prod_{j=0}^{n^i_m-1}\left(1-\bar{\varphi}_m\left(x^i_j\right)\right) & : \quad \text{otherwise} \end{cases}. \tag{4.46}$$

Here, n^i_m is the stage in z^i at which component m fails. Particularly, $n^i_m = n^i$ when the m-th component still works normally at the last stage of the cascading failure path. Component set $F^m(x^i_j) := F(x^i_j) \setminus \{m\}$ consists of all the elements in $F(x^i_j)$ except for m. Similarly, $\bar{F}^m(x^i_j) := \bar{F}(x^i_j) \setminus \{m\}$ is the component set including all the elements in $\bar{F}(x^i_j)$ except for m. According to (4.42), the sample weight that only CoFPF of component m changes is

$$w\left(z^i\right) = \frac{f\left(z^i\right)}{g\left(z^i\right)} = \frac{\Gamma\left(\bar{\varphi}_m, z^i\right)}{\Gamma\left(\varphi_m, z^i\right)}. \tag{4.47}$$

Substituting (4.47) into (4.43), the unbiased estimation of blackout risk is

$$\hat{R}_f\left(Y_0\right)=\frac{1}{N}\sum_{i=1}^{N}\frac{\Gamma\left(\bar{\varphi}_m, z^i\right)}{\Gamma\left(\varphi_m, z^i\right)}h\left(z^i\right)\delta_{\{h(z^i)\geq Y_0\}}. \tag{4.48}$$

Equation (4.48) enables an unbiased estimation of blackout risk when CoFPF changes, where the original samples are utilized, and no additional simulations are required.

4.4.3.2 Changing Multiple CoFPFs

The general case is that multiple CoFPFs change simultaneously. Invoking the expression of $f(z^i)$ in (4.45), there is

$$g\left(z^i\right)=\prod_{k\in K}\Gamma\left(\varphi_k, z^i\right) \tag{4.49}$$

$$f\left(z^i\right)=\prod_{k\in K_c}\Gamma\left(\bar{\varphi}_k, z^i\right)\cdot\prod_{k\in K_u}\Gamma\left(\varphi_k, z^i\right), \tag{4.50}$$

where K is the universal set of all components in the system; K_c is the set of components whose CoFPFs change; K_u is the set of others, i.e., $K = K_c \cup K_u$; $\bar{\varphi}_k$ is the new CoFPF of the k-th component.

According to (4.49) and (4.50), the sample weight is given by

$$w\left(z^i\right)=\prod_{k\in K_c}\frac{\Gamma\left(\bar{\varphi}_k, z^i\right)}{\Gamma\left(\varphi_k, z^i\right)}. \tag{4.51}$$

Substituting (4.51) into (4.43) yields

$$\hat{R}_f\left(Y_0\right)=\frac{1}{N}\sum_{i=1}^{N}\left(\prod_{k\in K_c}\frac{\Gamma\left(\bar{\varphi}_k, z^i\right)}{\Gamma\left(\varphi_k, z^i\right)}h\left(z^i\right)\delta_{\{h(z^i)\geq Y_0\}}\right). \tag{4.52}$$

Equation (4.52) is a generalization of Eq. (4.48). Equation (4.52) provides a mapping between the unbiased estimation of blackout risk and CoFPFs. When multiple CoFPFs change, the impartial estimate of blackout risk can be directly calculated using (4.52). It is computationally efficient since no additional cascading failure simulations are required, and only algebraic calculations are involved.

4.4.3.3 Unbiased Estimation of Blackout Risks

To clearly illustrate the algorithm, we first rewrite (4.52) in a matrix form as

$$\hat{R}_f(Y_0) = \frac{1}{N}\mathbf{L}\mathbf{F}_\mathbf{p}. \tag{4.53}$$

In (4.53), $\mathbf{L}$ is an N-dimensional row vector, where $\mathbf{L}_i = h(z^i)\delta_{\{h(z^i)\geq Y_0\}}/g(z^i)$. $\mathbf{F}_\mathbf{p}$ is an N-dimensional column vector, where $\mathbf{F}_{\mathbf{p}_i} = f(z^i)$.

Further define two $N \times k_a$ matrices $\mathbf{A}$ and $\mathbf{B}$, where k_a is the total number of all components, $\mathbf{A}_{ik} = \Gamma(\varphi_k, z^i)$, $\mathbf{B}_{ik} = \Gamma(\bar{\varphi}_k, z^i)$. According to (4.50), there is

$$\mathbf{F}_{\mathbf{p}_i} = \prod_{k\in K_c} \mathbf{B}_{ik} \cdot \prod_{k\in K_u} \mathbf{A}_{ik}. \tag{4.54}$$

Then, the algorithm is given as follows:

- **Step 1: Generating samples**. Based on the system and blackout model in consideration, generate a set of i.i.d. samples. Record the sample sets Z_g and Y_g, as well as the row vector $\mathbf{L}$.
- **Step 2: Calculating $\mathbf{F}_\mathbf{p}$**. Define the new CoFPFs for each component in K_c, and calculate $\mathbf{B}$ and $\mathbf{A}$. Then calculate $\mathbf{F}_\mathbf{p}$ according to (4.54). Particularly, instead of calculation, $\mathbf{A}$ can be saved in Step 1.
- **Step 3: Data analysis**. According to (4.53), estimate blackout risk for the changed CoFPFs.

A toy 4-bus system (see Fig. 4.8) illustrates the calculation steps to clarify the algorithm better. For simplicity, we number the lines and use symbols to represent CFPs (see Tables 4.2 and 4.3).

Step 1. First, generate samples with the simulation model.

Step 2. Then, we take the i-th element in $\mathbf{F}_\mathbf{p}$ as an example. Consider the path of the i-th cascading failure shown in Tab II. (At stage 1, lines 2 and 3 are tripped. Then, at stage 2, line 4 fails, and the blackout happens as a consequence. The load shedding is y^i.)

For simplicity, the original CFP of the k-th line at stage j is denoted by $p_{j,k}$, e.g., the original CFP of line 2 at stage 1 is $p_{1,2}$. Then, the probability of the i-th cascading failure is

$$\begin{aligned} g\left(z^i\right) =& p_{1,2}p_{1,3}\left(1-p_{1,1}\right)\left(1-p_{1,4}\right)\left(1-p_{1,5}\right) \\ & \cdots \times p_{2,4}\left(1-p_{2,1}\right)\left(1-p_{2,5}\right). \end{aligned}$$

And there is

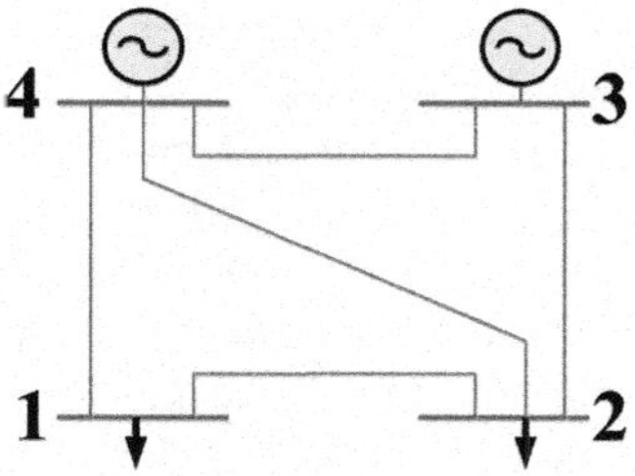

Fig. 4.8 Toy 4-bus system

Table 4.2 Line numbers

Line	Bus1–Bus2	Bus1–Bus4	Bus2–Bus3	Bus2–Bus4	Bus3–Bus4
Number	1	2	3	4	5

Table 4.3 One cascading path

Stage	Fault line	Original CFP
1	2	$p_{1,2}$
1	3	$p_{1,3}$
2	4	$p_{2,4}$

$$\mathbf{L}_i = \frac{y^i \delta_{\{y^i \geq Y_0\}}}{g\left(z^i\right)}$$

$$\mathbf{A}_{i4} = \left(1 - p_{1,4}\right) p_{2,4}.$$

Other elements in the i-th row of **A** can be obtained similarly.

On the other hand, suppose the CFPs of line 4 at stage 1 and stage 2 change to $p'_{1,4}$ and $p'_{2,4}$, respectively. Then $\mathbf{B}_{i4} = (1 - p'_{1,4})p'_{2,4}$. Other elements in the i-th row of **B** can be obtained similarly.

If we only consider the CFP of line 4 changes, then $\mathbf{F_{pi}} = p_{1,2}p_{1,3}(1-p_{1,1})(1-p'_{1,4})(1 - p_{1,5}) \times p'_{2,4}(1 - p_{2,1})(1 - p_{2,5})$. Other scenarios are similar.

Step 3. Based on the previous calculation, the blackout risk can be directly estimated using (4.53).

4.4.3.4 Some Implications

The proposed method has important implications in blackout-related analyses. Two typical examples are (1) efficient estimation of blackout risk under extreme weather conditions and (2) risk-based maintenance scheduling.

For the first case, as well-known, extreme weather conditions (e.g., typhoons) often occur for a short time but affect a wide range of components. The failure probabilities of related components may increase remarkably. In this case, the proposed method can be applied to fast evaluate the consequent risk regarding the weather forecast.

For the second case, since maintenance can considerably reduce CFP, the proposed method allows the efficient identification of the most effective candidate devices in the system for mitigating blackout risk. Specifically, suppose that one only considers the simultaneous maintenance of at most d_m components. Then the number of possible scenarios is up to $\sum_{d=1}^{d_m} C(k_a, d)$, which turns out to be intractable in a sizeable practical system ($C(k_a, d)$ is the number of d-combinations from k_a elements). Moreover, in each scenario, many cascading failure simulations are required to estimate blackout risk, which is highly time-consuming or even practically impossible. In contrast, with the proposed method, one only needs to generate the sample set in the base scenario. Then, blackout risks for other scenarios can be directly calculated using only algebraic calculations, which are very simple and computationally efficient.

4.4.4 Example

4.4.4.1 Setting

In this experiment, the numerical experiments are carried out on the IEEE 39-bus system. Moreover, the widely used OPA model is employed to generate the samples in Z_g. Notably, since the proposed method aims to quantify the influence of CFP on blackout risk concerning a deterministic initial state, the slow dynamic representing the load growth is omitted. Then, the primary sampling steps that simulate the propagation of cascading failures are summarized as follows:

- **Step 1: Data initialization**. Initialize the system data and parameters. Particularly, define specific CoFPFs for each component. The initial state is x_0.
- **Step 2: Sampling outages**. At stage j of the i-th sampling, according to the system state, x_j^i and CoFPFs simulate the component failures concerning the failure probabilities.
- **Step 3: Termination judgment**. If new failures happen in Step 2, recalculate the system state x_{j+1}^i at stage $j+1$ with the new topology and return to Step 2; otherwise, one sampling ends. The corresponding samples are $z^i = \left\{x_0, x_1^i \cdots x_j^i\right\}$ and y^i. If all N simulations are completed, the sampling process ends.

In this simulation model, the state variables X_j are chosen as all components' ON/OFF statuses and power flow on corresponding components at stage j. Meanwhile, the random failures of transmission lines and power transformers are considered. The employed CoFPF is

$$\varphi_k\left(s_k, \boldsymbol{\eta}_k\right) = \begin{cases} p_{\min}^k & : s_k < s_d^k \\ \frac{p_{\max}^k - p_{\min}^k}{s_u^k - s_d^k}\left(s_k - s_d^k\right) + p_{\min}^k & : \text{else} \\ p_{\max}^k & : s_k > s_u^k \end{cases}, \tag{4.55}$$

where s_k is the load ratio of component k and $\boldsymbol{\eta}_k = [p_{\min}^k, p_{\max}^k, s_d^k, s_u^k]$. Specifically, $p_{\min}^k$ denotes the minimum failure probability of component k when the load ratio is less than s_d^k; $p_{\max}^k$ denotes the maximum failure probability when the load ratio is larger than s_u^k. Usually, it holds $0 < p_{\min}^k < p_{\max}^k < 1$, which depicts a certain probability of hidden failures in protection devices.

It is worth noting that the simulation process with the specific settings mentioned above is a simple way to emulate the propagation of cascading failures, which is only utilized to demonstrate the proposed method. The proposed method can be applied when more realistic models and parameters are adopted.

4.4.4.2 Unbiasedness of the Blackout Risk Estimation

In this case, it is shown that the proposed method can achieve an unbiased estimation of blackout risk. We initially choose the parameters as $s_d^k = 1.0$, $s_u^k = 1.25$, $p_{\max}^k = 0.99$, $p_{\min}^k = 0.05$, $k \in K$ and carry out 2000 cascading failure simulations with the initial parameters. Then the sample set Z_g and related $\mathbf{L}$ are obtained. Afterward, a set of failure components K_c are randomly selected, including two components. Accordingly, we modify the parameters, $\boldsymbol{\eta}_k$, of their CoFPFs to $\bar{p}_{min}^k = p_{min}^k - 0.001$, where $k \in K_c$. The blackout risks are estimated using (4.53) regarding the new settings and various load shedding levels. For comparison, we regenerate 2000 samples under the new settings and estimate the blackout risks using (4.41). The results are given in Fig. 4.9.

Figure 4.9 illustrates that the estimations of blackout risk with the two methods are almost the same, which indicates the proposed method can achieve an unbiased estimate of blackout risk. Note that the proposed method requires no additional simulations, making it far more efficient than the traditional MCS and scalable for large-scale systems.

4.4.4.3 Parameter Changes in CoFPFs

In this case, the performance of the proposed method is evaluated when parameters $\boldsymbol{\eta}_k$ of some CoFPFs change. The sample set Z_g and associated $\mathbf{L}$ are based on the 2000 samples concerning the initial parameters. We consider two settings: (1) $Y_0 = 0$; (2) $Y_0 = 2000$. Statistically, it can be obtained that $\hat{R}_g(0) = 975.12$ and $\hat{R}_g(2000) = 647.21$, respectively.

Since there are 46 components in the system, we consider $k_a = 46$ different scenarios. Specifically, each CoFPF is changed individually by allowing $\bar{p}_{min}^k =$

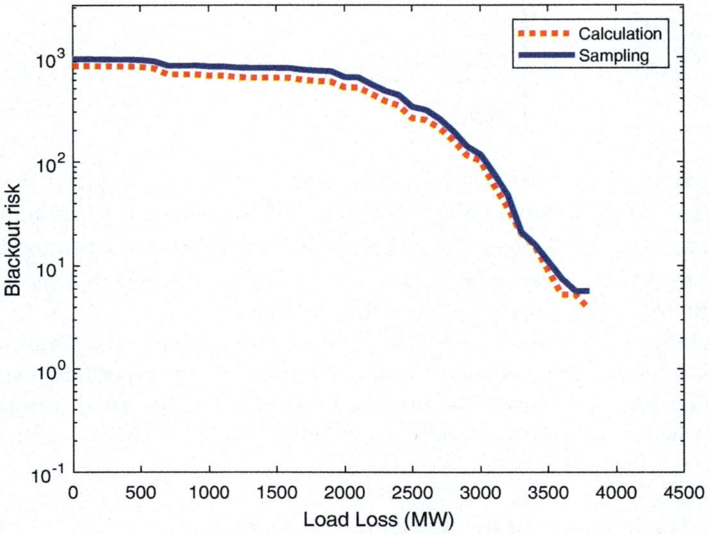

Fig. 4.9 Estimation of blackout risk with sampling and calculation

$p^k_{min} - 0.002$. Using the proposed method, blackout risks can be estimated quickly. Some results are presented in Table 4.4 ($Y_0 = 0$) and Table 4.5 ($Y_0 = 2000$). In Tables 4.4 and 4.5, particularly, the average computational times of unbiased estimation of blackout risk in each scenario are 0.00050 s and 0.00039 s, respectively.

Table 4.4 displays the top twelve scenarios having the lowest value of blackout risk as well as the average risk of all scenarios. Whereas decreasing failure probabilities of specific components can effectively mitigate blackout risk, others have little impact. For example, reducing p^k_{min} of component #14 results in a 1.8% reduction of blackout risk, while the average ratio of risk reduction is only 0.7%. This result implies that some critical components may play a core role in the propagation of cascading failures and the promotion of load shedding. The proposed method enables a scalable way to efficiently identify those components, which may facilitate the effective mitigation of blackout risk.

When considering $Y_0 = 2000$, which is associated with severe blackout events, it is interesting to see the most influential components in Table 4.5 are similar to those in Table 4.4. But in other cases, these influential components between different load losses may differ drastically. In other words, the impact of CoFPF varies with different load shedding levels, demonstrating the complex relationship between blackout risk and CoFPFs.

Then, we simultaneously lower p^k_{min} of two CoFPFs. Z_g and $\mathbf{L}$ are the same as before, and the number of scenarios is $C(k_a, 2) = C(46, 2)$. The calculated unbiased estimations concerning two Y_0 are listed in Tables 4.6 and 4.7. Unsurprisingly, the risk reduction ratios are more remarkable than the results in Table 4.4 and Table 4.5, where only one CoFPF decreases. Moreover, it should be noted that the

Table 4.4 BP estimation when parameters of CoFPFs change ($Y_0 = 0$)

Component index	Blackout risk	Risk reduction ratio
20	948.541	0.027
21	956.665	0.019
14	957.489	0.018
22	958.342	0.017
37	959.305	0.016
6	959.307	0.016
35	960.103	0.015
34	960.207	0.015
28	960.282	0.015
39	960.463	0.015
33	960.675	0.015
27	960.925	0.015
...	...	...
Mean value	968.205	0.007

Table 4.5 BP estimation when parameters of CoFPFs change ($Y_0 = 2000$)

Component index	Blackout risk	Risk reduction ratio
20	948.541	0.027
14	957.489	0.018
21	956.665	0.019
22	958.342	0.017
6	959.307	0.016
37	959.305	0.016
35	960.103	0.015
39	960.463	0.015
36	961.001	0.014
28	960.282	0.015
34	960.207	0.015
33	960.675	0.015
...	...	...
Mean value	640.918	0.010

pairs of components in Tables 4.6 and 4.7 are the combinations of the ones shown in Tables 4.4 and 4.5. But in other cases, the combinations are not directly composed of individual critical components, for which the relationship between blackout risk and CFP is complicated and nonlinear, resulting in difficulties in analyses.

4.5 An Application of Probabilistic Analytics to Blackout Risk Mitigation

This subsection introduces an application, i.e., using a dynamic thermal rating (DTR) sensor to mitigate cascading failure risk. However, traditional methods can-

Table 4.6 Risk reduction ratio when parameters of CoFPFs change ($Y_0 = 0$)

Component index	Blackout risk	Risk reduction ratio
[20,21]	930.753	0.045
[14,20]	931.562	0.045
[20,22]	932.393	0.044
[20,37]	933.250	0.043
[6,20]	933.308	0.043
[20,35]	934.050	0.042
[20,34]	934.151	0.042
[20,28]	934.214	0.042
[20,39]	934.413	0.042
[20,33]	934.611	0.042
[20,38]	934.860	0.041
[20,27]	934.862	0.041
...	...	...
Mean value	961.392	0.014

Table 4.7 Risk reduction ratio when parameters of CoFPFs change ($Y_0 = 2000$)

Component index	Blackout risk	Risk reduction ratio
[14,20]	608.784	0.059
[20,21]	609.984	0.058
[20,22]	611.609	0.055
[6,20]	611.794	0.055
[20,37]	613.822	0.052
[20,35]	613.822	0.052
[20,39]	613.822	0.052
[14,21]	613.976	0.051
[20,36]	613.982	0.051
[20,28]	614.076	0.051
[20,34]	614.145	0.051
[20,33]	614.171	0.051
...	...	...
Mean value	634.706	0.019

not optimally address the combinatorial optimization issue of sensor placement and may suffer from the dimensionality curse. Risk mitigation may result in the Braess paradox, where line updates increase failure risk. A submodular optimization-based DTR placement model is presented to address these challenges, considering the Braess paradox. Based on the models in Sects. 4.2 and 4.4, the failure risk is quickly measured. Then, a unique submodular optimization methodology is provided to reformulate the risk mitigation model, while a computationally efficient algorithm is developed with a provable approximation guarantee. Case results validate the benefits of the suggested models.

4.5.1 Preliminaries

4.5.1.1 DTR

DTR is an advanced technique that temporarily allows transmission lines to increase the maximum thermal rating. In detail, the placed DTR sensors can measure the conductor's ambient environment and line state data, which are then sent to the system operator to determine the new thermal rating dynamically. These measured data include local environment parameters such as line surrounding temperature and wind speed, as well as line states such as conductor temperature and line sag. DTR enables the system to take advantage of existing conductors' transformation capability, which postpones the costly and complex grid expansion, including tower erection, new line construction, and old-line updates. Another benefit of DTR is that it allows power systems to integrate more wind power, resulting in more significant emission reduction. It can also reduce the sensing error of system state and environment, allowing operators to design more accurate schedules. Furthermore, DTR can reinforce the power system reliability by decreasing load loss and not-served energy. In this chapter, we investigate the benefit of this promising technique in mitigating the risk of cascading failure when the power system is under heavy load or extreme weather.

There are mainly three international standards detailing the DTR composition and scheme: IEC, CIGRE, and IEEE, and some analyses indicate that the ratings computed by each standard differ by less than 10%. Then, in this chapter, we apply the CIGRE standard for DTR calculation:

$$Q_c + Q_r + M_c \frac{\mathrm{d}T}{\mathrm{d}t} = I^2 R + Q_s \tag{4.56}$$

$$\frac{\mathrm{d}T}{\mathrm{d}t} = \frac{I^2 R + Q_s - Q_c - Q_r}{M_c}, \tag{4.57}$$

where I^2R denotes the conductor heating from power flow, Q_s the solar radiation heating, Q_c the wind cooling, Q_r the radiation cooling, and $\mathrm{d}T/\mathrm{d}t$ the non-steady form related to temperature T, time t, and conductor heat capacity M_c.

Equation (4.56) is the non-steady-state form of heat-balance equation for DTR calculation, primarily applied to the schedule problems on a short-term time scale, assuming DTR has been installed in the system and containing the situation in which the conductor temperature changes rapidly. On the other hand, this chapter aims to mitigate the cascading failure risk using DTR placement, a type of planning issue with a long-term time scale. Namely, the effect of DTR in planning issues is evaluated from historical long-term weather data, specifically the average weather data over an appointed period. Regularly, the Drake aluminum cable steel reinforced (ACSR) reported that the conductor temperature could reach a steady value within sufficient time so that in the planning issue, the referred weather data for DTR can be regarded as collected from a static conductor ambient weather, meaning that it is feasible to apply the static heat-balance equation to the DTR placement, as described follows:

$$Q_c + Q_r = I^2 R + Q_s. \tag{4.58}$$

Specifically, the values of Q_s, Q_c, and Q_r are related to the ambient weather condition, such as wind speed V and surrounding temperature T (or average values), and their detailed formulas can be found in [61]. For convenience, we adopt $I^2 R$, the power as DTR value in the following risk mitigation model.

Even though DTR has many benefits for system operation, mass installment of DTR is expensive. Consequently, how to deploy DTR under a limited budget to optimally reduce system risk is a primary problem.

4.5.1.2 DTR Function in Cascading Failures

The cascading failure evolution depends on relay actions of power protection. When the line is overcurrent, the relay will cut off the line automatically, and the successive cutting procedure will result in cascading failure. Whether the line is overcurrent is identified based on a predefined threshold called the maximum allowed conductor current (or power). Traditionally, this threshold is default from a relatively conservative weather condition, e.g., the condition of 40°C ambient temperature and 0.61 m/s wind speed. However, the realistic environments for DTR placement are usually less harsh than the default weather. In other words, the value of maximum current (or power) set by DTR measured data is greater than the default, leading to a lower probability for relay action when the conductor still has sufficient capacity to sustain the higher power energy transform. Thus, DTR can assist power systems in using the remaining transformation capability in existing conductors, reducing the cascading failure risk.

Additionally, the DTR-based threshold in relay action differs from the real-time DTR value. Due to the automation of power protection, these security thresholds must be predefined, and the real-time DTR value is not practical and accurate for the system operator to determine the maximum transform. Instead, these real-time DTR data are suitable to aggregate for preparation of the threshold setting of the following hour or a longer period. Regarding risk mitigation, there are two scenarios to consider for DTR-based thresholds. The first scenario concerns planning, where the DTR has not yet been placed. In this case, the DTR-based threshold is calculated using long-term historical weather data to estimate the conductivity of the conductor. This helps system planners determine the optimal location for the DTR to reduce risk. The second scenario relates to scheduling problems, where the DTR is already in place within the system. In this case, the threshold is determined using weather data from both shorter historical periods measured by the DTR and weather predictions, such as data from the previous hour and forecasts for the next hour. In the two scenarios, the threshold values are all predefined to guide the relay action in system protection based on data for a period long enough to ignore the non-steady state in heat balance so that the static heat-balance equation (4.58) is more suitable for DTR calculation in the relay threshold setting.

4.5.1.3 Submodular Function

Submodularity is defined as the diminishing return property in some special set functions. Specifically, it expresses that the incremental "value," "cost," or "gain" of adding elements to a set A decrease when A grows larger. This property appears in a wide variety of fields, including economics, computer science, network analysis, and so on. Its definition is as follows.

Definition 2 Suppose set $A \subseteq B \subset N$ and set $v \in N \backslash B$, where N is the ground set. A function $f : 2^{|N|} \to \mathbb{R}$ is submodular if it satisfies

$$f(A \cup v) - f(A) \geq f(B \cup v) - f(B). \tag{4.59}$$

Starting with the definition, we have the following Lemmas, which provide a facilitated tool for determining when a function is submodular.

Lemma 1 *If* $\alpha_i \geq 0$ *and* $f_i : 2^{|N|} \to \mathbb{R}$ *is a submodular function, then so is* $\sum_i \alpha_i f_i$.

Lemma 2 *Any submodular function* f *can be represented as a sum of submodular functions* $\widetilde{f}$ *and a modular function* m*, i.e.,* $f = \widetilde{f} - m$.

The submodular function enables combinatorial optimization with some provable guarantees, offering an analytical tool for detailed analysis of the optimization process. It can also be resolved in polynomial time, making it suited for dealing with dimensionality curse problems and the latest online applications. These benefits make submodular function prominent in several fields of research.

4.5.2 DTR-Based Risk Mitigation Model

4.5.2.1 Modeling of Cascading Failures

Define $M^k = \{(FG_0^k, \ldots, FG_i^k, \ldots) : M^k \in M\}$ as the k-th cascading failure chain, in which $FG_i^k \in 2^N$ denotes the system line state of i-th failure generation, and M denotes the cascading failure chain database. The system's line set is S_L. Each failure chain in M is independently identically distributed (i.i.d). Assume that the single component failure probability follows the piecewise probability distribution of hidden failure. Owing to the piecewise structure, the single component failure probability $\varphi_e(FG_i^k)$ is approximated smoothly by a Sigmoid form:

$$\varphi_e(FG_i^k) = \mathrm{Pr}_e^{\min} + \frac{\mathrm{Pr}_e^{\max} - \mathrm{Pr}_e^{\min}}{1 + \exp\left[-\mu \frac{2P_e - (P_e^{\min} + P_e^{\max})}{P_e^{\min} + P_e^{\max}}\right]}, \tag{4.60}$$

where P_e, $P_e^{\min}$, $P_e^{\max}$, respectively, denote current power flow, minimum and maximum transmission capabilities of line e, $\mathrm{Pr}_e^{\min}$, $\mathrm{Pr}_e^{\max}$ the minimum and maximum of failure probability, respectively, and μ the approximate factor to avoid the vanishing gradient problem.

Suppose that the cascading failure chain is of Markov property. Then define f^k as the probability of a specific cascading failure chain M^k with $|M^k| = d$:

$$\begin{aligned} f^k =& f(FG_0^k) f(FG_1^k | FG_0^k) \cdots f(FG_i^k | FG_{i-1}^k) \\ & \cdots f(FG_d^k | FG_{d-1}^k), \end{aligned} \tag{4.61}$$

where $f(FG_i^k | FG_{i-1}^k)$ denotes the transition probability from failure generation $i-1$ to i, and $f(FG_0^k) \approx 1$ denotes the initial.

Specifically, for a single failure generation FG_i^k, its transition probability is

$$f(FG_i^k | FG_{i-1}^k) = \prod_{e \in S_{f_i}} \varphi_e(FG_i^k) \cdot \prod_{e \in \overline{S_{f_i}}} \left(1 - \varphi_e(FG_i^k)\right), \tag{4.62}$$

where S_{f_i} denotes the line set that functioned normally in FG_{i-1}^k but failed in FG_i^k, and $\overline{S_{f_i}}$ denotes the line set that functioned normally in FG_i^k.

4.5.2.2 Risk Model Considering Cascading Failures

In this part, the load loss and the probability of failure chain are combined to construct a risk model. First, define load loss induced by a single failure chain M^k as Y_k obtained from the cascading failure simulator.

Since cascading failure is treated as an infrequent but extreme event, which has less occurrence probability but can bring great damage to the whole power system, we define $\delta_{\{Y_k > Y_{ext}\}}$ as an indicator that selects the cascading failure chain in which the load loss exceeds Y_{ext}, the setting loss threshold. Specifically, $\delta_{\{Y_k > Y_{ext}\}}$ can be formulated as

$$\delta_{\{Y_k > Y_{ext}\}} = \begin{cases} 1 & : Y_k > Y_{ext} \\ 0 & : \text{others} \end{cases}. \tag{4.63}$$

Then, the original risk of a single failure chain M^k is proposed as

$$Risk_{\emptyset}^k = Y_k \cdot f^k \cdot \delta_{\{Y_k > Y_{ext}\}}. \tag{4.64}$$

Above all, the original risk of whole failure chains can be modeled as

$$RiskW_{\emptyset} = \mathbb{E}\left(Y \cdot \delta_{\{Y>Y_{ext}\}}\right) = \sum_{M^k \in M} Risk_{\emptyset}^k = \frac{1}{|M|} \sum_{M^k \in M} Y_k \cdot \delta_{\{Y_k > Y_{ext}\}}. \tag{4.65}$$

It can indicate that this kind of risk is the unbiased estimation, treated as a variant of Conditional Value at Risk.

4.5.3 DTR-Based Risk Mitigation Model

4.5.3.1 DTR Placement in a Single Line

Suppose DTR is placed in line e. Then, its failure probability in FG_i^k is given by

$$\varphi_e'(FG_i^k) = \mathrm{Pr}_e^{\min} + \frac{\mathrm{Pr}_e^{\max} - \mathrm{Pr}_e^{\min}}{1 + \exp\left[-\mu \frac{2P_e - \alpha(P_e^{\min} + P_e^{\max})}{\alpha(P_e^{\min} + P_e^{\max})}\right]}, \tag{4.66}$$

where α denotes the DTR improved parameter for power transmission threshold, enhancing the original to $\alpha P_e^{\max}$ related to ambient weather like wind power and temperature.

Then, the occurrence probability of M^k changes from f^k to g_e^k corresponding to DTR placement in line e. Given $d = |M^k|$, g_e^k can be formulated as

$$g_e^k = \prod_{i=1}^{d} \left[\prod_{j \in \overline{S_{f_i}^{non}}} \varphi_j(FG_i^k) \cdot \prod_{j \in S_{f_i}^{non}} \left(1 - \varphi_j(FG_i^k)\right) \right] \ldots \cdot \mathrm{H}^k(\varphi_e'), \tag{4.67}$$

where $\overline{S_{f_i}^{non}}$ denotes the no-DTR line set that functioned normally in FG_{i-1}^k but failed in FG_i^k, $S_{f_i}^{non}$ denotes the no-DTR line set that functioned normally in FG_i^k, and $\mathrm{H}^k(\varphi\prime_e)$ is defined as

$$\mathrm{H}^k(\varphi_e') = \begin{cases} \prod_{i=1}^{d} \left(1 - \varphi_e'(FG_i^k)\right) & : d_e > d \\ \varphi_e'(FG_{d_e}^k) \cdot \prod_{i=1}^{d_e - 1} \left(1 - \varphi_e'(FG_i^k)\right) & : \text{others}, \end{cases} \tag{4.68}$$

where d_e is the generation when line e fails. Otherwise, if $d_e > d$, there is no failure in line e during M^k.

4.5.3.2 DTR Placement in a Set of Lines

Letting S_{A} as the line set chosen to install DTR, we set the occurrence probability of M^k changes from f^k to g_{A}^k formulated as

$$g_{\mathrm{A}}^k = \prod_{i=1}^{d} \left[\prod_{j \in \overline{S_{f_i}^{\mathrm{non}}}} \varphi_j(FG_i^k) \cdot \prod_{j \in S_{f_i}^{\mathrm{non}}} \left(1 - \varphi_j(FG_i^k)\right) \right] \prod_{e \in S_{\mathrm{A}}} \mathrm{H}^k(\varphi_e'), \tag{4.69}$$

where $\mathrm{H}^k(\varphi\prime_e)$ is defined similarly as (4.68).

4.5.3.3 Important Sampling Weight Technique

DTR technology can be placed in transmission lines that set the dynamic thermal rating rather than the static. However, there is a problem that when some parameters such as line maximum capacity change, the related cascading failure chains may be different, affecting failure chain probability and risk index. However, regenerating the failure database based on the updated system will cause a huge computation burden, especially in optimization problems. To tackle this problem, the critical sampling weight technique is applied.

Given line set S_{A} placed with DTR, the underlying relationship between g_{A}^k and f^k can be expressed as

$$W_{\mathrm{A}-\emptyset}^k = \frac{g_{\mathrm{A}}^k}{f^k} = \prod_{e \in S_{\mathrm{A}}} \frac{\mathrm{H}^k(\varphi_e')}{\mathrm{H}^k(\varphi_e)}, \tag{4.70}$$

where φ_e denotes the original failure probability.

Furthermore, for two different line sets installed with DTR, S_{A} and S_{B}, the hidden relation between g_{A}^k and g_{B}^k can be formulated as

$$W_{\mathrm{A}-\mathrm{B}}^k = \frac{g_{\mathrm{A}}^k}{g_{\mathrm{B}}^k} = \prod_{e \in S_{\mathrm{A}} \backslash S_{\mathrm{B}}} \frac{\mathrm{H}^k(\varphi_{\mathrm{A}e}')}{\mathrm{H}^k(\varphi_{\mathrm{B}e})} \cdot \prod_{e \in S_{\mathrm{B}} \backslash S_{\mathrm{A}}} \frac{\mathrm{H}^k(\varphi_{\mathrm{A}e})}{\mathrm{H}^k(\varphi_{\mathrm{B}e}')}, \tag{4.71}$$

where $\varphi_{\mathrm{A}e}'$ and $\varphi_{\mathrm{B}e}'$ denote the failure probabilities of line e when DTR is placed in S_{A} and S_{B}, respectively, and $\varphi_{\mathrm{A}e}$ and $\varphi_{\mathrm{B}e}$ the original probabilities. Moreover, if $S_{\mathrm{B}} \subset S_{\mathrm{A}}$, (4.71) can be calculated as

$$W_{\mathrm{A}-\mathrm{B}}^k = \prod_{e \in S_{\mathrm{A}} \backslash S_{\mathrm{B}}} \frac{\mathrm{H}^k(\varphi_{\mathrm{A}e}')}{\mathrm{H}^k(\varphi_e)}, \tag{4.72}$$

where $\varphi_e = \varphi_{\mathrm{B}e}$. On the basis of the assumption in (4.72), it is easily obtained that

$$W^k_{\{A+v\}-\{B+v\}} = W^k_{A-B}, \tag{4.73}$$

where v denotes an arbitrary line set belonging to S_L.

On the other hand, in failure chain database M where $|M| = D$, the important sampling weight is not only related to component failure probability but also the load loss. Define load loss matrix as $\mathbf{Y} = (Y_1\delta_{\{Y_1>Y_{ext}\}}, \ldots, Y_D\delta_{\{Y_D>Y_{ext}\}})^T$, and the weight matrix compared with the original system as $\mathbf{W}_{A-\emptyset} = (W^1_{A-\emptyset}, \ldots, W^D_{A-\emptyset})^T$. Suppose that there are 2 different updated line sets S_A and S_B. Their sampling weight in database M can be defined as $\widetilde{W}_{A-B}$

$$\widetilde{W}_{A-B} = \frac{RiskW_A}{RiskW_B} = \mathbf{W}^T_{A-\emptyset}\mathbf{Y} \cdot \left(\mathbf{W}^T_{B-\emptyset}\mathbf{Y}\right)^{-1}. \tag{4.74}$$

4.5.4 Braess Paradox in Failure Risk Mitigation

The Braess paradox, known as a counter-intuitive phenomenon that improving or adding some components contrarily worsens the system performance, was initially discovered in transportation research. In the previous study, it was found that the mitigation of failure risk also exists in the Braess paradox, especially in random mitigation and static state mitigation strategies. In this part, the condition of the Braess paradox occurring is explored in failure risk mitigation.

For a specified cascading failure chain M^k with maximal generation d, suppose only line e is placed with DTR and fails in generation d_e. There are two situations for consideration:

1. If line e still functions normally when M^k ends, i.e., $d_e > d$, from (4.68), the sampling weight becomes

$$W^k_{e-\emptyset} = \prod_{i=1}^{d} \frac{1 - \varphi'_e(FG^k_i)}{1 - \varphi_e(FG^k_i)}. \tag{4.75}$$

   Due to $\varphi'_e(FG^k_i) < \varphi_e(FG^k_i)$ for each generation i, the ratio $\frac{1-\varphi'_e(FG^k_i)}{1-\varphi_e(FG^k_i)}$ would be larger than 1, so does the product of all generation ratios $\prod_{i=1}^{d} \frac{1-\varphi'_e(FG^k_i)}{1-\varphi_e(FG^k_i)}$. Therefore, the failure probability g^k_e with updated line e will be greater than the original failure probability f^k since $g^k_e = W^k_{e-\emptyset} f^k$, meaning that the Braess paradox occurs in this situation.
2. If line e fails in M^k, i.e., $d_e \le d$, it can obtain the sampling weight

$$W^k_{e-\emptyset} = \frac{\varphi'_e(FG^k_{d_e}) \cdot \prod_{i=1}^{d_e-1} (1 - \varphi'_e(FG^k_i))}{\varphi_e(FG^k_{d_e}) \cdot \prod_{i=1}^{d_e-1} (1 - \varphi_e(FG^k_i))}. \tag{4.76}$$

Based on (4.76), if there is $W^k_{e-\emptyset} > 1$, it means that g^k_e will be greater than f^k. Then, the Braess paradox happens in this situation. There are some helpful remarks about this situation.

Remark 6 In (4.76), we know the ratio $\frac{1-\varphi'_e(FG^k_i)}{1-\varphi_e(FG^k_i)} > 1$ for each failure generation, so the product of generation ratios line e functions normally, $\prod_{i=1}^{d_e-1} \frac{1-\varphi'_e(FG^k_i)}{1-\varphi_e(FG^k_i)}$, is more than 1. With d_e increases, $\prod_{i=1}^{d_e-1} \frac{1-\varphi'_e(FG^k_i)}{1-\varphi_e(FG^k_i)}$ will become larger. Consequently, for cascading failure propagating to a larger scale, updating some lines with higher survival probability during the failure period will increase the risk of failure propagation.

Remark 7 In (4.76), $\frac{\varphi'_e(FG^k_i)}{\varphi_e(FG^k_i)}$ is less than 1. To prevent the case that $W^k_{e-\emptyset} > 1$, $\frac{\varphi'_e(FG^k_i)}{\varphi_e(FG^k_i)}$ should be as small as possible. Define the deviation of probability as $\Delta\varphi_e = \varphi_e\left(FG^k_i\right) - \varphi'_e(FG^k_i)$. Thus, updating the line with smaller $\Delta\varphi_e$ will increase the probability of Braess paradox occurrence.

Remark 8 In (4.76), there is a threshold for the generation amount of line e to prevent the Braess paradox, $d^{thrd}_e \in \mathbf{N}^+$, satisfying both (4.77) and (4.78):

$$\frac{\varphi'_e(FG^k_{d^{thrd}_e-1})}{\varphi_e(FG^k_{d^{thrd}_e-1})} \leq \prod_{i=1}^{d^{thrd}_e-2} \frac{1 - \varphi_e(FG^k_i)}{1 - \varphi'_e(FG^k_i)} \tag{4.77}$$

$$\frac{\varphi'_e(FG^k_{d^{thrd}_e})}{\varphi_e(FG^k_{d^{thrd}_e})} > \prod_{i=1}^{d^{thrd}_e-1} \frac{1 - \varphi_e(FG^k_i)}{1 - \varphi'_e(FG^k_i)}. \tag{4.78}$$

Similarly, if the power system includes a line set S_{A} with DTR placement, it can obtain the sampling weight for M^k with maximal generation d as (4.70). This case is complicated since some updated lines would fail in M^k, while the others can survive. Define S^{fail}_{A} as the updated line set that failed in M^k, S^{surv}_{A} as the updated line set that survived in M^k, and $S_{\mathrm{A}} = S^{fail}_{\mathrm{A}} \cup S^{surv}_{\mathrm{A}}$. Then $W^k_{\mathrm{A}-\emptyset}$ can be expressed as

$$W^k_{\mathrm{A}-\emptyset} = \prod_{e \in S^{fail}_{\mathrm{A}}} \left(\frac{\varphi'_e(FG^k_{d_e}) \cdot \prod_{i=1}^{d_e-1} (1-\varphi'_e(FG^k_i))}{\varphi_e(FG^k_{d_e}) \cdot \prod_{i=1}^{d_e-1} (1-\varphi_e(FG^k_i))} \right)$$

$$\cdots \prod_{e \in S^{surv}_{\mathrm{A}}} \left(\prod_{i=1}^{d} \frac{1-\varphi'_e(FG^k_i)}{1-\varphi_e(FG^k_i)} \right), \tag{4.79}$$

where d_e denotes the failure generation of line e in M^k. If there is the Braess paradox, it has $W^k_{\mathrm{A}-\emptyset} > 1$.

To decrease the probability of the Braess paradox in this circumstance, we can adopt a DTR placement strategy based on Remark 9.

Remark 9 In (4.79), we have $\prod_{e \in S^{surv}_{\mathrm{A}}} (\prod_{i=1}^{d} \frac{1-\varphi'_e(FG^k_i)}{1-\varphi_e(FG^k_i)})$ is greater than 1. However, the term $\frac{\varphi'_{d_e}(FG^k_{d_e})}{\varphi_{d_e}(FG^k_{d_e})} \prod_{i=1}^{d_e-1} \frac{1-\varphi_e(FG^k_i)}{1-\varphi'_e(FG^k_i)}$ cannot be determined whether it is larger than 1. To decrease the probability of $W^k_{\mathrm{A}-\emptyset} > 1$, the updated lines should meet some of the following conditions:

1. Lines belonging to S^{fail}_{A}
2. Lines failed in the early generation of M^k
3. Lines with positive larger value of $\Delta\varphi_e = \varphi_e(FG^k_i) - \varphi'_e(FG^k_i)$

4.5.5 *Submodular Optimization of Risk Mitigation*

4.5.5.1 Optimization Construction

To establish the DTR placement strategy to mitigate cascading failure risk, define $S_{\mathrm{A}} \subseteq S_{\mathrm{L}}$ as an allocation of potential lines with DTR placement, and $C(S_{\mathrm{A}})$ as a linear function for the cost of placement S_{A}. Then we intend to find S_{A} that

$$\max_{S_{\mathrm{A}} \subseteq S_{\mathrm{L}}} -C(S_{\mathrm{A}}) + \gamma(RiskW_{\emptyset} - RiskW_{\mathrm{A}}), \tag{4.80}$$

where γ denotes an adjustment factor and $RiskW_{\emptyset}$ the original system risk. It can easily be seen that this objective function is non-monotone.

From Sect. 4.5.4, sometimes the Braess paradox occurs in risk mitigation, such that the objective function is not strictly submodular, i.e., the satisfaction of $W > 1$ will lead to invalidation of the submodular function definition. The detail is presented in Theorem 1.

Theorem 1 *Let the failure chain database M have $|M| = D$. Define 2 updated line sets as S_A and S_B, in which $S_\mathrm{A} \subseteq S_\mathrm{B} \subset S_\mathrm{L}$, and there is $v \notin S_\mathrm{B}$. Their sampling weight in each failure chain is $\mathbf{W_{B-A}} = (W^1_{\mathbf{B-A}}, W^2_{\mathbf{B-A}}, \ldots, W^D_{\mathbf{B-A}})^T$. If there are Braess paradox phenomena in mitigating some failure chains, the function* (4.80) *will not be satisfied with the submodular function definition.*

Proof Suppose that there is the Braess paradox happening in M^k when updating S_B compared with S_A so that

$$W^k_{\mathrm{B-A}} = \prod_{e \in S_\mathrm{B} \setminus S_\mathrm{A}} \frac{\mathrm{H}^k(\varphi'_{\mathrm{B}e})}{\mathrm{H}^k(\varphi_e)} > 1. \tag{4.81}$$

Then it has

$$\begin{aligned}
&(Risk^k_\emptyset - Risk^k_{\mathrm{A}+v}) - (Risk^k_\emptyset - Risk^k_\mathrm{A}) \\
&= Risk^k_\mathrm{A} - Risk^k_{\mathrm{A}+v} \\
&\leq W^k_{\mathrm{B-A}}(Risk^k_\mathrm{A} - Risk^k_{\mathrm{A}+v}) \\
&= Risk^k_\mathrm{B} - Risk^k_{\mathrm{B}+v} \\
&= (Risk^k_\emptyset - Risk^k_{\mathrm{B}+v}) - (Risk^k_\emptyset - Risk^k_\mathrm{B}),
\end{aligned} \tag{4.82}$$

which is violating Definition 2. It is noted that $\sum_{M^k \in M} Risk^k_\emptyset = RiskW_\emptyset$ and $\sum_{M^k \in M} Risk^k_\mathrm{A} = RiskW_\mathrm{A}$. Then, based upon Lemma 2, once there are Braess paradox phenomena in failure chains such that some sub-functions in accumulation term are not submodular, the function (4.80) is not a submodular function. □

4.5.5.2 Submodular Optimization Approach

The above analysis shows that the Braess paradox in some chains invalidates the submodular optimization definition. Thus, we propose a submodular optimization approach to reformulate the risk mitigation optimization. Specifically, for M^k, we define the following function:

$$G^k(S_\mathrm{B}) = Risk^k_\emptyset - Risk^k_\mathrm{B} + \max(w^k_{com} - 1, 0) \cdot Risk^k_\mathrm{A}, \tag{4.83}$$

where $w^k_{com} = \frac{Risk^k_\mathrm{B}}{Risk^k_\mathrm{A}}$ denotes the compared sampling weight between new updating set $S_\mathrm{B} \subseteq S_\mathrm{L}$ with previous updating set $S_\mathrm{A} \subset S_\mathrm{B}$ or $S_\mathrm{A} \supset S_\mathrm{B}$, where $||S_\mathrm{B}| - |S_\mathrm{A}|| = 1$. The property of function (4.83) is given in Theorem 2.

Theorem 2 *$G^k(S_\mathrm{B})$ is a submodular function no matter whether or not the Braess paradox occurs.*

Proof Based on (4.73), there is a component $v \in S_\mathrm{L} \backslash S_\mathrm{B}$ not causing Braess paradox, such that

$$\frac{Risk^k_{\mathrm{B}+v}}{Risk^k_{\mathrm{A}+v}} = \frac{Risk^k_\mathrm{B}}{Risk^k_\mathrm{A}} = w^k_{com}. \tag{4.84}$$

First, for $w^k_{com} = W^k_{\mathrm{B-A}} < 1$, there is no Braess paradox from the updated line set S_A to S_B for M^k. Based upon $\max\left(w^k_{com} - 1,\ 0\right) = 0$, there is

$$\begin{aligned}
&G^k(S_\mathrm{B} \cup v) - G^k(S_\mathrm{B}) \\
&= -Risk^k_{\mathrm{B}+v} + Risk^k_\mathrm{B} + 0 \cdot (Risk^k_{\mathrm{A}+v} - Risk^k_\mathrm{A}) \\
&< \frac{1}{w^k_{com}}(-Risk^k_{\mathrm{B}+v} + Risk^k_\mathrm{B}) \\
&= -Risk^k_{\mathrm{A}+v} + Risk^k_\mathrm{A} \\
&= G^k(S_\mathrm{A} \cup v) - G^k(S_\mathrm{A}).
\end{aligned} \tag{4.85}$$

Second, for $w^k_{com} = W^k_{B-A} \geq 1$, there is a Braess paradox from the updated line set S_A to S_B for M^k. According to $\max\left(w^k_{com} - 1,\ 0\right) = w^k_{com} - 1$, it has

$$\begin{aligned}
&G^k(S_\mathrm{B} \cup v) - G^k(S_\mathrm{B}) \\
&= -Risk^k_{\mathrm{B}+v} + Risk^k_\mathrm{B} + (w^k_{com} - 1) \cdot (Risk^k_{\mathrm{A}+v} - Risk^k_\mathrm{A}) \\
&= -Risk^k_{\mathrm{B}+v} + Risk^k_\mathrm{B} + (Risk^k_{\mathrm{B}+v} - Risk^k_\mathrm{B}) \\
&\quad \ldots - (Risk^k_{\mathrm{A}+v} - Risk^k_\mathrm{A}) \\
&= -Risk^k_{\mathrm{A}+v} + Risk^k_\mathrm{A} \\
&= G^k(S_\mathrm{A} \cup v) - G^k(S_\mathrm{A}).
\end{aligned} \tag{4.86}$$

In conclusion, we know that

$$G^k(S_\mathrm{A} \cup v) - G^k(S_\mathrm{A}) \geq G^k(S_\mathrm{B} \cup v) - G^k(S_\mathrm{B}), \tag{4.87}$$

which satisfies Definition 2. Thus, $G^k(S_\mathrm{B})$ is a submodular function immune to the Braess paradox, which completes the proof. □

In (4.83), for the term $\max(w_{com}^k - 1, 0) \cdot Risk_{\mathrm{A}}^k$, define a constant $\eta \in (0, 1]$ as an adjustment multiplier that will not violate the submodular function definition based upon the proof of Theorem 2. Then there is

$$G^k(S_{\mathrm{B}}) = Risk_{\emptyset}^k - Risk_{\mathrm{B}}^k + \eta \cdot \max(w_{com}^k - 1, 0) \cdot Risk_{\mathrm{A}}^k. \tag{4.88}$$

Via the sum of $G^k(S_{\mathrm{B}})$ for all $M^k \in M$ and the addition of linear cost function $C(S_{\mathrm{B}})$, one can obtain the objective function

$$\begin{aligned} F(S_{\mathrm{B}}) &= C(S_{\mathrm{B}}) + \gamma \sum_{M^k \in M} (Risk_{\emptyset}^k - Risk_{\mathrm{B}}^k) \\ &\quad \ldots + \eta\gamma \sum_{M^k \in M} [\max(w_{com}^k - 1, 0) \cdot Risk_{\mathrm{A}}^k] \\ &= C(S_{\mathrm{B}}) + \gamma(RiskW_{\emptyset} - RiskW_{\mathrm{B}}) + \eta\gamma \cdot BPI, \end{aligned} \tag{4.89}$$

where $BPI = \sum_{M^k \in M} [\max(w_{com} - 1, 0) \cdot Risk_{\mathrm{A}}^k]$ denotes the Braess paradox indicator, with a higher value indicating more side effects induced by the Braess paradox. The following theorem shows the property of function (4.89).

Theorem 3 *The function $F(S_{\mathrm{B}})$ defined in* (4.89) *is of submodularity.*

Proof The linear cost function $C(S_{\mathrm{B}})$ can be seen as a modular function. And the sum of $G^k(S_{\mathrm{B}})$ for all $M^k \in M$ is essentially the sum of a group of submodular functions. Then, based on Lemmas 1 and 2, it immediately concludes that $F(S_{\mathrm{B}})$ is submodular. □

According to the overall submodular optimization approach, the reformulated risk mitigation model is of submodularity regardless of whether the Braess paradox occurs throughout the risk mitigation process, ultimately enabling the proposed model to benefit from submodular optimization.

4.5.5.3 Estimation Error Analysis

The optimization result generally includes estimation error since the failure chain database is generated from a cascading failure simulator. Define the absolute error of $RiskW_{\emptyset}$ as ε, the confidence level as τ, and the realistic value of $RiskW_{\emptyset}$ as $\overline{RiskW_{\emptyset}}$. Then, it can obtain

$$\Pr\left(\frac{-\varepsilon}{\sigma/\sqrt{D}} \le \frac{\overline{RiskW_{\emptyset}} - RiskW_{\emptyset}}{\sigma/\sqrt{D}} \le \frac{\varepsilon}{\sigma/\sqrt{D}}\right) = \tau, \tag{4.90}$$

where $|M| = D$, and σ denotes the standard deviation of $RiskW_{\emptyset}$ for convenience. Then, according to the Central Limit Theorem, it has

$$\varepsilon = \Phi^{-1}\left(\frac{\tau+1}{2}\right)\cdot \sigma/\sqrt{D}, \tag{4.91}$$

where $\Phi^{-1}(\cdot)$ is the inverse function of standard normal distribution. Therefore, assuming $\varepsilon > 0$, the realistic submodular function of cascading failure risk mitigation can be built as $\overline{F}(S_{\rm B})$

$$\begin{aligned}
\overline{F}(S_{\rm B}) &= C(S_{\rm B}) + \gamma(\overline{RiskW_{\emptyset}} - \overline{RiskW_{\rm B}}) + \eta\gamma\cdot BPI \\
&= C(S_{\rm B}) + \gamma(RiskW_{\emptyset} + \varepsilon - RiskW_{\rm B} - \varepsilon \widetilde{W}_{\rm B-\emptyset}) \\
&\quad \ldots + \eta\gamma\cdot BPI \\
&= C(S_{\rm B}) + \gamma(RiskW_{\emptyset} - RiskW_{\rm B}) + \eta\gamma\cdot BPI \\
&\quad \ldots + \gamma\varepsilon(1 - \widetilde{W}_{\rm B-\emptyset}) \\
&= F(S_{\rm B}) + \gamma\varepsilon(1 - \widetilde{W}_{\rm B-\emptyset}),
\end{aligned} \tag{4.92}$$

which is also non-monotone.

4.5.6 Risk Mitigation Solving Algorithm

The problem (4.92) is a combinatorial optimization originally with at least $O(2^n)$ computation complexity for candidate number n. As n grows, the amount of calculation increases exponentially required to traverse all candidate combinations to obtain the optimal solution, known as an NP-hard problem suffering from the notorious "*curse-of-dimension*." To get a sub-optimal solution with an acceptable computation time, it is appropriate to design a solving method with reduced computation complexity, often lowering the original complexity $O(2^n)$ to a polynomial-time one [62]. In other words, as compared with the traditional traversal process, the method with polynomial-time complexity can complete the same combinatorial optimization in less time. Through the idea of computation complexity reduction, we propose a risk mitigation solving algorithm based on a modified double greedy algorithm to deal with this submodular function, as presented in Algorithm 1.

In terms of performance analysis of Algorithm 1, define variable $S_{opt}^i = (S_{\rm opt} \bigcup S_x^i) \bigcap S_y^i$, where $S_{\rm opt}$ is the realistic optimal set. According to Algorithm 1, it is noted that $S_x^0 = \emptyset$, $S_y^0 = S_{\rm L}$, $S_{opt}^0 = S_{\rm opt}$, and $S_x^n = S_y^n = S_{opt}^n$. We have the following theorem for the relation among objective function values at S_x^i, S_y^i, and S_{opt}^i at each iteration.

Algorithm 1 Risk mitigation solving algorithm

Input: Failure chain database M, load loss vector Y and related threshold Y_{ext}, line set S_{L}, line capacity maximum vector $P_{\max}$, DTR in-proved parameter α, objective submodular function $\overline{F}(S)$.
Output: Line set S_x^n installed with DTR.
1) $S_x^0 \leftarrow \emptyset$ and $S_y^0 \leftarrow S_{\mathrm{L}}$
2) Set $n \leftarrow |S_{\mathrm{L}}|$
3) **for** i=1 to n **do**
4) Select $u_i \in S_{\mathrm{L}}$
5) $S_x^i \leftarrow S_x^{i-1} + u_i$ and $S_y^i \leftarrow S_y^{i-1} - u_i$
6) $a_i \leftarrow \max\{0, \overline{F}(S_x^i) - \overline{F}(S_x^{i-1})\}$
7) $b_i \leftarrow \max\{0, \overline{F}(S_y^i) - \overline{F}(S_y^{i-1})\}$
8) Draw $RanNum_i \sim Unif(0, 1)$
9) **if** $RanNum_i < \frac{a_i}{a_i+b_i}$ **then**
10) $S_x^i \leftarrow S_x^{i-1} + u_i$ and $S_y^i \leftarrow S_y^{i-1}$
11) **else do**
12) $S_x^i \leftarrow S_x^{i-1}$ and $S_y^i \leftarrow S_y^{i-1} + u_i$
13) **end if**
14) **end for**
15) Return S_x^n (or equivalently S_y^n)

Theorem 4 *For every $i \in [1, n]$,*

$$\mathbb{E}\left[F(S_{opt}^{i-1}) - F(S_{opt}^{i})\right] \leq \frac{1}{2}\mathbb{E}\left[F(S_x^i) - F(S_x^{i-1}) + F(S_y^i) - F(S_y^{i-1})\right]. \tag{4.93}$$

By Theorem 4, the following corollary can be obtained.

Corollary 1 *The following inequality equation holds for every $i \in [1, n]$.*

$$\begin{aligned}\overline{F}(S_{opt}^{i-1}) - \overline{F}(S_{opt}^{i}) \leq \frac{1}{2}\Big[&F(S_x^i) - F(S_x^{i-1}) + F(S_y^i) - F(S_y^{i-1}) \\ &\ldots + \gamma\varepsilon(\widetilde{W}_{x^{i-1}-\emptyset} - \widetilde{W}_{x^i-\emptyset} + \widetilde{W}_{y^{i-1}-\emptyset} - \widetilde{W}_{y^i-\emptyset})\Big].\end{aligned} \tag{4.94}$$

Note that functions $\overline{F}(S)$ and $F(S)$ in this part implicitly contain an unbiased estimation of the mean from formulas (4.65) and (4.92). It is easy to prove Corollary 1, which we omit here. Then, Corollary 1 leads to Theorem 5, establishing the approximation guarantee for the proposed non-monotone submodular function.

Theorem 5 *Let $S_x^n \subseteq S_{\mathrm{L}}$ be the line set selected by Algorithm 1 and S_{opt} be the optimal solution to $\{\max \overline{F}(S) : S \subseteq S_{\mathrm{L}}\}$. Then Algorithm 1 can obtain a solution achieving an approximation guarantee of 1/2 with an estimation error, i.e.,*

$$F(S_x^n) \geq \frac{1}{2}\overline{F}(S_{\mathrm{opt}}) - \gamma\varepsilon\left(\frac{3}{4} + \frac{1}{4}\widetilde{W}_{\mathrm{L}-\emptyset}\right). \tag{4.95}$$

Proof By summing the inequality (4.94) by i from 1 to n, it has

$$\sum_{i=1}^{n}[\overline{F}(S_{opt}^{i-1}) - \overline{F}(S_{opt}^{i})] \leq \frac{1}{2}\sum_{i=1}^{n}[F(S_x^i) - F(S_x^{i-1}) + F(S_y^i) - F(S_y^{i-1})$$
$$\ldots + \gamma\varepsilon(\widetilde{W}_{x^{i-1}\_\emptyset} - \widetilde{W}_{x^i\_\emptyset} + \widetilde{W}_{y^{i-1}\_\emptyset} - \widetilde{W}_{y^i\_\emptyset})]. \tag{4.96}$$

Via cancellation of the same items in (4.96), it leads to

$$\overline{F}(S_{opt}^0) - \overline{F}(S_{opt}^n) \leq \frac{1}{2}[F(S_x^n) - F(S_x^0) + F(S_y^n) - F(S_y^0)$$
$$\ldots + \gamma\varepsilon(\widetilde{W}_{x^0\_\emptyset} - \widetilde{W}_{x^n\_\emptyset} + \widetilde{W}_{y^0\_\emptyset} - \widetilde{W}_{y^n\_\emptyset})]. \tag{4.97}$$

Based upon the relations among S_x^i, S_y^i, and S_{opt}^i in Algorithm 1 and $\widetilde{W}_{x^0\_\emptyset} = 1$, there is

$$\begin{aligned}\overline{F}(S_{\text{opt}}) &\leq \frac{1}{2}[F(S_x^n) - F(S_x^0) + F(S_y^n) - F(S_y^0)\\ &\quad \ldots + \gamma\varepsilon(1 - \widetilde{W}_{x^n\_\emptyset} + \widetilde{W}_{y^0\_\emptyset} - \widetilde{W}_{y^n\_\emptyset})]\\ &\quad \ldots + F(S_x^n) + \gamma\varepsilon(1 - \widetilde{W}_{x^n\_\emptyset})\\ &\leq \frac{1}{2}[4F(S_x^n) - F(S_x^0) - F(S_y^0)\\ &\quad \ldots + \gamma\varepsilon(3 - 3\widetilde{W}_{x^n\_\emptyset} + \widetilde{W}_{y^0\_\emptyset} - \widetilde{W}_{y^n\_\emptyset})]\\ &\leq 2F(S_x^n) + \gamma\varepsilon(\frac{3}{2} + \frac{1}{2}\widetilde{W}_{y^0\_\emptyset}).\end{aligned} \tag{4.98}$$

By rearranging terms in (4.98) and $S_y^0 = S_{\text{L}}$, it yields (4.95) and completes the proof. □

From Theorem 5, the solution of Algorithm 1 for the proposed non-monotone submodular function can achieve a roughly 1/2 approximation guarantee. Note that this guarantee can be regarded as the minimum ratio of the searching solution to the real optimum, whose numerator is the lower bound for risk mitigation optimization, even though the actual optimal solution is practically impossible to achieve. However, the approximation guarantee cannot quantitatively determine how good the final search function value is or compare the two algorithms' performance from their guarantees. In practice, this submodular structure and corresponding solving algorithm often yield results closer to the realistic optimum than the $1/2-$approximation value. Instead, it is acceptable to claim qualitatively that the algorithms with a higher approximation guarantee usually have a better objective value than those with a lower guarantee.

The calculation efficiency of algorithm one is given in the theorem below.

Theorem 6 *Let* $d_{max} = \max\{|M^k| : M^k \in M\}$, *D be the number of unidentical chains in M, and* $n = |S_L|$. *The optimization of Algorithm 1 runs in* $O(nDd_{max})$ *time, i.e., the solution can be found in polynomial time.*

Proof The runtime of Algorithm 1 is decided by a double greedy searching process and the recalculation of submodular risk function $\overline{F}(S)$ in terms of the selected updated line set. For the greedy searching stage, the algorithm needs $O(n)$ evaluations for S_L. Recalculation of $\overline{F}(S)$ requires $O(D)$ computations for the failure chains, each of which entails solving at most $O(d_{max})$ risk computation at each failure chain. Thus, the computational complexity of Algorithm 1 is $O(nDd_{max})$, which completes the proof. □

Theorem 6 implies that the dedicated solving algorithm can handle the non-monotone submodular optimization issue in polynomial time. However, solving in polynomial time does not mean a fast computation time. When the parameters in $O(nDd_{max})$ rise, this algorithm still takes longer to solve the problem than the cases with smaller parameters. This study compares computation efficiency with the original computation complexity $O(2^n)$ in tackling the identical problem. The proposed method's scalability is more evident when dealing with large-scale cases than the traversal process with $O(2^n)$ complexity. Therefore, as a result of the reduction of computation complexity, this polynomial-time calculation efficiency is critical for dealing with large-scale instances that are prone to the curse of dimensionality, as well as online applications that require a specific level of accuracy at any given iteration.

4.5.7 Example

4.5.7.1 Important Sampling Weight Approximation

A comparative test is implemented in the IEEE 39-bus system to assess the effectiveness of important sampling weight. After lines updating, the Monte Carlo (MC) method is traditionally applied to regenerate the cascading failure chain database using a cascading failure simulator [33], also used to produce the original database. Specifically, this failure simulator is based on DC power flow on the MATPOWER toolkit, considering the aspects of load shedding and generator tripping, island operation, and hidden failures. On the other hand, the Importance Sampling (IS) approach is utilized to obtain a new database straight from the original. The set database contains 5000 simulations, and DTR is placed on line 4. The results are presented below.

When load loss is less than 3000 MW, the cumulative distribution functions of MC and IS approaches are practically identical, as seen in Fig. 4.10. For load loss of more than 3000 MW, the SIS technique can be more focused on searching for this kind of severe failures. Table 4.8 provides that the random bias of average power loss between two approaches is approximately 22.7, and the approximation ratio is

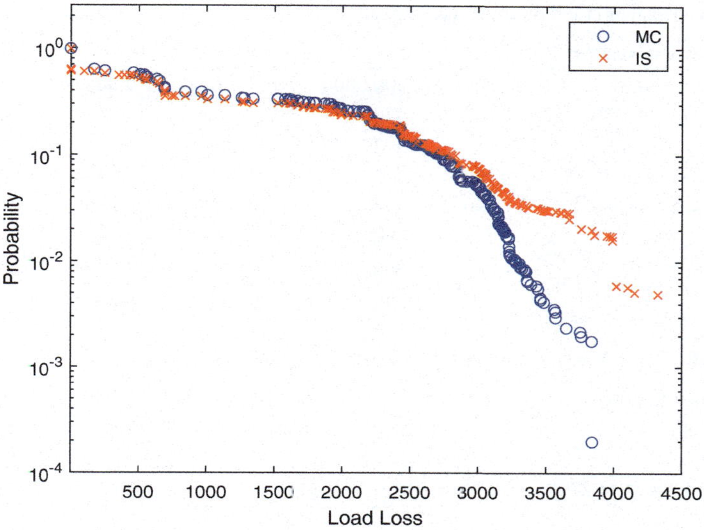

Fig. 4.10 The cumulative distribution functions of MC and IS methods

Table 4.8 Performance comparison between MC and IS methods

Database with $N = 5000$	MC	IS
Average power loss (MW)	1015.9	993.2
Time (s)	391.4	10.1

97.8%. However, the computation time for MC is 391.4 s, significantly larger than 10.1 s of the SIS, demonstrating that the SIS method is more time-efficient.

4.5.7.2 Impacts of Weather and System Factors on DTR Risk Mitigation

The calculation of DTR depends on the ambient weather conditions. Thus, we compare the effect of different weather conditions on DTR risk mitigation. To determine the suitable threshold for relay action, it is better to apply the average worst weather condition on critical line spans to calculate the DTR value.

Experiments are simulated on the IEEE 39-bus system. Specifically, set $cost = 0.35$ as the cost for unit improved line capacity from DTR, $\gamma = 1$, $\eta = 0.8$, and $Y_{ext} = 2000$ for submodular function. And database M contains 5000 simulations. Assume that the default threshold rating is from the condition that $T = 40°\text{C}$ and $V = 0.61\,\text{m/s}$, meaning that $\alpha = 1.0$. Then, we select the average worst weather data in different critical line spans from weather databases MRCC and NOAA to calculate α.

Table 4.9 shows the results of weather effect on DTR risk mitigation, revealing that all DTR strategies have lower $RiskW$ values than the default value of 657.646, indicating that DTR can help to mitigate system risk on a large scale. Note that the

Table 4.9 Weather effect on DTR risk mitigation

Average worst weather	α	F	$RiskW$	$Cost$	BPI
$T = 40$, $V = 0.61$	1.0	200.000	657.646	0.000	0.000
$T = 36.8$, $V = 1.50$	1.108	470.387	217.007	196.473	32.775
$T = 34$, $V = 2.47$	1.153	512.775	144.308	217.463	21.125
$T = 33.6$, $V = 2.31$	1.180	535.028	118.719	221.254	21.695
$T = 32.8$, $V = 2.65$	1.189	593.445	115.999	160.160	14.948
$T = 32.0$, $V = 3.21$	1.201	586.002	94.771	184.207	9.167
$T = 30.8$, $V = 2.59$	1.212	554.008	74.406	241.849	15.770
$T = 31.4$, $V = 2.91$	1.252	540.178	150.512	188.856	27.375

Table 4.10 Generation and load distribution effect on DTR risk mitigation

Spatial distribution	F	$RiskW$	$Cost$	BPI
$20 \rightleftharpoons 27, 8 \rightleftharpoons 12$	1000.680	161.925	274.729	34.594
$31 \rightleftharpoons 24$	604.720	122.177	216.516	15.840
$16 \rightleftharpoons 20, 22 \rightleftharpoons 2, 31 \rightleftharpoons 8$	570.058	130.792	148.349	35.757

DTR placement schemes change under different average worst weathers, which are not listed due to space limitations. And it is found that there is the best condition for system risk mitigation. In this case, the best situation is that α is approximately in the range [1.19, 1.21], which can achieve the best values of $F = 593.445$, $RiskW = 94.771$, and $BPI = 9.167$, meaning that it can decrease the risk mostly and suffer the least from Braess paradox. When α decreases from 1.19, the values of $RiskW$ and BPI will increase. The reason is that the relatively bad weather curtails the DTR effect, so there is only a limited threshold increment. On the other hand, the same situation happens when α increases from 1.21. Because the excessive DTR effect will exacerbate the imbalance of system failure distribution, i.e., when a line improves significantly with DTR, drawing more power to pass through, it will transfer the risk to other lines that without DTR placement but having hidden failure link with it, resulting in a worse Braess paradox, as concluded in research [33].

Moreover, the generation and load pattern can affect the failure database, influencing the DTR effect on risk mitigation. From the previous research, the simultaneous changes of generation and load sizes in the steady state have less impact on critical lines, i.e., on cascading failure evolution, meaning that this kind of change would not significantly alter the DTR placement scheme. However, the spatial distribution of generation and load plays a vital role in the DTR effect. The reason is that different generation and load spatial distributions would have varied power flow distributions, which will alter line outage distribution (failure database). Thus, given $\alpha = 1.2$, we choose three different generation and load distributions (e.g., $20 \rightleftharpoons 27$ means that the parameters in buses 20 and 27 are exchanged) in IEEE 39-bus system to compare the DTR risk mitigation.

The result is shown in Table 4.10. Note that DTR placement schemes change under different spatial distributions, which are not listed here to save space. It can be seen that when generation and load spatial distribution differ, the final DTR risk mitigation effect will be different. In this case, distribution 1 has the largest F value, indicating that its original risk is more significant and DTR can provide considerable benefit to risk mitigation. Yet, it still suffers from the Braess paradox a lot represented by $BPI = 34.594$. Other distributions also suggest that DTR can significantly impact risk mitigation, but all with higher BPI values. It reminds us that the Braess paradox is common in risk mitigation, and one should be aware of it.

In summary, the DTR indeed can assist in mitigating the risk. For the weather factor, there is a suitable region for a DTR improved ratio that achieves a better risk mitigation effect. Too much or too little α would curtail the risk mitigation effect, requiring the operator to thoroughly examine the realistic weather condition and propose a suitable DTR placement scheme. On the other hand, varied generation and load spatial distributions also lead to different DTR placements. But note that each district usually has relatively constant worst weather conditions and generation and load spatial distribution so that a uniform DTR placement scheme can be functional.

4.5.7.3 Performance Comparison with Different Placement Strategies

(1) Performance comparison with traditional placement strategies

This part aims to evaluate the performance of the proposed strategy, namely the **double greedy strategy (DG)**, with that of other strategies using traditional assessment indexes. Four traditional assessment index-based strategies are introduced:

- **Random line strategy (RL)**. Randomly select lines to place DTR.
- **Failure rate strategy (FR)**. Rank lines in decreasing order from failure number in the database, and place DTR in the highest ranked lines.
- **Largest power flow strategy (LPF)**. Rank lines in decreasing order from initial power flow, and place DTR in the highest ranked lines.
- **Largest hidden failure strategy (LHF)**. Using the $N-1$ security test, rank lines in decreasing order from hidden failure probabilities, and place DTR in the highest ranked lines.

Experiments are simulated on the IEEE 39-bus system, and the settings are the same as in Sect. 4.5.7.2 besides fixed $\alpha = 1.2$. Results are shown in Table 4.11 and Table 4.12.

Table 4.11 lists DTR placement plans from several strategies. The line numbers they chose are within the range between 5 and 9. The DG strategy selects fewer lines to place DTR, while the LPF strategy selects the largest number.

For objective function value, Table 4.12 illustrates DG strategy has the highest value, $F_{\text{DG}} = 586.002$, compared with other traditional strategies, such as $F_{\text{RL}} = 421.198$ of RL, $F_{\text{FR}} = 432.965$ of FR, and $F_{\text{LHF}} = 519.372$ of LHF. Especially for the LPF strategy, its objective function is -187.126, due to $Cost_{\text{LPF}} = 610.291$, considerably higher than other strategies. Additionally, when it comes to failure risk and placing cost, DG strategy's $RiskW_{\text{DG}} = 94.771$ and $Cost_{\text{DG}} = 184.20$ are all

Table 4.11 DTR placement plans of different strategies in 39-bus system

Strategy	Lines with DTR placement
DG	7, 10, 12, 13, 24
RL	3, 6, 8, 23, 26, 30, 36
FR	7, 8, 9, 18, 19, 20, 24, 30
LPF	10, 14, 20, 33, 35, 37, 39, 41, 46
LHF	3, 9, 10, 13, 18, 19

Table 4.12 Performance comparison among different strategies in IEEE 39-bus system

Strategy	F	$RiskW$	$Cost$	BPI
DG	586.002	94.771	184.207	9.167
RL	421.198	257.253	233.090	67.368
FR	432.965	174.279	269.067	23.331
LPF	−187.126	461.957	610.291	34.345
LHF	519.372	138.058	208.299	10.103
No DTR	200.000	657.646	0.000	0.000

the best when compared to others, suggesting that it can achieve better performance with less burden. Although FL, FR, and LHF can all reduce system failure risk on a large scale, their cost-effectiveness is worse than DG's.

Furthermore, when compared to other strategies, $BPI_{\text{DG}} = 9.167$ is the lowest in the Braess paradox index, implying that there is less Braess paradox effect on the system following DG strategy execution. The paradox effect on LHF is likewise curtailed. However, for $BPI_{\text{RL}} = 67.368$, $BPI_{\text{FR}} = 23.331$, and $BPI_{\text{LPF}} = 34.345$, those effects are remarkable, meaning that in some failure propagation scenarios, these DTR placement strategies will have considerable side effect on risk mitigation. In short, this test suggests that the Braess paradox significantly influences the assessment index-based strategies, greatly reducing their mitigation performance. The proposed DG strategy outperforms traditional assessment index-based strategies in risk mitigation, cost-effectiveness, and the Braess paradox effect in this submodular optimization.

(2) Performance comparison among different searching strategies

The performances of different searching strategies for this submodular optimization are compared in this part. Popular searching strategies for solving non-monotone submodular maximization include:

- **Traditional greedy strategy (TG)**. Select lines with the highest objective function gain iteratively, terminated on a modest negative tolerance value. Note that the time needed for TG to deal with non-monotone problems will be more costly than the monotone.
- **Deterministic greedy strategy (DMG)**. This searching technique is based on the squeeze theorem perspective, with a nearly 1/3 approximation guarantee.
- **Implicit enumeration strategy (IE)**. Select lines with a maximal objective function to gain while introducing constraints to reduce searching space. Note that the ground set suffers from the dimension curse as it grows.

Table 4.13 DTR placement plans of different searching methods in IEEE 39-bus system

Strategy	Lines with DTR placement
DG	7, 10, 12, 13, 24
TG	7, 9, 10, 13, 16, 18, 24
DMG	6, 10, 13, 16, 23, 24, 26
IE	3, 13, 16, 23, 24

Table 4.14 Performance comparison among different searching methods in IEEE 39-bus system

Strategy	*F*	*RiskW*	*Cost*	*BPI*	*Computation time (s)*
DG	586.002	94.771	184.207	9.167	133.25
TG	576.948	56.495	230.332	7.660	428.32
DMG	541.010	74.464	255.671	16.873	132.05
IE	570.769	140.593	167.067	25.978	1517.62

Experiments are applied to the IEEE 39-bus system. The parameters are the same as in the previous comparison experiment. Results are provided in Tables 4.13 and 4.14.

Table 4.13 lists the line selection from different searching strategies. It can be seen that all strategies choose lines 13 and 24. Table 4.14 presents that all the objective function values of strategies are greater than 530 compared to assessment index-based strategies. And DG strategy has the highest value, $F_{\text{DG}} = 586.002$, while $F_{\text{TG}} = 576.948$ and $F_{\text{IE}} = 570.769$ also perform better. Although $RiskW_{\text{TG}} = 56.495$ and $RiskW_{\text{DMG}} = 74.464$ are less than DGs, these strategies have higher costs, 230.332 and 255.671, compared to $Cost_{\text{DG}} = 184.207$. In addition, we can see that for the Braess paradox index, BPIs of DG and TG are relatively lower than others, implying that they suffer less from the Braess paradox. However, DMG and IE suffer greatly from the Braess paradox phenomenon, as seen by $BPI_{\text{DMG}} = 16.873$ and $BPI_{\text{IE}} = 25.978$.

Furthermore, for computation times shown in Table 4.14, it can be observed that DG and DMG spend less time, i.e., $Time_{\text{DG}} = 133.25$ s and $Time_{\text{DMG}} = 132.05$ s, respectively, while DMG obtains an inferior result. Even if the performances are close to that of DG, the computation times for TG and IE are longer, i.e., $Time_{\text{TG}} = 428.32$ s and $Time_{\text{IE}} = 1517.62$ s.

In summary, it can be observed that searching-based strategies are superior to traditional assessment index-based strategies. For objective function value, the proposed method has the highest value as it can achieve a better balance between risk mitigation and cost burden. It also significantly reduces the Braess paradox effect, whereas other traditional strategies suffer greatly from it. Moreover, the DG strategy takes less time to obtain a near-optimal solution than TG and IE. When dealing with larger-scale cases, this time efficiency is quite important.

4.6 Summary and Conclusions

This chapter introduces probabilistic models for cascading failures. Specifically, an analytic model is formulated to simulate cascading failures and analyze the blackout risk, utilizing the Markov property and sequential importance sampling. Furthermore, a characteristic model is devised to establish the relationship between blackout risk and component failure probability. Both models are then employed to address complicated cascading failure risk mitigation problems, demonstrating their distinct advantages and benefits. Numerical experiments reveal some major findings listed as follows:

1. Compared to the traditional Monte Carlo simulation strategy, the proposed analytic strategy can efficaciously improve the computational efficiency and decrease the estimation variance of blackout probability/risk, improving the capability of capturing rare events in cascading failure simulations.
2. The characteristic model can accurately assess the relationship between blackout risk and component failure probability, allowing for a quick calculation of the blackout risk when the component failure probability functions change.
3. These two proposed models can provide a convenient and efficient way to design a risk mitigation strategy for cascading failures.

The developed models and analytics have been deployed for the blackout risk analysis and mitigation of large-scale power systems, including the IEEE 300-bus standard systems and a real 1122-bus system. The results primarily verify their applicability. Please refer to [27, 31, 63, 64] for details. As the monitoring technology and external conditions prediction develop, the proposed methodologies are promising for realistic cascading failure analysis and relevant prevention strategy design.

References

1. I. Dobson, Estimating the propagation and extent of cascading line outages from utility data with a branching process. IEEE Trans. Power Syst. **27**(4), 2146–2155 (2012)
2. M. Vaiman, K. Bell, Y. Chen, B. Chowdhury, I. Dobson, P. Hines, M. Papic, S. Miller, P. Zhang, Risk assessment of cascading outages: methodologies and challenges. IEEE Trans. Power Syst. **27**(2), 631 (2012)
3. U.O. Handbook, Union for the co-ordination of transmission of electricity, UCTE, 20 July (2004)
4. D. Cornforth, Long tails from the distribution of 23 years of electrical disturbance data, in *2009 IEEE/PES Power Systems Conference and Exposition* (IEEE, 2009), pp. 1–8
5. J. Qi, Utility outage data driven interaction networks for cascading failure analysis and mitigation. IEEE Trans. Power Syst. **36**(2), 1409–1418 (2020)
6. H. Guo, C. Zheng, H.H.-C. Iu, T. Fernando, A critical review of cascading failure analysis and modeling of power system. Renew. Sust. Energ. Rev. **80**, 9–22 (2017)
7. P. Hines, J. Apt, S. Talukdar, Large blackouts in North America: historical trends and policy implications. Energy Policy **37**(12), 5249–5259 (2009)

8. K. Zhou, I. Dobson, Z. Wang, A. Roitershtein, A.P. Ghosh, A Markovian influence graph formed from utility line outage data to mitigate large cascades. IEEE Trans. Power Syst. **35**(4), 3224–3235 (2020)
9. B.A. Carreras, D.E. Newman, I. Dobson, North American blackout time series statistics and implications for blackout risk. IEEE Trans. Power Syst. **31**(6), 4406–4414 (2016)
10. J. Bialek, E. Ciapessoni, D. Cirio, E. Cotilla-Sanchez, C. Dent, I. Dobson, P. Henneaux, P. Hines, J. Jardim, S. Miller, Benchmarking and validation of cascading failure analysis tools. IEEE Trans. Power Syst. **31**(6), 4887–4900 (2016)
11. W.-K. Ching, M.K. Ng, Markov chains, in *Models, Algorithms and Applications* (Springer, New York, USA 2006)
12. M. Rahnamay-Naeini, M.M. Hayat, Impacts of operating characteristics on sensitivity of power grids to cascading failures, in *2016 IEEE Power and Energy Society General Meeting (PESGM)* (IEEE, 2016), pp. 1–5
13. Q. Huang, L. Shao, N. Li, Dynamic detection of transmission line outages using hidden Markov models. IEEE Trans. Power Syst. **31**(3), 2026–2033 (2015)
14. Z. Ma, C. Shen, F. Liu, S. Mei, Fast screening of vulnerable transmission lines in power grids: a PageRank-based approach. IEEE Trans. Smart Grid **10**(2), 1982–1991 (2017)
15. P.D. Hines, I. Dobson, P. Rezaei, Cascading power outages propagate locally in an influence graph that is not the actual grid topology. IEEE Trans. Power Syst. **32**(2), 958–967 (2016)
16. J. Qi, K. Sun, S. Mei, An interaction model for simulation and mitigation of cascading failures. IEEE Trans. Power Syst. **30**(2), 804–819 (2014)
17. I. Dobson, B.A. Carreras, D.E. Newman, A loading-dependent model of probabilistic cascading failure Probab. Eng. Inf. Sci. **19**(1), 15–32 (2005)
18. I. Dobson, B.A. Carreras, V.E. Lynch, D.E. Newman, Complex systems analysis of series of blackouts: cascading failure, critical points, and self-organization. Chaos: Interdisciplinary J. Nonlinear Sci. **17**(2), 026103 (2007)
19. H. Wu, I. Dobson, Analysis of induction motor cascading stall in a simple system based on the cascade model. IEEE Trans. Power Syst. **28**(3), 3184–3193 (2013)
20. H. Dong, L. Cui, System reliability under cascading failure models. IEEE Trans. Reliab. **65**(2), 929–940 (2015)
21. K.B. Athreya, P.E. Ney, *Branching Processes* (Springer, Berlin, Heidelberg, 1972)
22. S.K. Baek, H.A.T. Kiet, B.J. Kim, Family name distributions: master equation approach. Phys. Rev. E **76**(4), 046113 (2007)
23. T. Aldemir, A survey of dynamic methodologies for probabilistic safety assessment of nuclear power plants. Ann. Nucl. Energy **52**, 113–124 (2013)
24. I. Dobson, B.A. Carreras, D.E. Newman, A branching process approximation to cascading load-dependent system failure, in *Proceedings of the 37th Annual Hawaii International Conference on System Sciences, 2004* (IEEE, 2004), pp. 10–pp
25. I. Dobson, K.R. Wierzbicki, B.A. Carreras, V.E. Lynch, D.E. Newman, An estimator of propagation of cascading failure, in *Proceedings of the 39th Annual Hawaii International Conference on System Sciences (HICSS'06)*, vol. 10 (IEEE, 2006), pp. 245c–245c
26. J. Chen, J.S. Thorp, I. Dobson, Cascading dynamics and mitigation assessment in power system disturbances via a hidden failure model. Int. J. Electr. Power Energy Syst. **27**(4), 318–326 (2005)
27. Q. Long, J. Liu, F. Liu, Y. Hou, Submodular optimization of dynamic thermal rating for cascading failure risk mitigation considering Braess paradox. IEEE Trans. Syst. **38**(4), 3605–3620 (2022)
28. I. Dobson, B.A. Carreras, V.E. Lynch, D.E. Newman, An initial model for complex dynamics in electric power system blackouts, in *HICSS* (2001)
29. S. Mei, Y. Ni, G. Wang, S. Wu, A study of self-organized criticality of power system under cascading failures based on AC-OPF with voltage stability margin. IEEE Trans. Power Syst. **23**(4), 1719–1726 (2008)
30. M. Alvi, *A Manual for Selecting Sampling Techniques in Research* (2016). https://mpra.ub.uni-muenchen.de/70218/

31. J. Guo, F. Liu, J. Wang, J. Lin, S. Mei, Toward efficient cascading outage simulation and probability analysis in power systems. IEEE Trans. Power Syst. **33**(3), 2370–2382 (2017)
32. R.Y. Rubinstein, P.W. Glynn, How to deal with the curse of dimensionality of likelihood ratios in Monte Carlo simulation. Stoch. Model. **25**(4), 547–568 (2009)
33. Q. Long, Z. Ma, F. Liu, S. Mei, Y. Hou, Analyzing patterns transference and mitigation of cascading failures with interaction graphs, in *2021 IEEE PES Innovative Smart Grid Technologies Europe (ISGT-Europe)* (2021)
34. P. Rezaei, P.D. Hines, M.J. Eppstein, Estimating cascading failure risk with random chemistry. IEEE Trans. Power Syst. **30**(5), 2726–2735 (2014)
35. J. Kim, J.A. Bucklew, I. Dobson, Splitting method for speedy simulation of cascading blackouts. IEEE Trans. Power Syst. **28**(3), 3010–3017 (2012)
36. J. Shortle, Efficient simulation of blackout probabilities using splitting. Int. J. Electr. Power Energy Syst. **44**(1), 743–751 (2013)
37. S.-P. Wang, A. Chen, C.-W. Liu, C.-H. Chen, J. Shortle, J.-Y. Wu, Efficient splitting simulation for blackout analysis. IEEE Trans. Power Syst. **30**(4), 1775–1783 (2014)
38. O.A. Ansari, S.M. Mazhari, Y. Gong, C.Y. Chung, Short-term reliability evaluation of generating systems using fixed-effort generalized splitting, in *2020 IEEE Power and Energy Society General Meeting (PESGM)* (IEEE, 2020), pp. 1–5
39. J. Liu, *Monte Carlo Strategies in Scientific Computing* (Springer, New York, USA, 2008)
40. J.A. Bucklew, J. Bucklew, *Introduction to Rare Event Simulation*, vol. 5 (Springer, New York, USA, 2004)
41. E. Tomasson, L. Söder, Improved importance sampling for reliability evaluation of composite power systems. IEEE Trans. Power Syst. **32**(3), 2426–2434 (2016)
42. J. Huang, Y. Xue, Z.Y. Dong, K.P. Wong, An efficient probabilistic assessment method for electricity market risk management. IEEE Trans. Power Syst. **27**(3), 1485–1493 (2012)
43. J. Thorp, A. Phadke, S. Horowitz, S. Tamronglak, Anatomy of power system disturbances: importance sampling. Int. J. Electr. Power Energy Syst. **20**(2), 147–152 (1998)
44. Q. Chen, L. Mili, Composite power system vulnerability evaluation to cascading failures using importance sampling and antithetic variates. IEEE Trans. Power Syst. **28**(3), 2321–2330 (2013)
45. A. Doucet, N. De Freitas, N.J. Gordon et al., *Sequential Monte Carlo Methods in Practice*, vol. 1 (Springer, New York, USA, 2001)
46. M. Perninge, F. Lindskog, L. Soder, Importance sampling of injected powers for electric power system security analysis. IEEE Trans. Power Syst. **27**(1), 3–11 (2011)
47. Y. Wang, C. Guo, Q. Wu, A cross-entropy-based three-stage sequential importance sampling for composite power system short-term reliability evaluation. IEEE Trans. Power Syst. **28**(4), 4254–4263 (2013)
48. J. Qiu, Z.Y. Dong, J.H. Zhao, Y. Xu, Y. Zheng, C. Li, K.P. Wong, Multi-stage flexible expansion co-planning under uncertainties in a combined electricity and gas market. IEEE Trans. Power Syst. **30**(4), 2119–2129 (2014)
49. J. Yan, H. He, Y. Sun, Integrated security analysis on cascading failure in complex networks. IEEE Trans. Inf. Forensics Secur. **9**(3), 451–463 (2014)
50. T. Wang, Q. Long, X. Gu, W. Chai, Information flow modeling and performance evaluation of communication networks serving power grids. IEEE Access **8**, 13,735–13,747 (2020)
51. R. Yao, X. Zhang, S. Huang, S. Mei, X. Li, Q. Zhu et al., Cascading outage preventive control for large-scale AC-DC interconnected power grid, in *2014 IEEE PES General Meeting| Conference & Exposition*. (IEEE, 2014), pp. 1–5
52. R. Yao, K. Sun, F. Liu, S. Mei, Management of cascading outage risk based on risk gradient and Markovian tree search. IEEE Trans. Power Syst. **33**(4), 4050–4060 (2017)
53. M.J. Eppstein, P.D. Hines, A "random chemistry" algorithm for identifying collections of multiple contingencies that initiate cascading failure. IEEE Trans. Power Syst. **27**(3), 1698–1705 (2012)
54. H. Ren, I. Dobson, B.A. Carreras, Long-term effect of the $N-1$ criterion on cascading line outages in an evolving power transmission grid. IEEE Trans. Power Syst. **23**(3), 1217–1225 (2008)

55. C. Vellaithurai, A. Srivastava, S. Zonouz, R. Berthier, CPIndex: cyber-physical vulnerability assessment for power-grid infrastructures. IEEE Trans. Smart Grid **6**(2), 566–575 (2014)
56. A.E. David, B. Gjorgiev, G. Sansavini, Quantitative comparison of cascading failure models for risk-based decision making in power systems. Reliab. Eng. Syst. Saf. **198**, 106877 (2020)
57. Z. Wang, A. Scaglione, R.J. Thomas, A Markov-transition model for cascading failures in power grids, in *2012 45th Hawaii International Conference on System Sciences* (IEEE, 2012), pp. 2115–2124
58. C.P. Robert, G. Casella, G. Casella, *Monte Carlo Statistical Methods*, vol. 2 (Springer, 1999)
59. S. Mei, F. He, X. Zhang, S. Wu, G. Wang, An improved OPA model and blackout risk assessment. IEEE Trans. Power Syst. **24**(2), 814–823 (2009)
60. J. Qi, S. Mei, F. Liu, Blackout model considering slow process. IEEE Trans. Power Syst. **28**(3), 3274–3282 (2013)
61. J. Iglesias, G. Watt, D. Douglass, V. Morgan, R. Stephen, M. Bertinat, D. Muftic, R. Puffer, D. Guery, S. Ueda et al., *Guide for Thermal Rating Calculations of Overhead Lines* (CIGRE, Paris, France, 2014)
62. A. Krause, D. Golovin, Submodular function maximization. Tractability **3**, 71–104 (2014)
63. J. Guo, F. Liu, J. Wang, M. Cao, S. Mei, Quantifying the influence of component failure probability on cascading blackout risk. IEEE Trans. Power Syst. **33**(5), 5671–5681 (2018)
64. F. Liu, J. Guo, X. Zhang, Y. Hou, S. Mei, Mitigating the risk of cascading blackouts: a data inference based maintenance method. IEEE Access **6**, 39,197–39,207 (2018)

Chapter 5
Modeling Cascading Failures in Power Systems: Quasi-Steady-State Models and Dynamic Models

Eduardo Cotilla-Sanchez

Nomenclature

QSS	Quasi-Steady-State.
$p.u.$	per unit.
N	Number of buses in the system.
NM	Number of branches in the system (transmission lines and transformers).
N_G	The set of all generator buses in the system $N_G \subset n = \{1, 2, \ldots, N\}$.
N_D	The set of all demand/load buses in the system $N_D \subset n = \{1, 2, \ldots, N\}$.
MW	Megawatts.
C	Set of contingencies $C = \{c_1, c_2, \ldots\}$.
m_i	Cascading simulator model i .
R	Relative agreement of cascading path.
$r_{g,i}$	Generator equivalent series resistance.
$X_{d,i}$	Direct axis synchronous reactance for the generator at Bus i. $X'_{d,i}$ is the transient reactance. X_q refers to the quadrature axis reactance.
M	Generator inertia constant.
D	Generator damping constant.
$\tilde{V}_i$	Complex voltage at Bus i: $\tilde{V}_i = \|V_i\|e^{j\theta_i}$.

E. Cotilla-Sanchez (✉)
Kelley Engineering Center, Oregon State University, Corvallis, OR, USA
e-mail: ecs@oregonstate.edu

K. Sun (ed.), *Cascading Failures in Power Grids*, Power Electronics and Power Systems, https://doi.org/10.1007/978-3-031-48000-3_5

5.1 Modeling Cascading Failures in Power Systems: Quasi-Steady-State Models and Dynamic Models

5.1.1 Introduction

In essence, cascading failures emerge when outage mechanisms interact and show dependency patterns. Because modern grids incorporate new devices with new modes of becoming unstable, and these contribute to increasing the complexity of cascades, it becomes very important to understand how to place an upper bound on the modeling needs that one should implement when performing cascading analysis.

Generally for advanced simulation of cascading outages, it is understood that we need the ingredients of dynamics and protection. However, in a majority of current approaches to simulating cascading, the efforts to model dynamics and protection are either shorthanded or so detailed that results are difficult to interpret and benchmark [8]. The challenges on choosing appropriate timescales between quasi-steady-state and dynamic simulations, or other solutions, such as quasi-dynamic implementations, are also well justified in [14, 22]. An intended contribution of this chapter is to discuss and compare example simulators and experiments that have comparable protection elements and tunable dynamics in an open-source platform.

Recently, Dai et al. [5] show that security dispatches with multiple constraints can increase cascading outage risk due to overload of critical lines. They also compare a QSS simulator and a dynamic simulator for their experiments. Similarly, counter-intuitive relationships between system loading levels and cascading risk were observed by [19] using the QSS simulator that we compare in this work with a closely related dynamic simulator. This work shows that increased risk can be observed at lower demand levels.

One of the major challenges in the simulation is how to handle islands. This is necessary, and while some QSS simulators explicitly address separation and how to account for those (lines or load) losses, some of the dynamic simulators present challenges upon stability of islands. There are efforts to study the post-disturbance stability guarantees [16], and research groups have also explored control strategies to reconnect at appropriate times and mitigate further separation and damage [15].

Along with relatively recent benchmarking efforts, [2], the recommended standards to implement within these proposed benchmarks have also evolved and increasingly include additional types of relay mechanisms in cascading studies. Back to the PRC-023-2 standard, from 2012, one out of three mechanisms (overloads) could be implemented with a QSS simulator; however, the remainder of mechanisms requires some level of dynamic modeling. Newer standards and mechanisms, for example, inverter-interfaced renewable generation, will continue to push the complexity of required studies without a clear path to equitable compare across the methodologies used by different utilities, for example.

More recent benchmarking studies, e.g., [11] by the IEEE Cascading Failure Working Group, have broadly implemented analysis with multiple simulators, including research-grade (sometimes open source) and commercial-grade methodologies. The QSS simulation templates in [11] are used as baseline in this chapter for the proposed experiments in the remainder of this chapter.

5.1.2 Metrics to Benchmark Experiments with QSS and Dynamic Simulators

In addition to common cascade and resulting blackout statistics seen in previous benchmark studies, we adapt a metric of cascade path similarity, first introduced in [7] (Evaluating the Impact of Modeling Assumptions for Cascading Failure Simulation), the *Relative Agreement of Cascading Path*, R, defined as follows:

$$R(m_1, m_2) = \frac{1}{|C|} \sum_{i=1}^{|C|} \frac{|A_i \cap B_i|}{|A_i \cup B_i|}, \tag{5.1}$$

where m_1 and m_2 are the two cascading simulators being compared, A_i and B_i are the sets of dependent events for each cascade, and C is the set of cascades under consideration. In summary, for all experiments developed in this chapter, we measure and record the following statistics at the end of each cascade simulation, for each simulator:

- Blackout size: Total unserved load in per unit MW
- Cascade size: The number of line failures and their sequence
- Relative agreement of cascading path: R, defined as in Eq. 5.1

Similar to these, as well as other related metrics, are regularly used to assess risk derived from cascading outages or common mode failures [17, 18]. Real datasets derived from transmission owners are particularly valuable and help calibrate simulator models. In [17], the authors cluster and analyze data from the NERC TADS (Transmission Availability Data System), obtaining useful results about the timing of the outages. They also highlight the opportunity to use an underlying network model so that electrical distance for the clusters of outages can be included in the analysis.

5.1.3 A QSS Model Example

The quasi-steady-state (QSS) example model *dcsimsep* [12] is used here to investigate assumptions of models that rely solely on the steady-state operation of the system. There are no generator or load dynamics in the model, and the only departures from steady-state conditions come from relay-based switching of overloaded transmission lines.

The *dcsimsep* simulator was originally developed at the University of Vermont, led by Prof. Paul Hines and with contributions by Dr. Pooya Rezaei, Prof. Maggie Eppstein, as well as the author of this chapter. While the horizon timeline is simplified into QSS and the power flow implemented uses the dc approximation, the simulator features a relatively sophisticated separation (islanding) scheme and the possibility to interface with external solvers to improve the chances of finding a feasible dispatch solution by balancing generation and demand.

Here we focus the configuration of the simulation on the QSS group of parameters to better evaluate similarities and differences with respect to the dynamic simulator. For example, a tunable aspect in *dcsimsep* is the *pseudo*-time allowed for generator rampings, that is, discretizing the next QSS timestep where the power flow is calculated by performing a generator output projection before advancing the solver to the next steady-state timestep. By starting with a minimum fixed value, e.g., 60 seconds, of ramping that is allowed before resorting to load shedding to balance the island, we can later reproduce as similar behavior as possible in the dynamic simulator by tuning the machines' inertia to equivalent values.

5.1.4 A Dynamic Model Example

By choosing a dynamic model example, *COSMIC* [4], that is built upon the same code baseline as the QSS example, we try to focus the experiments in the remainder of this chapter on their direct similarities and differences, while minimizing baseline noise coming from possible discrepancies across different families of simulators and implementations on substantially different platforms.

The dynamics of a power system can be described with a set of hybrid differential (**f**) and algebraic (**g**) equations [3, 21], where hybrid refers to the addition of a set of equations (**h**) that represent discrete events (for example, a relay trip). The resulting hybrid differential–algebraic system of equations (DAE) is given by

$$\begin{aligned} \frac{d\mathbf{x}(t)}{dt} &= \mathbf{f}(t, \mathbf{x}(t), \mathbf{y}(t), \mathbf{z}(t)) \\ \mathbf{0} &= \mathbf{g}(t, \mathbf{x}(t), \mathbf{y}(t), \mathbf{z}(t)) \\ \mathbf{0} &> \mathbf{h}(t, \mathbf{x}(t), \mathbf{y}(t), \mathbf{z}(t)), \end{aligned} \tag{5.2}$$

where $\mathbf{x}(t)$ is a vector of continuous state variables linked to the differential equations, $\mathbf{y}(t)$ is a vector of continuous state variables linked to the algebraic equations, and $\mathbf{z}(t)$ is a vector of discrete state variables. There are different options in *COSMIC* to solve the differential portion of the system, for example, Matlab's standard functions *ode15s* and *ode23t* [20], which feature an interface to compute a semi-explicit DAE system with a direct approach. The most challenging part of this approach is the initialization of the integration, in particular, right after discrete events. At these breaking points, *COSMIC* uses a nonlinear solver to calculate where to "reconnect" the upcoming state vectors with the appropriate family of differential equations that match the current memory of the system. At this point in the time horizon, the options to implement the nonlinear solver in *COSMIC* are similar to the capabilities of *dcsimsep*, and one can customize the function *fsolve* to use external algorithms and settings that work better with the particular power system at hand. To speed up simulation by default, *COSMIC* uses a *trust region* method with *dogleg* implementation that only requires one linear solution per iteration.

A relatively large simulation horizon is computationally expensive for dynamic simulators, and typically, there are developed strategies to speed up without compromising on the mechanisms that are characterized. In the case of the *COSMIC* simulator, a relatively simple approach is to use a variable step size. Here, for the purposes of these experiments and comparative exercise with the QSS simulator, we choose a 10 minute simulation horizon, sufficient to resolve the time constants of the relays. We implement several types of relays:

- Branch temperature relays–overcurrent relays: In order to detect the overload of transmission lines, at each timestep, the simulator updates the temperature portion of the integer state variables ($\mathbf{z}_{temp,i}$) after comparing the current branch temperatures from the differential state vector ($\mathbf{x}(t)$) with the maximum temperature limit for each branch (calculated from the standard IEEE rate line limits) [3].
- Under-voltage and under-frequency relays: After detecting instantaneous off-nominal values for voltage or frequency at each bus (or generator bus), the simulator will apply a relay delay and trip if necessary. In the experiments for this chapter, we use a threshold of 0.87 p.u. for under-voltage and 0.985 p.u. for under-frequency.
- Distance relays: Here we also configure a relatively simple Zone 1 distance relay, tuned at 0.9 p.u.

Another feature implemented in *COSMIC* to accelerate the simulation is the calculation of an equivalent generator set of dynamics for those buses where multiple units exist. We aggregate inertia and capacity as well as tuning the equivalent machine controls. Adapted from [3], the description of the machine dynamics for each generator in *COSMIC* can be summarized by the following elements:

The base equation related to machine dynamics is the standard second-order swing equation, describing rotor speed for a generator connected to Bus i:

$$M\frac{d\overline{\omega_i}}{dt} = P_{m,i} - P_{g,i} - D(\overline{\omega_i} - 1), \tag{5.3}$$

where M is the machine inertia constant, $P_{m,i}$ is the mechanical power input, $P_{g,i}$ is the electrical power output, and D is the machine damping constant.

The rotor angle is given by the equation:

$$\frac{d\delta_i(t)}{dt} = 2\pi f_0(\overline{\omega_i} - 1), \tag{5.4}$$

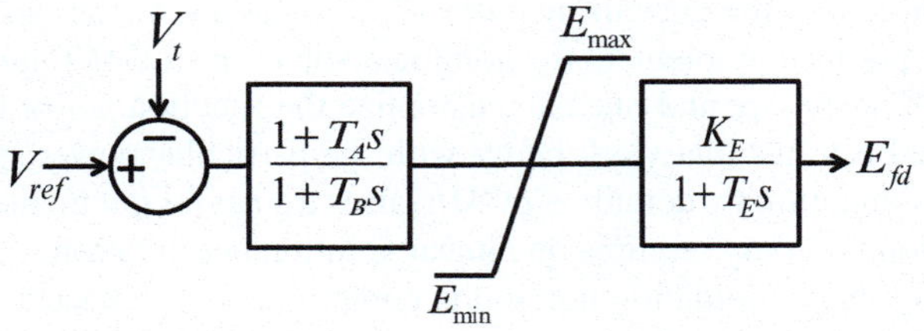

Fig. 5.1 Diagram of the machine exciter model implemented in the *COSMIC* dynamic simulator

and *COSMIC* allows the choice between a standard angle reference or a center-of-inertia reference. This feature is useful to compare and validate across dynamic simulators.

The transient open circuit voltage magnitude is calculated with the differential equation:

$$\frac{d\left|E'_{a,i}(t)\right|}{dt} = -\left|E'_{a,i}\right| \frac{X_{d,i}}{T'_{do,i} X'_{d,i}} + \left(\frac{X_{d,i}}{X'_{d,i}} - 1\right) \frac{|V_i(t)|}{T'_{do,i}} cos(\delta_{m,i}(t)) + \frac{E_{fd,i}}{T'_{do,i}}, \tag{5.5}$$

where $X_{d,i}$ and $X'_{d,i}$ are the direct axis generator synchronous and transient reactances, respectively, $T'_{do,i}$ is the direct axis transient time constant, and $E_{fd,i}$ is the machine exciter output (see Eq. 5.6). The three equations described above describe the basic physical properties of the generator, adding up to a 3rd-order differential equation system. In order to complete the machine model, we add two additional sets of equations describing the machine exciter and the machine governor.

The machine exciter equations define two of the differential variables in $\mathbf{x}(t)$:

$$\begin{aligned} \frac{dE_{fd}}{dt} &= \frac{1}{T_E}\left(K_E.\text{sigm}\left(\left(1-\frac{T_A}{T_B}\right)E_1 + \frac{T_A}{T_B}(V_{ref}-V_t)\right) - E_{fd}\right) \\ \frac{dE_1}{dt} &= \frac{1}{T_B}\left(V_{ref} - V_t - E_1\right), \end{aligned} \tag{5.6}$$

where T_A, T_B, and K_E are exciter time constants, and sigm(·) we write as a differentiable sigmoidal function that acts as a limiter between E_{min} and E_{max}. We implement this function in a way that is similar to *rail limiters* whereby the smooth joints in between linear segments are encoded as differentiable cubic splines. Figure 5.1 illustrates this simplified exciter configuration.

Similarly, we use a similar differentiable rail limiter technique to complete the dynamic model of the generator with a machine governor, describing the mechanical forcing given a deviation in the machine speed:

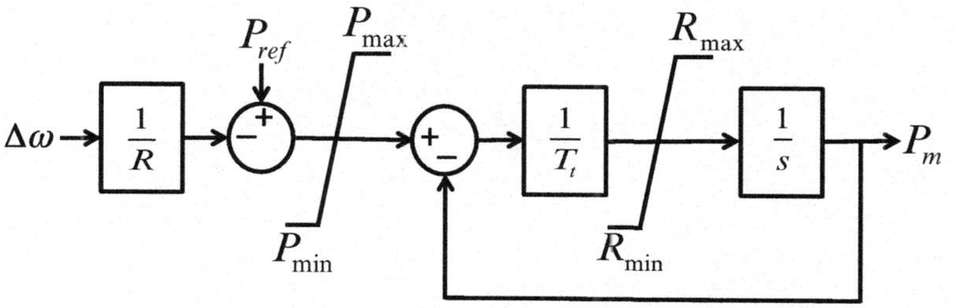

Fig. 5.2 Diagram of the machine governor model implemented in the *COSMIC* dynamic simulator

$$\frac{dP_m}{dt} = \text{sigm}\left(\frac{1}{T_t}(\text{sigm}\left(P_{ref} - \frac{1}{R}\Delta\omega\right)P_2 - P_m)\right), \tag{5.7}$$

where R and T_t are the droop and time constants, respectively. Figure 5.2 describes the interactions among the governor variables and the rail limiters for $P_{\min}$ versus $P_{\max}$ and $R_{\min}$ versus $R_{\max}$.

5.1.5 Benchmark Experiments

For the experiments in this chapter, we focus on the test system IEEE RTS-96 [9], which was also proposed as a good benchmarking network by [10] after some initial adjustments to be initially dispatched as $N-1$ secure. We keep the same modifications for this work. For each cascading failure simulator, we subjected the test grid to $N-2$ and $N-3$ line contingencies. We enumerate all the $N-2$ contingencies (up to a total of 7140 runs) and then simulate an equal number of $N-3$ contingencies, randomly sampled, and resulting on a total of 14,280 runs for all $N-2$ and $N-3$ contingencies.

For the *dcsimsep* simulator, we implement the baseline redispatch mechanisms included in the code package, without additional emergency control activated [6, 13, 19]. In terms of protection, we focus on the mechanisms of separation, load shedding, and line overload relays. At the end of each QSS simulation epoch, we record the initial exogenous events, the dependent endogenous events (including lines tripped), and total load lost.

For the *COSMIC* simulator, we also implement the baseline redispatch mechanisms, in this case an ac algebraic equations resolve that "reconnects" the new algebraic state after a discrete event with the necessary dynamics from the generator machines, exciters, and governors. This is handled as a standard Differential–Algebraic Equation (DAE) solution across discrete events [21], and for consistency, we do not implement additional controls as it is the case in recent applications for this simulator [1]. It is important to note that one of the main relay mechanisms in *COSMIC*, the temperature relay, is an aggregate relay mechanism inspired by the

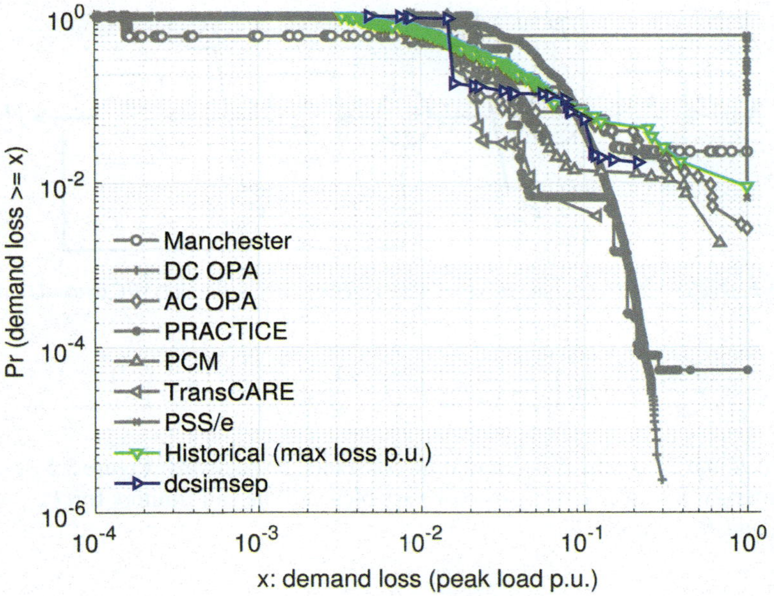

Fig. 5.3 Loss of load distribution for all $N-2$ and equal size set of random $N-3$ contingencies. Comparison between *dcsimsep*, normalized historical data reference, and benchmark QSS simulators from [11]

line overload relay in *dcsimsep*. In order to keep similar rate overload constants for this work, we extended the dynamic simulation horizon in *COSMIC* to 10 minutes, so that we are able to observe those cascades that have an equivalent time horizon in the QSS implementation. Besides the temperature/overload relay, we also implement separation, under-frequency and under-voltage load shedding, distance relays, overcurrent relays, and generator off-nominal frequency trips. At the end of each dynamic simulation epoch, we also record the initial exogenous events, the dependent endogenous events, and total load lost.

5.1.5.1 Size of Blackouts: Load Distributions

The first metric we analyze from the described experiments is the size of the blackouts observed, measured in MW of load not met. We plot these results as a complementary cumulative distribution function (CCDF) that allows us to compare the trends across simulators, and in particular, the areas corresponding to small, medium, or large blackouts, depending on the x-axis region that we focus on. For example, a trace with a heavy tail corresponds with a relatively large probability of observing large blackouts.

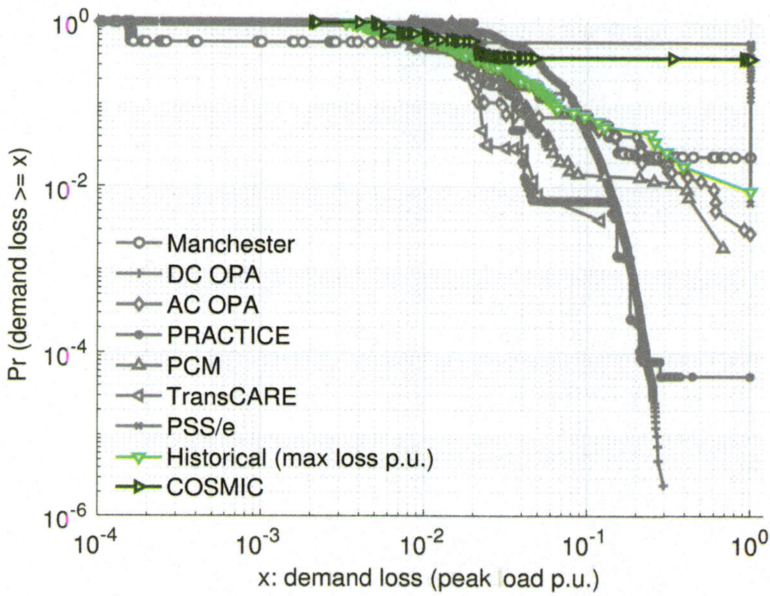

Fig. 5.4 Loss of load distribution for all $N-2$ and equal size set of random $N-3$ contingencies. Comparison between *COSMIC*, normalized historical data reference, and benchmark QSS simulators from [11]

In Fig. 5.3 we can observe the new loss of load distribution obtained with the QSS simulator *dcsimsep* (blue trace) and how it compares with those QSS simulators benchmarked in [11]. Distribution results suggest that *dcsimsep* produces blackouts similar in size with the mid-range group of those in the benchmark set (all grayed out traces for clarity). Also from [11], we highlight (light green trace) the historical data reference, adjusted in per unit to the size of the maximum blackout in the RTS-96. For medium–large blackouts, the *dcsimsep* trace appears similar to the historical reference, a bit under for certain blackout sizes, but overall with a comparable slope.

A recommendation from this analysis is that in order to produce larger blackout sizes with an open-source simulator such as *dcsimsep*, one could adjust the separation mechanisms. Effectively we would be fine-tuning the definition of *when* to consider an island separated as a brownout, continuing to recursively re-balance smaller areas, versus considering that the cascade is finished from the point of view of the original connected components in the network, because we cannot reliably dispatch such small areas. This is relevant as we increase the number of microgrids in our electrical network, as well as other modalities of distributed control, we need to reflect the mechanisms for separation and re-connection accordingly in the simulator, both QSS and dynamic [15].

Analogously, in Fig. 5.4 we can observe the new loss of load distribution obtained with the dynamic simulator *COSMIC* (dark green trace) in the context of the

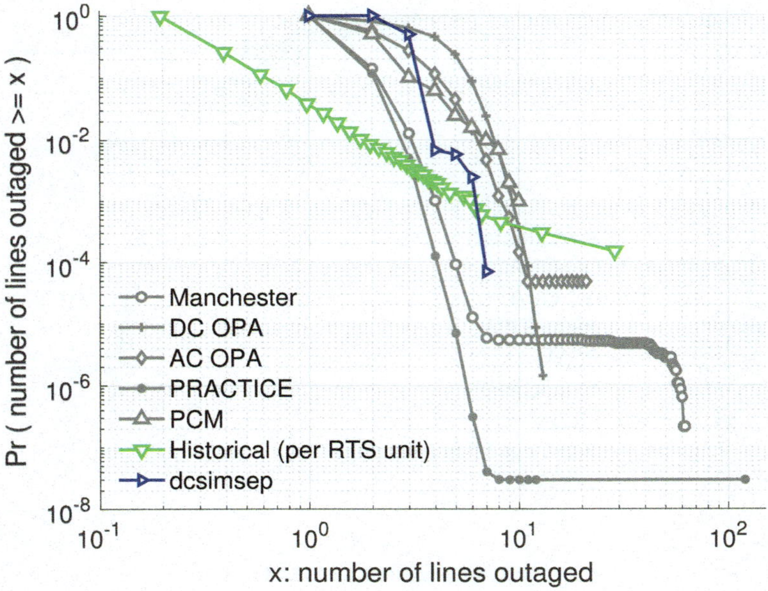

Fig. 5.5 Lines tripped distribution for all $N-2$ and equal size set of random $N-3$ contingencies. Comparison between *dcsimsep*, normalized historical data reference, and benchmark QSS simulators from [11]

datasets from [11]. In this case, small and medium blackout sizes follow fairly well the historical trend and benchmark QSS simulators. For large blackout sizes, the relatively heavier tail captures the behavior of possible dynamic instability for networks that are largely disturbed and tend to produce full outages (although this is not exclusive to dynamic simulators, as the trace corresponding to PSS/e shows, with some higher probabilities for the maximum size. That being said, the trace corresponding to *COSMIC* reflects the appearance of very large blackouts that are not full system collapse, and this certainly merits further exploration of the individual cascades so that it sheds light on the particular mechanism differentiating from the QSS trends in this benchmark ensemble.

5.1.5.2 Line Distributions

The second metric we analyze for this set of experiments is the distribution of transmission lines that were outaged. It is important to note that we include in this section all the lines that were removed from service in the $N-2$ and $N-3$ cascades, either by the initial events or by the subsequent dependent relay trips. In the next section, we will explore the differences between the two subsets (initial or dependent outages) in terms of line criticality by means of frequency of appearance.

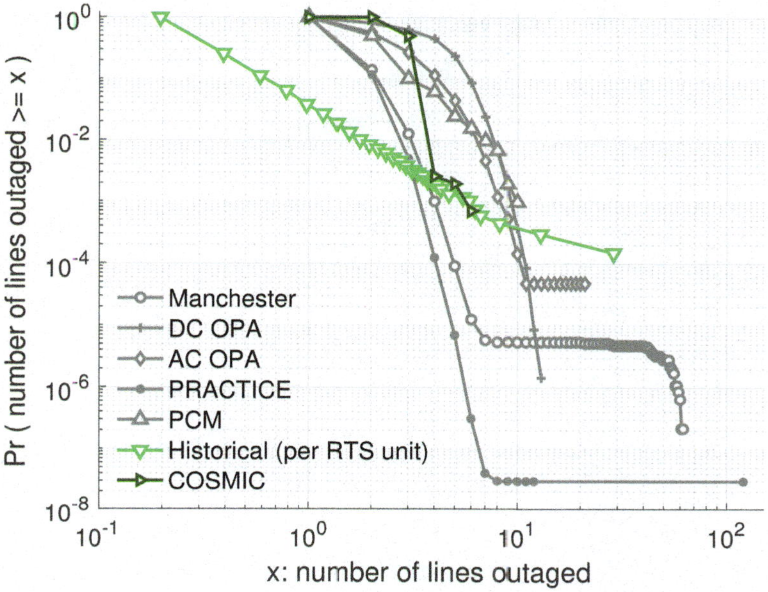

Fig. 5.6 Lines tripped distribution for all $N - 2$ and equal size set of random $N - 3$ contingencies. Comparison between *COSMIC*, normalized historical data reference, and benchmark QSS simulators from [11]

In Fig. 5.5, we compare the distribution of the number of lines outaged for *dcsimsep* with the historical reference and simulator benchmarks from [11]. It is encouraging that the trace shape for the medium–large size cascades (4,5,6, or 7 lines) approaches the historical trend.

We also note the difference with other simulators that obtain a much larger number of transmission lines tripped. As we discussed earlier with the threshold to consider the system separation as a terminating event or part of the cascade, it would be important also to define what is an upper bound on lines outaged for a given system size where one would expect the cascading simulation to still be meaningful. In this experiment's scale, for a system size like the RTS-96 with 120 lines, it is difficult to interpret how the simulation mechanisms continue in a similar regime after the relays tripped line 100*th* out of 120, for example. It appears that the overall distribution of lines cascaded is most similar between the *dcsimsep* and *PCM* QSS simulators.

For the dynamic simulation comparison, in Fig. 5.6, we see a similar trend for the lines outaged distribution in the *COSMIC* simulator, whereby the medium size cascades matched most closely the historical trend among all the benchmarks.

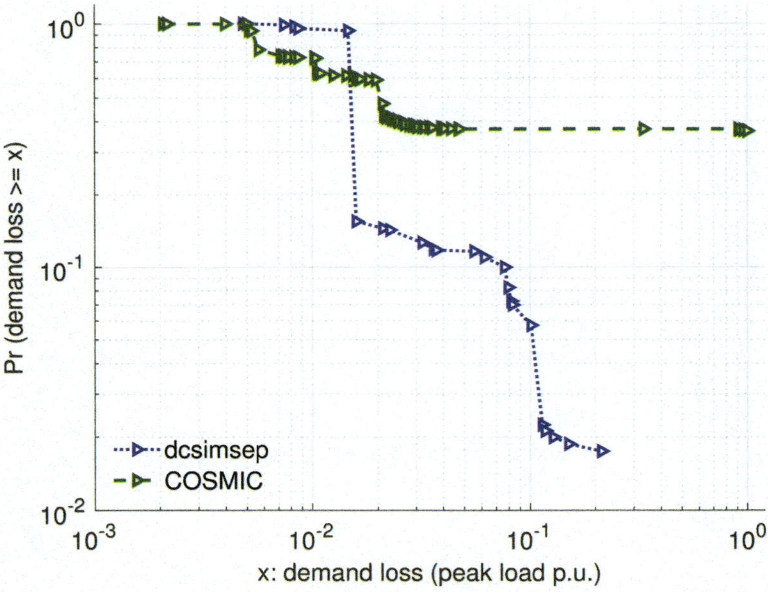

Fig. 5.7 Side-by-side comparison between *dcsimsep* and *COSMIC* simulators for the distribution of demand lost

5.1.5.3 Summary of Statistical Similarities and Differences Between *dcsimsep* and *COSMIC*

Here we compare side by side, according to the previous metrics in this section, our QSS example simulator, *dcsimsep*, and our dynamic example simulator, *COSMIC*. First for distribution of load lost, in Fig 5.7, we see higher similarities for small blackouts; however, for medium and large blackouts, the distributions for demand loss start to diverge. It is worth to note here that as we have discussed so far, we prioritized in this batch of experiments to equalize the overload relays to be able to compare the trajectories of the cascades between both the QSS and dynamic simulators. Another approach for future work could be to maintain as much similarity as possible with load shedding relays as well, although this is best achieved by working in some additional dispatch control, a feature that both example simulators implement.

Figure 5.8 suggests that the similarities between the lines tripped distributions are stronger and follow a very similar pattern for medium-sized cascades. The output in *dcsimsep* produced a few longer cascades, but the probability of those quickly falls.

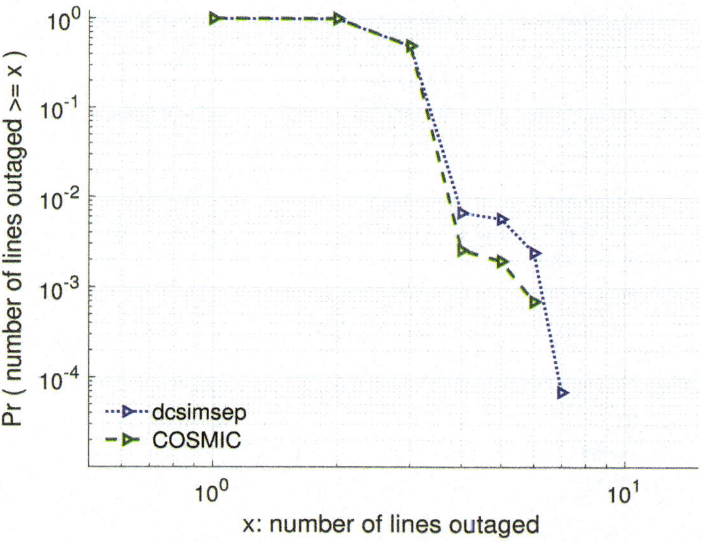

Fig. 5.8 Side-by-side comparison between *dcsimsep* and *COSMIC* simulators for the distribution of lines outaged

5.1.5.4 Cascade Sequence Benchmarks

In this section, we turn to explore the similarities and differences of the cascading sequences for both the QSS and dynamic simulators, with the same set of $N-2$ and $N-3$ experiments on the IEEE RTS-96 network. First, we measure the relative agreement of cascading path, R, among all the simulated initiating events, and then solely grouped by $N-2$ or $N-3$ types. Table 5.1 shows that the agreement between the simulators is substantially higher when analyzing the $N-2$ initiating events, whereas R decreases on the $N-3$ subset. The number of cascades that fall in one type versus the other is approximately a 25%/75% split.

5.1.5.5 Rank of Top 5 Critical Components Involved in Initial Outages

Now we turn to discuss the involvement of common initiating sets of lines in subsequent outages. From the set of $N-2$ and $N-3$ experiments, we count and rank those lines that more frequently appear as initial outage at the beginning of a cascade. In Table 5.2, we compare the results for *dcsimsep* on the left panel with *COSMIC* on the right panel. The five rows of data across both panels represent the Top 5 branches (including those that are tied in one of the five given positions), the line identifier, and the frequency of appearance in the full set of simulation epochs. After highlighting in boldface the line numbers that are common to both simulators, we can observe a very high degree of similarity, whereby 100% of those initial

Table 5.1 Relative agreement of cascading path for *dcsimsep* (m_1) and *COSMIC* (m_2) among all simulated initiating events and the $N-2$, $N-3$ subsets

	$R(m_1, m_2)$	$\|C\|$
$N-2$ and $N-3$	0.2686	103 cascades
$N-2$	0.5641	26 cascades
$N-3$	0.1688	77 cascades

Table 5.2 Rank of Top 5 critical components and their frequency of appearance within initial outages that cause cascades

dcsimsep			*COSMIC*		
Rank	Line number	Frequency	Rank	Line number	Frequency
Top 1	**30**	×18	Top 1	**27**	×8
Top 2	**7**	×15	Top 2	**7, 30**	×7
Top 3	**27, 31**	×14	Top 3	**28, 31, 69**	×6
Top 4	**26, 66, 69**	×12	Top 4	25, **29**, **66**, 119	×5
Top 5	**28, 29, 67**	×11	Top 5	**26**, **67**, 101	×4

contingencies that are Top 5 for *dcsimsep* appear on the panel corresponding to the *COSMIC* simulator, with only 3 lines that are not a commonality appearing on the dynamic simulator results.

5.1.5.6 Rank of Top 5 Critical Components Involved in Subsequent Outages

Similarly to the previous subsection, we now discuss the involvement of common subsequent sets dependent lines outages, with respect to both simulators, from the set of $N-2$ and $N-3$ experiments. In Table 5.3, we compare again the results for *dcsimsep* on the left panel with *COSMIC* on the right panel. In this case, after highlighting in boldface the line numbers that are common to both simulators, we also obtain a very high overall similarity on both sets, with an 80% agreement on the lines that appear on the Top 5 critical elements for both *dcsimsep* and *COSMIC*. Only lines 11 and 25 appeared in one or the other simulator but not both. It is important to note that overall, the set of dependent critical outaged lines is also more concentrated in fewer number of lines with higher frequency of appearance in cascading sequences, for both the QSS and the dynamic simulator. This suggests that any mitigation measures that focus on the propagating lines are likely to help reduce the overall risk, independently of whether one uses the QSS or the dynamic simulator version for this test case.

Table 5.3 Rank of Top 5 critical components and their frequency of appearance within initial outages that cause cascades

dcsimsep			*COSMIC*		
Rank	Line number	Frequency	Rank	Line number	Frequency
Top 1	**119**	×58	Top 1	**118**	×26
Top 2	11	×16	Top 2	**119**	×26
Top 3	**30**	×16	Top 3	**30**	×8
Top 4	**41**	×16	Top 4	**41**	×7
Top 5	**118**	×14	Top 5	25	×6

5.1.6 Conclusions and Future Work

In this chapter, we have presented a discussion of similarities and differences between quasi-steady-state (QSS) and dynamic models of cascading outages. We discuss new findings obtained from side-to-side comparison experiments between QSS and dynamic simulators that stem from related branches of open-source codes. We also positioned this analysis with respect to previous results obtained by recent benchmarking studies that included both research-grade and commercial simulator software. For future work, we recommend the concurrent use of multiple simulation fidelities in cascading outage studies, as well as the inclusion, early in the workflow, of real datasets for calibration and validation of the relay mechanisms in the models being considered.

References

1. F. Alanazi, J. Kim, E. Cotilla-Sanchez, Load oscillating attacks of smart grids: vulnerability analysis. IEEE Access **11**, 36538–36549 (2023). ISSN: 2169-3536. https://doi.org/10.1109/ACCESS.2023.3266249. https://ieeexplore.ieee.org/document/10098782/ (visited on 08/29/2023)
2. J. Bialek et al., Benchmarking and validation of cascading failure analysis tools. IEEE Trans. Power Syst. **31**(6), 4887–4900 (2016). ISSN: 08858950. https://doi.org/10.1109/TPWRS.2016.2518660
3. E. Cotilla-Sanchez, Big data and energy systems: efficient computational methods for the dynamic analysis of electric power infrastructure. PhD thesis. University of Vermont, 2012
4. E. Cotilla-Sanchez, *ecotillasanchez/cosmic*, Aug. 2023. https://github.com/ecotillasanchez/cosmic (visited on 08/30/2023)
5. Y. Dai et al., Risk assessment and mitigation of cascading failures using critical line sensitivities. IEEE Trans. Power Syst., 1–12 (2023). ISSN: 0885-8950, 1558-0679. https://doi.org/10.1109/TPWRS.2023.3305093. https://ieeexplore.ieee.org/document/10219000/ (visited on 08/27/2023)
6. M.J. Eppstein, P.D.H. Hines, A "random chemistry" algorithm for identifying collections of multiple contingencies that initiate cascading failure. IEEE Trans. Power Syst. **27**(3), 1698–1705 (2012). ISBN: 9781479913039. ISSN: 08858950. https://doi.org/10.1109/TPWRS.2012.2183624

7. R. Fitzmaurice, E. Cotilla-Sanchez, P. Hines, Evaluating the impact of modeling assumptions for cascading failure simulation, in *IEEE Power and Energy Society General Meeting*. ISSN: 19449925.2012. ISBN: 978-1-4673-2727-5. https://doi.org/10.1109/PESGM.2012.6345378
8. A.J. Flueck et al., Dynamics and protection in cascading outages, in *2020 IEEE Power & Energy Society General Meeting (PESGM)* (IEEE, Montreal, QC, Canada, Aug. 2020), pp. 1–5 ISBN: 978-1-72815-508-1. https://doi.org/10.1109/PESGM41954.2020.9281823. https://ieeexplore.ieee.org/document/9281823/ (visited on 08/30/2023)
9. C. Grigg, P. Wong, The IEEE reliability test system—1996 a report prepared by the reliability test system task force of the application of probability methods subcommittee. IEEE Trans. Power Syst. **14**(3), 1010–1020 (1999). ISBN: 0885-8950. ISSN: 08858950. https://doi.org/10.1109/59.780914
10. P. Henneaux et al., A two-level probabilistic risk assessment of cascading outages. IEEE Trans. Power Syst. **31**(3), 2393–2403 (2016)
11. P. Henneaux et al., Benchmarking quasi-steady state cascading outage analysis methodologies, *2018 IEEE International Conference on Probabilistic Methods Applied to Power Systems (PMAPS)* (IEEE, Boise, ID, June 2018), pp. 1–6. ISBN: 978-1-5386-3596-4. https://doi.org/10.1109/PMAPS.2018.8440212. https://ieeexplore.ieee.org/document/8440212/ (visited on 08/04/2023)
12. P. Hines, *DCSIMSEP*, June 2023. https://github.com/phines/dcsimsep (visited on 08/30/2023)
13. P. Hines, E. Cotilla-Sanchez, S. Blumsack, Do topological models provide good information about electricity infrastructure vulnerability? Chaos **20**(3) (2010). arXiv:1002.2268. ISBN: 1054-1500. ISSN: 10541500. https://doi.org/10.1063/1.3489887
14. W. Ju, K. Sun, R. Yao, Simulation of cascading outages using a power flow model considering frequency. IEEE Access (2018). IEEE, p. 1. https://doi.org/10.1109/ACCESS.2018.2851022
15. C. Lassetter, E. Cotilla-Sanchez, J. Kim, A learning scheme for microgrid reconnection. IEEE Trans. Power Syst. **33**(1), 691–700 (2018). ISSN: 0885-8950, 1558-0679. https://doi.org/10.1109/TPWRS.2017.2709741. http://ieeexplore.ieee.org/document/7935511/ (visited on 05/16/2020)
16. L. Niu et al., *A Hybrid Submodular Optimization Approach to Controlled Islanding with Post-Disturbance Stability Guarantees*. arXiv:2302.10308 [cs, eess, math], Feb. 2023. http://arxiv.org/abs/2302.10308 (visited on 02/27/2023)
17. M. Papic, S. Ekisheva, E. Cotilla-Sanchez, A risk-based approach to assess the operational resilience of transmission grids. Appl. Sci. **10**(14), 4761 (2020). ISSN: 2076-3417. https://doi.org/10.3390/app10144761. https://www.mdpi.com/2076-3417/10/14/4761 (visited on 07/11/2020)
18. M. Papic et al., Multiple outage challenges to transmission grid resilience, in *2019 IEEE Power & Energy Society General Meeting* (*PESGM*) (IEEE, Atlanta, GA, USA, Aug. 2019), pp. 1–5. ISBN: 978-1-72811-981-6. https://doi.org/10.1109/PESGM40551.2019.8973606. https://ieeexplore.ieee.org/document/8973606/ (visited on 02/03/2020)
19. P. Rezaei, P.D.H. Hines, M.J. Eppstein, Estimating cascading failure risk with random chemistry. IEEE Trans. Power Syst. **30**(5), 2726–2735 (2015). arXiv:1405.4213. ISBN:9781467380409. ISSN: 08858950. https://doi.org/10.1109/TPWRS.2014.2361735
20. L.F. Shampine, M.W. Reichelt, J.a. Kierzenka, Solving Index-1 DAEs in MATLAB and simulink. SIAM Rev. **41**(3), 538–552 (1999). ISSN: 0036-1445. https://doi.org/10.1137/S003614459933425X
21. J. Song et al., Dynamic modeling of cascading failure in power systems. IEEE Trans. Power Syst. **31**(3), 2085–2095 (2016). arXiv:1411.3990. https://doi.org/10.1109/TPWRS.2015.2439237
22. R. Yao et al., A multi-timescale quasi-dynamic model for simulation of cascading outages. IEEE Trans. Power Syst. **31**(4), 3189–3201 (2016). ISSN: 08858950. https://doi.org/10.1109/TPWRS.2015.2466116

Chapter 6
Multi-timescale Simulation, Risk Assessment, and Mitigation of Cascading Outages

Rui Yao, Kai Sun, Feng Liu, Shengwei Mei, and Shaowei Huang

6.1 Multi-timescale Quasi-dynamic Simulation of Cascading Outages

Cascading outages in electric power grids are dependent outage processes triggered by one or a set of initial faults. Cascading outages gradually impact and deteriorate transmission systems and may cause large blackouts as well as massive losses [1]. Since in recent years power system operation is facing more uncertainty and stress, the simulation, analysis, prediction, and mitigation of cascading outages have attracted more interests from both academia and industry [2]. Cascading outages are complex processes involving various dynamics in quite different timescales. The relay protection [3] and emergency control (e.g., load shedding [4]) usually take tens of milliseconds to seconds, while the load variation is much slower, in timescale of hours. Also, there are processes with timescales in between, such as overhead line outages caused by overheat and tree contact [5], and generator outages

The work presented in this chapter was done when Rui Yao was with Tsinghua University, Beijing, China, and the University of Tennessee, Knoxville, USA.

R. Yao (✉)
Google LLC, Mountain View, CA, USA
e-mail: ruiyao@ieee.org

K. Sun
University of Tennessee, Knoxville, Knoxville, TN, USA
e-mail: kaisun@utk.edu

F. Liu · S. Mei · S. Huang
The State Key Laboratory of Power Systems, Department of Electrical Engineering, Tsinghua University, Beijing, China
e-mail: lfeng@tsinghua.edu.cn; meishengwei@tsinghua.edu.cn; huangsw@tsinghua.edu.cn

K. Sun (ed.), *Cascading Failures in Power Grids*, Power Electronics and Power Systems, https://doi.org/10.1007/978-3-031-48000-3_6

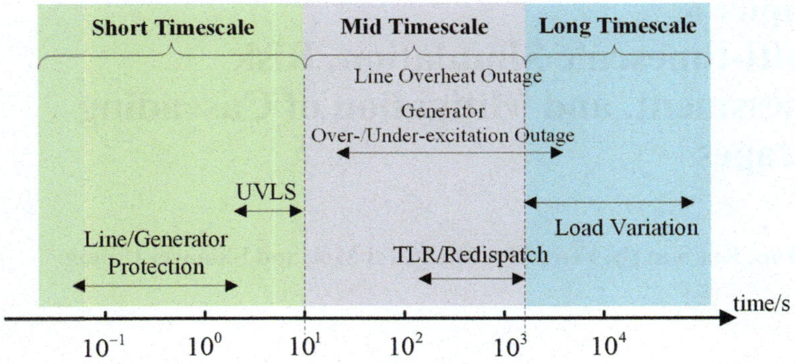

Fig. 6.1 Timescales of dynamics in cascading outages

caused by over-excitation or under-excitation [1]. These slow outage processes are in the timescales of minute to hour depending on the extent of stress and many other random factors. Also, transmission loading relief (TLR) and re-dispatch undertaken by operators against overloading generally last for 10–30 minutes or even longer. The conventional cascading outage simulation methods cannot reflect the multi-timescale characteristics of cascading outage process. Only with a proper methodology treating different timescales and representing time in cascading outage, can the model reasonably simulate interactions among related dynamics and obtain practical results with time information. Also, the representation of time facilitates practical application of cascading outage simulation, e.g., to assess and optimize the time performance with control actions.

Figure 6.1 shows timescales of typical dynamics involved in cascading outages [1, 6, 7]. To utilize quasi-dynamic concept in simulation, the dynamics should be categorized by timescales so that in cascading outage simulation, faster processes can be simulated between neighboring slower process transitions. To establish multi-timescale cascading outage simulation, the related dynamics are grouped into 3 categories:

(1) Short-term process. This includes overloading or faults directly causing branch and generator protections, as well as emergency load shedding, which usually last a few seconds.
(2) Mid-term process. This category includes overhead line outages caused by overheat and tree contact and generator outages by over-excitation or under-excitation, named as "Mid-term Random Outage" (MTRO). This timescale also includes re-dispatch operation taken by operators. The mid-term process usually lasts for minutes, and often with notable uncertainty.
(3) Long-term process. This process refers to variation of load, which is slow and continuous throughout the entire process of cascading outages. The categorization of timescales is expected to facilitate cascading outage simulation in that when simulating processes of shorter timescales, the states and parameters of longer timescale processes can be regarded as constant.

Fig. 6.2 Interactions among timescales

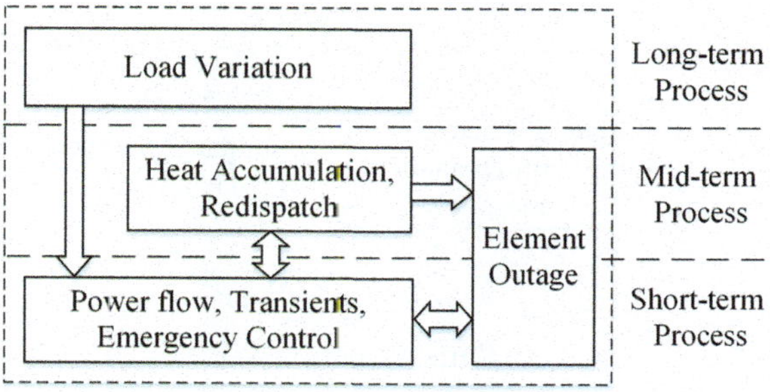

6.1.1 *Quasi-dynamic Framework: Simulate Interactions Among All Timescales*

To realize the multi-timescale simulation of cascading outages, it is necessary to figure out the dominant physical characteristics in each timescale and analyze their interactions. Figure 6.2 illustrates the interactions among timescales. In cascading outages, load variation is the slow driving force changing system states. The variations of system states directly influence the loading of elements and the heat accumulation, thus directly or slowly causing elements outages. The outages in turn render transients and redistribution of power flow pattern. The risky system states or transients may trigger re-dispatch operation or emergency control. The simulation procedures can be designed using the decomposition of timescales, and interactions among all the related processes should be realized.

The interactions between mid-term and long-term dynamics account for most of the time in the cascading outage process. Considering the related dynamics as shown in Fig. 6.1, the quasi-dynamic simulation procedure can be regarded as a loop of MTROs and re-dispatch operations accompanied with long-term system state changes (e.g., load variations). Here select a reasonable time interval Δt_{Mid} of the mid-term process, and discretize the long-term process with Δt_{Mid}; then the simulation is performed as a loop including simulation of the mid-term process during interval Δt_{Mid} and updating long-term states to the next interval, as shown in Fig. 6.3. Besides, the interactions between the mid-term and short-term processes can be described using the quasi-dynamics method. Short-term processes are checked and simulated once they are triggered.

Basically, Δt_{Mid} should not be shorter than the timeframe of re-dispatch operation, which is generally longer than 10 minutes. And Δt_{Mid} should not be too long, in order to avoid large load variation during each interval. A recommendation is that for cascading outage simulation involving human dispatchers, Δt_{Mid} should be between 15 minutes and 1 hour. Although this equal-interval discretization and simulation differs from the fact that actual random outage events may occur at any time, in the perspective of risk assessment, such an equal-interval discretization of time reflects the average effect of all possible cascades occurring at arbitrary time during the interval. Thus the statistical validity of this approach is kept thanks to

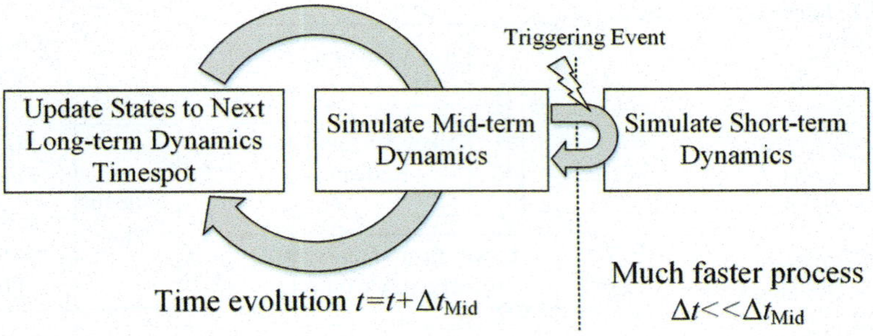

Fig. 6.3 Quasi-dynamic multi-timescale simulation framework

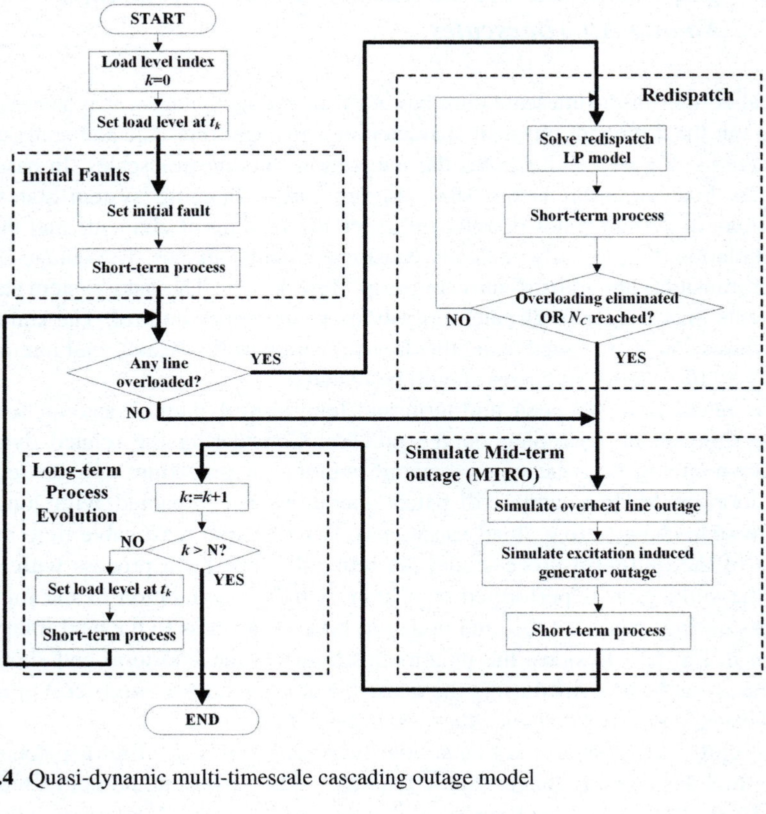

Fig. 6.4 Quasi-dynamic multi-timescale cascading outage model

random outage sampling on time interval Δt_{Mid}, which also adds flexibility to the selection of Δt_{Mid}.

Figure 6.4 shows the overall procedure of proposed cascading outage model. The model consists of several loops representing the evolution of cascading outages in different timescales. Also, related dynamics are modeled in the cascading outage simulation procedure. The detailed procedure and related dynamic modeling methods are described as following.

6.1.2 Detailed Modeling of Timescales

6.1.2.1 Long-Timescale Processes

The long-timescale processes extend over hour levels and mainly include load-level changes. In the simulation model, assume that the simulation process starts at time T_0, then set the load level at T_0 as the initial system condition before the initial outage, and for each Δt_{Mid}, set the system loading to the corresponding level until the simulation of the whole cascading outage process ends.

6.1.2.2 Mid-Timescale Processes

The mid-timescale processes are mainly thermal-driven. And under limited modeling details, the outage occurrence can be modeled as Poisson process. Here we consider line outages due to conductor overheat and generator outages due to over- and under-excitation.

For generator outages due to over- and under-excitation, when generator operates near its reactive power limits, there is increasing chance to outage. Moreover, deep over-excitation or under-excitation causes generator voltage deviation and possibly triggers generator voltage protection. We use time to failure (TTF) as the key metric, so for a generator, there is a TTF $\tau_G(V, Q)$ dependent on its reactive power output Q and its output voltage V.

For line outage due to overheat, similarly we can define a TTF metric τ_L, which can be determined by the line loading rate and multiple external factors, e.g., the environmental temperature, wind speed, and heights of objects on ground.

In the mid-timescale simulation, the observation time window is $\tau_D = \Delta t_{\text{Mid}}$, and thus we can sample the outages of generators and lines using the TTF metrics. For generators, their outage probabilities are

$$p_{\text{G}} = 1 - \mathrm{e}^{-\tau_D/\tau_G(V,Q)}. \tag{6.1}$$

And for lines, their outage probabilities are

$$p_{\text{L}} = 1 - \mathrm{e}^{-\tau_D/\tau_L}. \tag{6.2}$$

6.1.2.3 Short-Timescale Processes

The short-timescale processes, e.g., relay protection and emergency load shedding, are much faster than the processes of other timescales, so they are hardly intervened by other dynamics in cascading outages. Therefore, in the proposed cascading outage model, the simulation of short-timescale process can be wrapped as an individual submodules:

(1) Simulation of cascading protection actions:
If severe overloading is detected on any line, protection is triggered and the line is tripped. In this cascading outage model, a loading rate threshold β_{Li} is set (practically, 1.8–2.4) for each line [3]. If loading rate of line i exceeds β_{Li}, then cut the line quickly. Similarly, generators are equipped with voltage protection with over-voltage threshold $v_{\mathrm{G}}^{\max}$ and under-voltage threshold $v_{\mathrm{G}}^{\min}$. If voltage of a generator bus goes beyond the limits, then the generator is cut off by protection.
(2) Simulation of load shedding:
Load shedding is a commonly used emergency control scheme in order to relieve system from collapse. In cascading outage simulation based on power flow, the divergence of power flow often corresponds to instability or severe stress. Therefore, in cascading outage simulation, once power flow fails to converge, load shedding is performed for up to a preset number of rounds N_{U} until power flow converges.
(3) Simulation of dynamic or transient processes:
Power system is essentially a dynamic system, and after initial disturbance, there will be dynamic and transient processes. These processes can be simulated by dynamic simulation, which traces the whole process until system collapse or reaching steady state. Figures 6.5, 6.6 demonstrate the short-term processes after triggering outages simulated with dynamic simulation. As Fig. 6.5 shows, the system finally reaches a steady state and the short-term process ends. It can be seen that before each outage, the transients caused by the previous outage have basically faded away, so the simulation of such a short-term process based on power flow is reasonable. Figure 6.6 demonstrates a process of fast cascades that leads to system collapse (instability). In other cases, the system may also undergo oscillations, frequency or transient instabilities, etc. Under these circumstances, the process has high nonlinearity, and the analysis only using power flow may be inadequate.
(4) Overall short-timescale simulation process:
Figure 6.7 demonstrates the flowchart of short-timescale process simulation. The simulation can be performed in two ways: with power flow or dynamic simulation. Choosing power flow or dynamic simulation is a trade-off between efficiency and accuracy, depending on the application requirements. Dynamic simulation is able to more accurately reflect the interactions among system states, protection actions, and emergency controls. But its high requirement of computational resources and difficulty in dealing with uncertainty are major issues in practice.

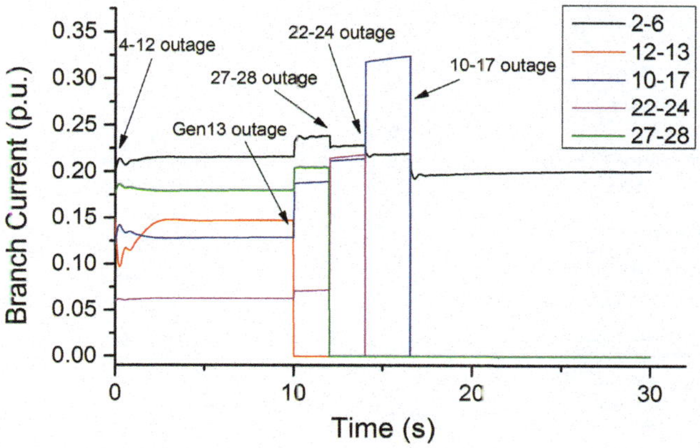

Fig. 6.5 Transient process in IEEE-30 bus system (stable case)

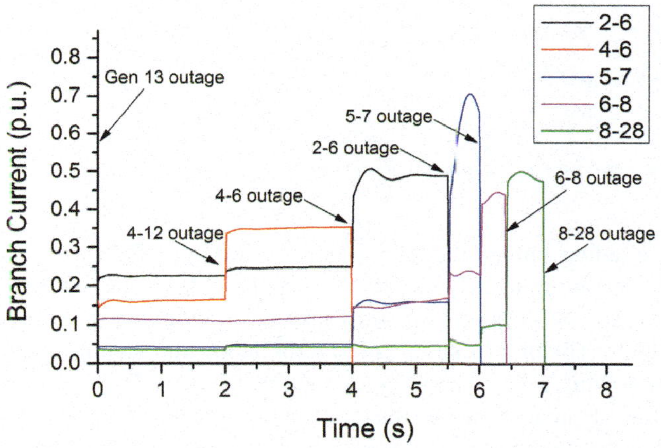

Fig. 6.6 Transient process in IEEE-30 bus system (unstable case)

6.1.3 Simulating Cascading Outages in US–Canada Northeast Grid Model

The proposed model is utilized to study the US–Canada Northeast power grid. This case study uses a reduced model of the system having all transmission lines of 230kV and above, and all generator buses. There are 410 buses, 882 branches, and 200 generators in total. The system load is 162,121.5 MW.

The proposed model is utilized to simulate possible patterns of outage evolutions. We simulate cascading outages of up to 2 hours and $\tau_D = 0.5$h. The reliability parameters of generators and branches are generated based on literatures [1, 6, 8, 9]. Figure 6.8 shows one of the cascading outage processes simulated using the

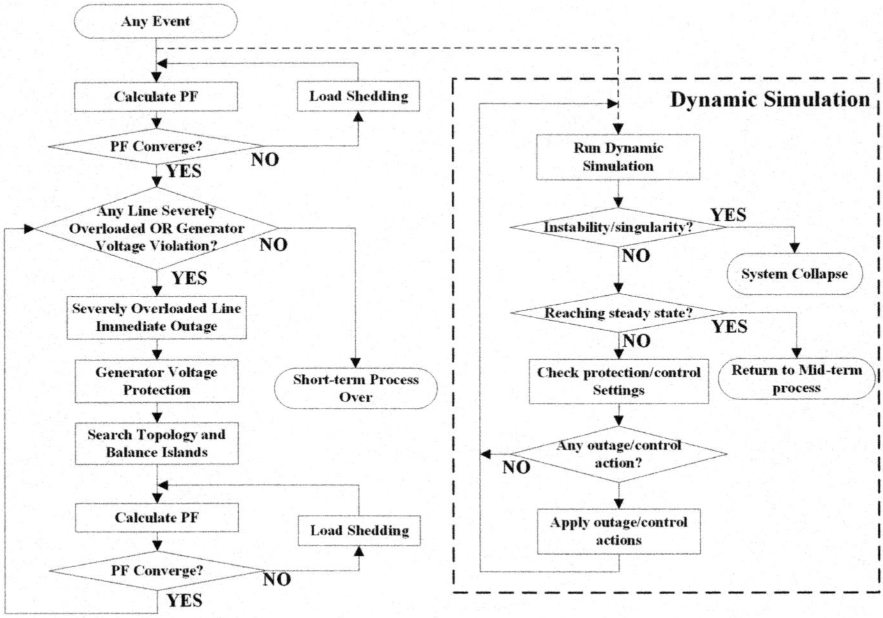

Fig. 6.7 Short-term process simulation procedure

proposed cascading outage model. The result indicates that the cascading outage process lasts for around 1.5 hours. It starts by the outage of Erie-Perry 345kV line and then develops with the loss of East lake generation and three more Ohio area lines in 1 hour. Till then, transmission paths along the south of Lake Erie have been cut, following which the outages then accelerate. The loss of several lines at Ohio–Pennsylvania border further narrows channels of power supply to Lake Erie south shore area. The outages also develop westward, forcing power flow to detour, causing more outages in Michigan and power flow reverse from Ontario. Such a pattern of transitions in power flow under cascading outages resembles that during the August 14th, 2003, blackout just before the final fast cascades stage, as illustrated in Fig. 6.9c. The simulated outages develop quickly at 1.5 hours after initial outages, finally separate Michigan from Ontario, and cut off almost all channels to Michigan but a shallow neck at west Michigan. The power supply to the Lake Erie south shore is also limited to only a few branches from southwest. The Michigan and Lake Erie south areas then experience severe stress, with high risk facing fast outages and blackout. The pattern of power flow is also similar to that before the 2003 blackout (Fig. 6.9d), and the stressed areas in simulation have large overlaps with the areas affected by the 2003 blackout.

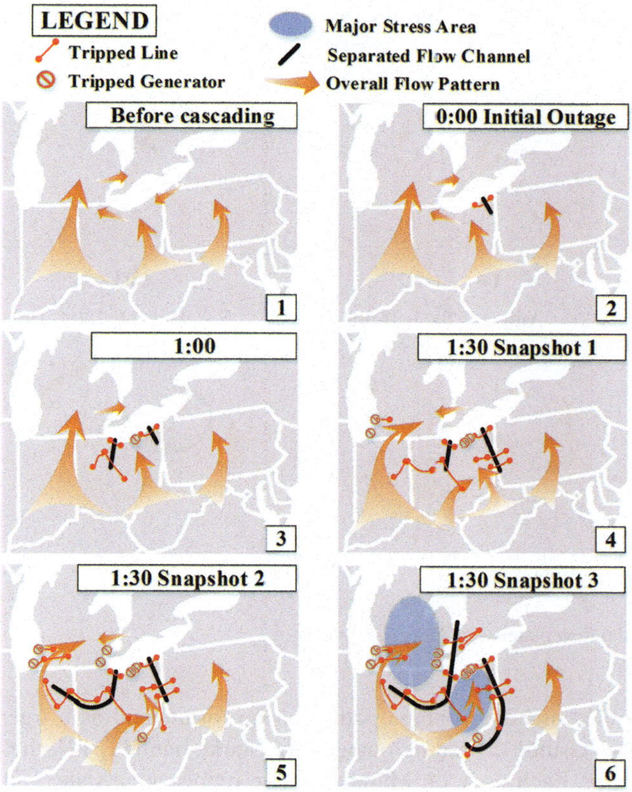

Fig. 6.8 A simulated cascading process in US–Canada Northeast system

6.2 Markovian Tree Model of Cascading Outages

6.2.1 Markovian Tree Model

In recent years, power system operation is facing more uncertainty and stress brought by increasing load, variable renewable energy, and natural disasters. The simulation, analysis, prediction, and mitigation of cascading outages have attracted more interests from both academia and industry [10]. It is important to model possible paths, especially the more probable ones and/or the ones with highest consequences. To characterize the severity of cascading outages, the concept of risk is usually used. Risk is defined as the expected consequence (in the amount of load loss or economic loss) and calculated as probability times the consequence. And thus, people can evaluate the severity of cascading outages by using the risk metrics.

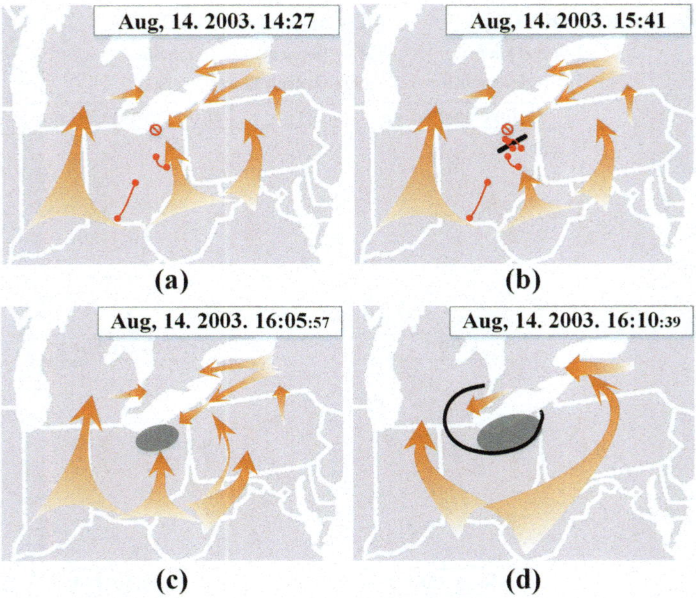

Fig. 6.9 Actual flow patterns in Aug. 2003 blackout

All possible cascading outage paths can be formulated as a tree structure, as figure shows. Assume cascading outages are Markovian [11, 12], the tree is then a Markovian tree. Each node on Markovian tree represents a state, and each branch on Markovian tree represents a random outage.

To represent the time elapse in simulation, we categorize events in cascading outages by timescales, as shown in Fig. 6.1. Then a quasi-dynamic simulation framework is utilized, which inserts the simulation of shorter timescales between adjacent longer-timescale events (Fig. 6.3). Thus multi-timescale simulation is realized, and approximate time elapse is provided, which reasonably reflects the actual characteristics of cascading outages. Since reasonable modeling and simulation of cascading outages is a prerequisite of practical risk assessment, next we will show how to reformulate the quasi-dynamic multi-timescale model as Markovian tree as a foundation of the novel risk assessment method.

We name this model as "quasi-dynamic model" in the rest of the chapter. The mid-timescale random outages have uncertainties, causing distinct cascading outage sequences. If we merge same states of all the possible cascading outage sequences from the beginning state, a tree structure like above will be formulated as Fig. 6.10 shows. Each node on Markovian tree represents a state, and each branch on Markovian tree represents a mid-timescale random outage. Similar to the quasi-dynamic model, every mid-timescale transition corresponds to a time elapse τ_D.

The labelling of states on a Markovian tree is determined as follows. The beginning state is level-0 node, and subsequent states are labelled sequentially

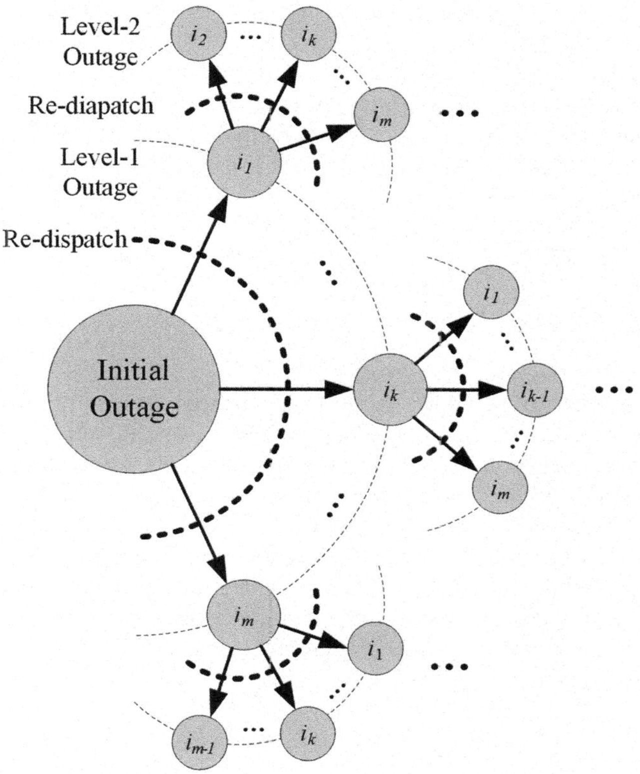

Fig. 6.10 Markovian tree representation of cascading outage paths

as level-1 nodes, level-2 nodes, etc. Note that not every state transition has a corresponding outage, and it is possible that no outages occur during the interval. If we label elements with positive integers, then a state is coded as the sequence of labels of failed elements from the initial outage up to the current level (no-outage labelled as 0), i.e., $(i_{k_1} i_{k_2} \cdots i_{k_n})$. Here only branch outage events are considered, but the modeling and simulation of generator outages are very similar.

The costs of cascading outages come from the dispatch control, load shed by power balancing when system separates, or load shed by emergency control measures on each level of the Markovian tree. The selection of the definition of the cost also has flexibility: the cost can be the load or the energy loss at each level of outage, or the economic loss caused by the outages. In this chapter, the cascading outage simulation is based on DC power flow, and the cost of an outage C on Markovian tree is the sum of load loss from re-dispatch C_R and load loss from network balancing C_B . Every state $(i_{k_1} i_{k_2} \cdots i_{k_n})$ corresponds to a non-negative cost denoted as $C(i_{k_1} i_{k_2} \cdots i_{k_n})$.

$$C(i_{k_1}i_{k_2}\cdots i_{k_n}) = C_R(i_{k_1}i_{k_2}\cdots i_{k_n}) + C_B(i_{k_1}i_{k_2}\cdots i_{k_n}). \tag{6.3}$$

The probability of any event $i_{k_{n+1}}$ (corresponding to state $(i_{k_1}i_{k_2}\cdots i_{k_{n+1}})$) depends on its previous state $(i_{k_1}i_{k_2}\cdots i_{k_n})$, so the conditional probability of event $i_{k_{n+1}}$ is denoted as $\Pr(i_{k_{n+1}}|i_{k_1}i_{k_2}\cdots i_{k_n})$. With these terms, take the risk assessment of expected load loss as example, and the risk is expressed as

$$\begin{aligned} R = C_0 &+ \sum_{k_1}\Pr(i_{k_1})C(i_{k_1}) + \sum_{k_1}\Pr(i_{k_1})\sum_{k_2}\Pr(i_{k_2}|i_{k_1})C(i_{k_1}i_{k_2}) \\ &+ \sum_{k_1}\Pr(i_{k_1})\sum_{k_2}\Pr(i_{k_2}|i_{k_1})\sum_{k_3}\Pr(i_{k_3}|i_{k_1}i_{k_2})C(i_{k_1}i_{k_2}i_{k_3}) + \cdots . \end{aligned} \tag{6.4}$$

Here each interval τ_D allows at most one element outage. To enforce this, the interval τ_D in the Markovian tree should be chosen that the probability of two or more random outages in τ_D should be negligible. It is found through tests that generally τ_D as 3–15 minutes can satisfy practical needs.[1]

The requirement of single outages on the Markovian tree renders different definitions of probability. In the quasi-dynamic model, outage events are sampled independently, while in the Markovian tree model, the outage probability is defined as "the probability that the outage is the first to occur." Assume that the occurrence of outages follows Poisson process, where the outage rate of element i is λ_i. Then in the Markovian tree model, the outage probability of each element in interval τ_D:

$$\Pr_i^{\text{MT}} = \frac{\lambda_i}{\sum_i \lambda_i}\left(1 - \mathrm{e}^{-\sum_j \lambda_j \tau_D}\right). \tag{6.5}$$

It can be seen that (6.5) the outage probability of element i is not only dependent on the outage rate of itself λ_i, but also on outage rates of other elements. The probability that there is no outage in interval τ_D is

$$\Pr_0 = \mathrm{e}^{-\sum_j \lambda_j \tau_D}. \tag{6.6}$$

6.2.2 Modeling of Grid Dispatch Behavior

Re-dispatch is categorized as a mid-timescale process. When overload occurs, dispatchers adjust generators or dump loads to relieve the overload. Ideally, re-dispatch is modeled as an optimization problem, and the optimal solution is instantly

[1] This framework can be easily extended to allow more than one element outage, especially when some outage combinations have non-negligible probabilities within τ_D. A side effect is that the tree will become wider. Reducing the τ_D will help limit the necessary tree width but will increase the tree depth, so there is a trade-off.

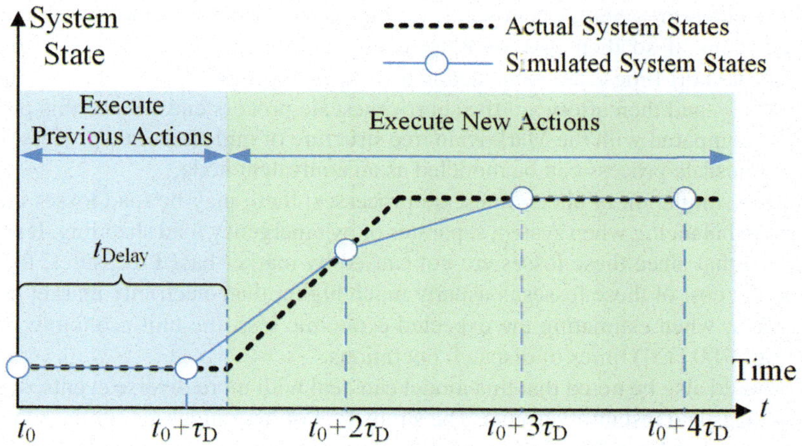

Fig. 6.11 Illustration of re-dispatch simulation

applied as the new state. However, re-dispatch takes time: when overloading occurs, the system needs time to acquire data, analyze system conditions, and reflect the data to operators; the operators also need time to judge and make decisions before taking actions. Moreover, due to the generation ramping speed constraints, it also takes time from the beginning of actions till the fulfillment of re-dispatch objectives. Therefore, the re-dispatch is a process with a time delay Δt_{Delay} and ramping.

As shown in Fig. 6.11, when an overloading event occurs at t_0, the re-dispatch action for the event is not immediately started. During interval $t_0 < t < t_0 + \Delta t_{Delay}$, the system may be taking actions dealing with previous events or there is no action at all. Considering the time-delay nature of re-dispatch, a queue of re-dispatch commands is prepared in simulation. As an event occurs, add the corresponding command to the queue and wait until the action is due. The latest command meeting the beginning time is offered from the queue and starts executing. The command in action is kept until re-dispatch is finished or it is replaced by a new command.

6.2.3 Modeling of Fast Cascade Processes

The processes considered in the Markovian tree model illustrated in Fig. 6.10 are the random outages caused by thermal and mechanical processes (namely, mid-timescale processes), such as line outages caused by overloading. But there may also be fast cascade processes such as outages directly triggered by protection relays and actions of emergency control. These processes usually finish in several seconds and are much shorter than processes in other timescales, and these processes follow strict preset logics.

In simulation, when system states change, check whether short-timescale processes occur, if so then first simulate them. As illustrated in Fig. 6.5, event i_{k_m} triggers a short-timescale event denoted as $i_{l_1|k_m}$, and then consequently triggers event $i_{l_2|k_m}$, and then afterward the short-timescale process ends. Since this process is short compared with the Markovian tree structure of mid-timescale processes, the short-timescale process can be modeled as an equivalent node.

In the simulation of short-timescale processes, there may be load losses caused by island balancing when system separates or by emergency load shedding. It should be noted that since these losses are not caused by market-based measures, the unit economic cost of these losses is usually much higher than electricity market prices. Therefore, when estimating the expected economic loss, the unit economic cost is μ_F (e.g., 100 [13]) times of dispatch operations.

It should also be noted that this model can deal with more diverse events, such as bus outages and instability events. The simulation of bus outages is similar to that of short-timescale branch/generator outages. The instability events usually need time-domain simulation [14, 15], which can also be incorporated into this simulation model.

6.2.4 An Illustrative Example of Markovian Tree

To better illustrate the mechanism of cascading outage simulation with Markovian tree, here an example is provided. As Fig. 6.6 shows, the cascading outages start with initial outage(s) at time $t = 0$, and then the initial outage(s) might trigger short-timescale events and re-dispatch operation. Since short-timescale events are much faster than the re-dispatch and mid-timescale outages, the short-timescale events should be the first to be simulated. After outages and short-timescale events, the re-dispatch operation and mid-timescale outage in an interval τ_D are simulated (Fig. 6.12).

Because any element may outage in an interval τ_D, so there are various possible directions of cascading outage development, corresponding to the forked structure in every interval shown in Fig. 6.6. The direction of cascading outage development is denoted with the index of the failed element. Note that there might be no outages during an interval, and in this case, the path is indexed as 0. So during the interval τ_D after any state, there is a forked structure of possible cascading outage directions. Therefore, all the possible cascading outage paths can be collected as a tree structure as Fig. 6.6 demonstrates.

In each cascading outage simulation with Markovian tree, a complete cascading outage path is simulated from the initial outages till the terminal of the path. On the Markovian tree structure, the simulated path is a linked list of nodes on the tree starting from the root node (initial outages) to the terminal. In Fig. 6.6, the filled nodes constitute a cascading outage path. The path means that at around 10 minutes after initial outages, the element 2 fails, and at around 15 minutes the element 3 fails.

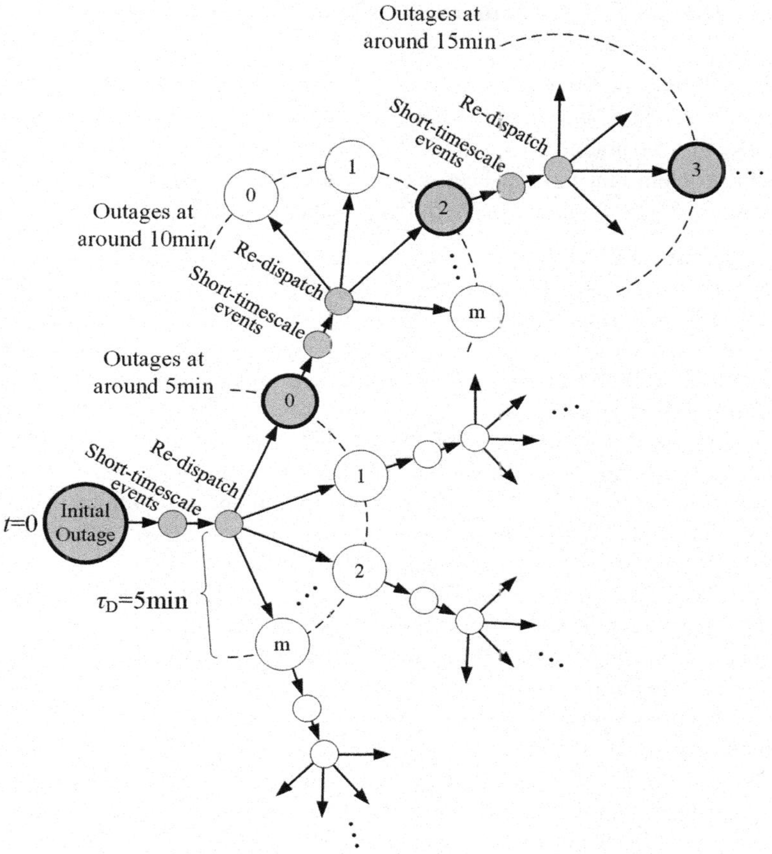

Fig. 6.12 An illustrative example of Markovian tree

It should be noted that due to the dependency in a cascading outage sequence, each post-outage state should be updated rather than regarding the outages as independent combinations. And since there are possibly re-dispatch operations, emergency control measures, and dynamic processes after each outage, the outcomes of different orders of outages from the same outage combination are probably different, and the risks of these different outage sequences are probably different. So in this chapter the outage sequences in different orders are treated as different and are simulated respectively. It is also remarkable that the proposed Markovian tree model maintains a good multi-timescale modeling of cascading outages, which is a prerequisite of accurate risk assessment. So the Markovian tree model plays as a bridge between reasonable simulation of cascading outages and realization of efficient risk assessment.

6.2.5 Discussion on Probability Quantification

The probabilities of outages are derived according to the model [14] that considers the time elapse of heat accumulation and some environmental factors. Yet it should be noted that in practice, estimating the probabilities of cascading outages is a difficult problem that has not been well-solved. The difficulties of estimating the probability of cascading outages mainly lie in the following factors:

(1) There are various kinds of causes to outages in a whole process of cascading outages. A power system consists of many kinds of elements, and any element may outage during the process of cascading outages due to various causes [16]. A line may outage due to sagging (2003 US–Canada blackout), annealing, lightning (2009 Brazil–Paraguay blackout), mis-operation (2011 Arizona–California outages), improper protection setting (2006 UCTE disturbance), power system oscillation (1996 WSCC outages), ice and snow (2008 South China blackout), geomagnetic storm (1989 Hydro-Quebec blackout), etc. A generator may fail due to over-current, over-/under-excitation, over-/under-frequency, etc. Also there are other kinds of elements or components that may outage, such as underground cables, transformers (2005 Moscow blackout), HVDC, control center (2003 US–Canada blackout) and communication infrastructure (1988 Hydro-Quebec blackout), etc. The power system itself also demonstrates complex dynamics in multiple timescales during cascading outage. Therefore, the cascading outage is a process involving complex and dependent system behavior and a variety of outages of different kinds of elements caused by system dynamics and/or external factors. To comprehensively study all kinds of events in cascading outages, it is necessary to acquire an adequate amount of data and logs in operation. However, currently the power systems still face lack of data, which hinders credible modeling and verification of possible kinds of element outages.
(2) The modeling of each kind of outage event is difficult due to various influencing factors. Currently, providing probabilistic model of each kind of outage mechanism is difficult because of (a) the difficulty in establishing the probabilistic model considering all the important influencing factors and (b) the lack of sources to collect data of those factors in practice. Take the outage of overhead line caused by sagging and tree contact as example, and the IEEE Standard 738 [17] points out that there are various factors that influence the steady-state sag of overhead lines, including the current on conductor, the type of conductor, ambient temperature, wind speed, wind direction, season, the time of day, sunshine illumination, etc. Moreover, the outage caused by tree contact also depends on the height of vegetation under the overhead line. We can see that not only electric-side variables but also many environmental factors influence the line outage event, while till now the line outage models in the existing literatures [14, 18–21] are intuitive but not accurate enough, and there is not yet credible probabilistic model of such outage event with comprehensive consideration of the influencing factors [22]. Moreover, in practice, it is difficult to monitor or

predict all these environmental factors accurately, making the model hard to utilize in application.

From the above, it can be concluded that the estimation of outage probability requires more accurate modeling of the physical process of various kinds of outages, which asks for enhanced study of element reliability, modeling of environmental factors as well as verification of models with field tests and observations during operation. Also, concentration should be placed on enhancing the situational awareness of power systems to enable the utilization of more accurate models, especially for the monitoring and management of environmental data.

6.3 Tree Search for Efficient Risk Assessment

6.3.1 Searching, Instead of Sampling

One of the most challenging problems of risk assessment is the limitation of calculation speed brought by the huge number of possible cascade paths and the lack of computation resources. One commonly utilized approach of risk assessment is the Monte Carlo method, which repeatedly creates samples of events based on their real probabilities until risk converges. However, Monte Carlo method requires large numbers of samples to converge, especially for rare events [10]. Since the convergence of sampling-based methods relies on the variance of sampling, various variance reduction methods are proposed, e.g., importance sampling [23], cross-entropy [24], stratification [25], etc. These methods can accelerate computation by several times compared to Monte Carlo method. However, the improvements are limited, and risk assessment still requires huge computation.

The selection of serious contingency patterns is another approach for risk assessment. Various contingency or state combination selection techniques have been proposed [26–28]. However, these methods treat outages as independent of each other, which neglects the fundamental "cascading" nature of cascading outages. Reasonable risk assessment requires updating the state as well as probability of element outages during simulation, which makes the techniques utilized in contingency selection methods ineffective.

Correctly capturing the dependency within sequences of outages is a prerequisite of reasonable and practical risk assessment, so the simulation of cascading outages is inevitable in risk assessment. An effective way of efficiency enhancement is to reduce the invocation of simulation. In sampling-based methods, a lot of time is wasted in duplicated sampling of same cascade paths. Actually upon knowing the conditional probabilities on each level of cascading outages (which is feasible in most existing cascading outage models), the efficiency of risk assessment can be significantly improved by simulating the cascading outage path and directly estimating risk indices using the probability, which avoids duplicated simulation of same paths.

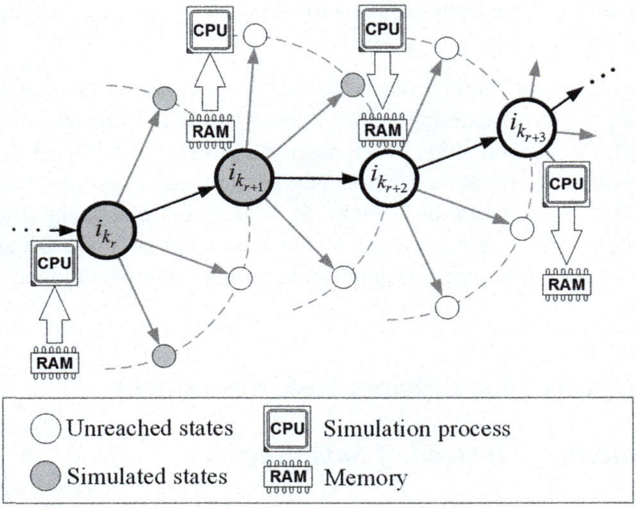

Fig. 6.13 Markovian tree search with caching

Following the details proposed in Sect. 6.2, simulation of cascading outages can be realized on the Markovian tree as equivalent to the model in [14], and risk assessment can be carried out by sampling on the Markovian tree. Yet sampling duplicates simulation of same cascade paths. To enhance efficiency, the simulated cascade paths can be recorded and avoided in further simulation. Thus risk assessment becomes searching on the Markovian tree.

The risk (6.4) can be regarded as the sum of risk terms of all the single states on the Markovian tree. Therefore, the risk assessment based on the Markovian tree can be regarded as simulating new states on the Markovian tree and adding new corresponding terms onto (6.4). Since the terms in (6.4) are non-negative, the risk is expected to keep increasing until reaching a value R, which is the cascading outage risk of the system.

Figure 6.13 illustrates the mechanism of risk assessment with Markovian tree search. The nodes with bold borderline constitute a partial cascading outage path. The nodes with gray color denote states that have been simulated in previous searches. In current search, these states are directly retrieved from the memory. The nodes with white color are states that have not been reached. The events of these states are simulated based on Section 6.2.1, the risk terms corresponding to these states are added to the risk metric, and then the states are stored in the memory.

6.3.2 Convergence Criteria of Risk Assessment

In the risk assessment with Markovian tree search, the theoretical value of risk is obtained when all the possible cascading outage paths are exhausted and simulated.

If there are N elements at the initial state, and K_D levels on the Markovian tree, then the number of all possible cascade paths on Markovian tree is

$$N_T = \sum_{i=0}^{K_D} \frac{N!K_D!}{(N - K_D + i)!(K_D - i)!i!}. \tag{6.7}$$

From (6.7), we can see that the number of possible cascade paths is huge for common-sized systems so that it is practically impossible to exhaust all the paths. Therefore, in practice, we can only simulate a portion of the paths. In risk assessment, as cascade paths are simulated, the value of risk keeps growing and gradually approaches the theoretical value, and the total probability of simulated cascade paths also approaches 1. Thus the criteria for the convergence of risk assessment are proposed as satisfying the following two conditions:

(1) The value of risk is stable (e.g., the growth of risk in the past 5000 searches is less than 0.1% of current risk value).
(2) The total probability of simulated cascade paths exceeds a certain threshold (e.g., 0.97).

Usually, the distribution of risks among all possible cascade paths is extremely non-uniform that risk is concentrated in a small portion of cascading outage paths. Observing the risk (6.4), since risk assessment is to add terms onto (6.4), the searching for states with larger risk terms in priority will achieve faster convergence of risk. To accelerate the convergence of the risk, a strategy of searching should be proposed, which guides searching to the paths with major contribution to the risk. Next we will establish strategies that guide searching to such states.

6.4 Risk Estimation and Forward–Backward Search Algorithm

6.4.1 Risk Estimation Index

When performing the tree search, it is important to lead the search direction to the cascading outages paths with high risk value, so the major portion of the risk can be identified with a least number of searches. Take the partial Markovian tree in Fig. 6.14 to study the searching strategy. Assume that searching has reached "*" state (labelled as $(i_{k_1} \cdots i_{k_{r-1}})$) and is about to select a next-level event i_{k_r} (hollow nodes pointed by solid line arrows) to simulate. The strategy should let the increment of risk of the selected path be as large as possible, so the first task is to estimate the risks of all the subsequent states with acceptable computation complexity.

Here a risk estimation index $\rho_{i_{k_r}|i_{k_1}\cdots i_{k_{r-1}}}$ (simply denoted as $\rho_{i_{k_r}}$ since all studied subsequent states in this section have the same previous events $i_{k_1} \cdots i_{k_{r-1}}$) is

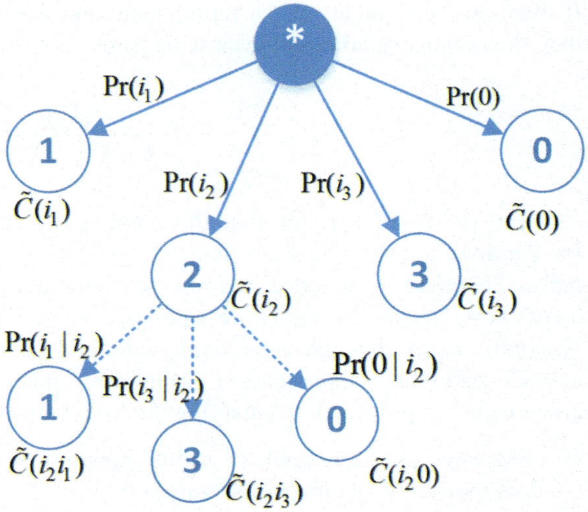

Fig. 6.14 Partial Markovian tree when search starts

established, and probabilities for searching are determined using $\rho_{i_{k_r}}$. The index consists of the following three parts:

1. Risk of System Separation $\rho^{\alpha}_{i_{k_r}}$. If the outage causes system to separate into islands, then there will be costs for generator dispatch and load shedding.
2. Risk of Overloading $\rho^{\beta}_{i_{k_r}}$. If the outage causes other components to overload, there will be costs for re-dispatch and/or emergency control actions.
3. Secondary Risk $\rho^{\gamma}_{i_{k_r}}$. This refers to the costs and losses not directly caused by this outage but in subsequent outage paths.

And the risk estimation index (REI) is calculated as

$$\rho_{i_{k_r}} = w_{\alpha}\rho^{\alpha}_{i_{k_r}} + w_{\beta}\rho^{\beta}_{i_{k_r}} + w_{\gamma}\rho^{\gamma}_{i_{k_r}}, \tag{6.8}$$

where $= w_{\alpha}$, $= w_{\beta}$, and $= w_{\gamma}$ are weights. Next sub-sections will explain how each component is calculated.

6.4.1.1 Risk Index Term of System Separation

If the outage of a branch causes the grid to separate, then the branch is called a cut branch of the grid. Cut branches can be identified with complexity of $O(|\mathbf{E}|)$, where $|\mathbf{E}|$ is the number of connected branches. Denote the admittance matrix of the grid as $\mathbf{Y}$, and then its Penrose–Moore pseudo-inverse ($(\cdot)^{+}$) uniquely exists, denoted as $\mathbf{Z} = \mathbf{Y}^{+}$.

A branch $i_{k_r} = u, v$ is a cut branch if and only if

$$Y_{uv}^{-1} - 2Z_{uv} + Z_{uu} + Z_{vv} = 0. \tag{6.9}$$

Considering numerical errors, set a sufficiently small threshold ε (e.g., 10^{-10}), and if

$$|Y_{uv}^{-1} - 2Z_{uv} + Z_{uu} + Z_{vv}| < \varepsilon, \tag{6.10}$$

then the branch is identified as a cut branch. If $i_{k_r} = \{u, v\}$ is a cut branch and it fails, the separated two parts of the system will have unbalanced power $\pm F_{uv}$, which needs power balancing and generates cost. Therefore, the cost of system separation caused by cut branch i_{k_r} outage is estimated as

$$\sigma_{i_{k_r}}^{\alpha} = \begin{cases} 2|F_{uv}|, & |Y_{uv}^{-1} - 2Z_{uv} + Z_{uu} + Z_{vv}| < \varepsilon \\ 0, & \text{otherwise.} \end{cases} \tag{6.11}$$

And the risk of system separation is estimated as

$$\rho_{i_{k_r}}^{\alpha} = \Pr(i_{k_1} \cdots i_{k_r}) \rho_{i_{k_r}}^{\alpha}. \tag{6.12}$$

6.4.1.2 Risk Index Term of Overloading

After the outage of non-cut branches, the power flow will re-distribute throughout the system and may cause overloading on other elements, leading to costs generated by re-dispatch or emergency control actions. The influence of a branch outage on other branches can be quantified by the power transfer distribution factor (PTDF). The PTDF of a non-cut branch $\{u, v\}$ to any other branch $\{p, q\}$ is

$$\delta_{uv}^{pq} = -\frac{Z_{up} + Z_{vq} - Z_{uq} - Z_{vp}}{1 + Y_{uv}(Z_{uu} + Z_{vv} - 2Z_{uv})} Y_{pq}. \tag{6.13}$$

The power flow on $\{p, q\}$ after the outage of $\{u, v\}$ is

$$F_{pq}^{*uv} = F_{pq} + \delta_{uv}^{pq} F_{uv}. \tag{6.14}$$

And the extent of overloading on branch $\{p, q\}$ is

$$\pi_{pq}^{*uv} = \max\{|F_{pq}^{*uv}| - F_{pq}^{\max}, 0\}. \tag{6.15}$$

Define the overloading index of branch $\{u, v\}$ as

$$\sigma_{i_{k_r}}^{\beta} = \sum_{\{p,q\} \in E} \pi_{pq}^{*uv}, i_{k_r} = \{u, v\}, \tag{6.16}$$

and define the estimation of overloading risk as

$$\rho_{i_{k_r}}^{\beta} = \Pr(i_{k_1} \cdots i_{k_r}) \rho_{i_{k_r}}^{\beta}. \tag{6.17}$$

For any studied state, about $|\mathbf{E}|^2$ PTDF values and post-outage flows on branches are needed to calculate, so the estimation of overloading risk has the complexity of $O(|\mathbf{E}|^2)$.

6.4.1.3 Secondary Risk

Considering Fig. 6.14, when selecting the next-level states of "*" state, the risk of subsequent states of next-level states should also be accounted for. This risk is called a secondary risk in this chapter. Since the secondary risk is hard to analytically quantify, a rough estimation is given here.

First calculate the power flow F_{pq}^{*uv} of all the connected lines $\{p, q\}$ after outage of branch $i_{k_r} = \{u, v\}$, and calculate the corresponding probabilities of outage $\Pr_{pq}^{*uv}$ during the next interval using (6.5)–(6.6). According to the overloading extent of F_{pq}^{*uv}, give an estimation of the cost $\tilde{C}_{pq}^{*uv}$ (it is difficult to accurately analyze, so in this chapter $\tilde{C}_{pq}^{*uv}$ is set as 1% of system load), and then the secondary risk of $i_{k_r} = \{u, v\}$ is

$$\rho_{i_{k_r}}^{\gamma} = \Pr(i_{k_1} \cdots i_{k_r}) \frac{\sum \Pr_{pq}^{*uv} \tilde{C}_{pq}^{*uv}}{|\mathbf{E}_{i_{k_1} \cdots i_{k_r}}|}, \tag{6.18}$$

where $\mathbf{E}_{i_{k_1} \cdots i_{k_r}}$ is the set of connected branches at state $(i_{k_1} \cdots i_{k_r})$, and $|\mathbf{E}_{i_{k_1} \cdots i_{k_r}}|$ is the number of connected branches.

If a next-level state has no outage, i.e., $i_{k_r} = 0$, then system separation risk and overloading risk are both considered as 0, but the secondary risk may be non-zero. In this case, if approximately regarding the system state at $i_{k_r} = 0$ the same as that of $i_{k_{r-1}}$, then the secondary events can be seen as shifting the next-level events of $i_{k_{r-1}}$ to i_{k_r}. If the probability of $i_{k_r} = 0$ is $\Pr_0$, then the corresponding secondary risk can be defined as

$$\rho_0^{\gamma} = \mu \frac{\Pr_0}{|\mathbf{E}_{i_{k_1} \cdots i_{k_{r-1}}}|} \sum_{i_{k_r}} \rho_{i_{k_r}}, \tag{6.19}$$

where $\mu \leq 1$ is a discount factor considering that risk will be reduced by control schemes in the system.

For a studied state, the complete estimation of secondary risks needs about $|\mathbf{E}|^2$ times of $\Pr_{pq}^{*uv}$ and $\tilde{C}_{pq}^{*uv}$ calculation, so the complexity of secondary risk estimation is $O(|\mathbf{E}|^2)$.

6.4.1.4 Computation Complexity of Risk Estimation Index

The above derivation of risk estimation index in this chapter only considers branch outages, while the methodology of establishing risk estimation index can be used for the risk estimation of other kinds of events. Since in power flow model, the system state after bus/generator outage can be similarly derived using distribution factor, so the risk estimation index of bus and generator outages can be similarly established. However, the instability events have significantly different mechanisms from the derivations above; therefore, the severity estimation of instability events may be different from the above analysis.

The overall complexity of risk estimation index is $O(|\mathbf{E}|^2)$, while in the simulation of each level of cascading outages, the update of matrices is $O(|V|^2)$ to $O(|V|^3)$, and the complexity of re-dispatch is bounded by $O(|\mathbf{E}|^{3.5})$. Therefore, the complexity of risk estimation index is much lower than the simulation, so the calculation of risk estimation index does not notably affect the overall efficiency.

6.4.2 Forward Searching Using Risk Estimation Index

As shown in Fig. 6.14, if a new state (labeled with asterisk) is reached on the Markovian tree, then all the subsequent states and paths are new. The risk estimation indices of next-level states are calculated, and probabilities for selecting these states can be obtained using risk estimation indices. If the index is thought as accurate reflection of risks, then the optimal search strategy is to guide to the path with the highest risk estimation index value, which is a deterministic strategy.

$$\Pr_{i_{k_r}}^{calc} = \begin{cases} 1, \, i_{k_r} = \arg\max_i \{\rho_i\} \\ 0, \text{otherwise} \end{cases}. \tag{6.20}$$

However, the risk estimation index may have error, so it is essential to have randomness in searching. Another strategy is random search with equal probability

$$\Pr_{i_{k_r}}^{calc} = \frac{1}{|\mathbf{E}_{i_{k_1}\cdots i_{k_{r-1}}}| + 1}. \tag{6.21}$$

The searching strategies of (6.20) and (6.21) represent two extremes: deterministic search vs. pure random search. To keep the merits of both approaches, the strategy can be selected in between. Introduce a parameter $\lambda \geq 0$ and set probability

$$\Pr_{i_{k_r}}^{calc} = \frac{(\rho_{k_r})^{\lambda}}{\sum_{\xi \in \mathbf{E}_{i_{k_1}\cdots i_{k_{r-1}}} \cup \{0\}} (\rho_{\xi})^{\lambda}}. \tag{6.22}$$

For $\lambda = 0$, Eq. (6.22) is equivalent to (6.20), and for $\lambda \rightarrow +\infty$, Eq. (6.22) approaches to (6.21).

During the simulation of cascading outages, the matrices $\mathbf{Y}$, $\mathbf{Z}$ need to be updated. If a set of branches $\{i_k\}$ are removed from the grid, then the admittance matrix can be updated with

$$\mathbf{Y}' = \mathbf{Y} - \mathbf{M}_{\{i_k\}}\mathrm{diag}(\mathbf{y}_{\{i_k\}})\mathbf{M}^{\mathrm{T}}_{\{i_k\}}, \tag{6.23}$$

where $M_{\{i_k\}}$ is a $|\mathbf{V}| \times |\{i_k\}|$ matrix. Each of its column $M_{\{i_k\}}$ corresponding to a branch $i_k = \{u, v\}$ satisfies $M_{\{i_k\},u} = 1$, $M_{\{i_k\},v} = -1$, and all other entries are 0. $\mathrm{diag}(\mathbf{y}_{\{i_k\}})$ is a matrix with branch admittances of $\{i_k\}$ as its diagonal elements. The update of matrix $\mathbf{Y}$ (6.23) can be finished with a very small amount of calculation, with complexity of $O(|\{i_k\}|)$. The update of $\mathbf{Z}$ can be done with Woodbury matrix identity:

$$\mathbf{Z}' = \mathbf{Z} + \mathbf{Z}\mathbf{M}_{\{i_k\}}\mathbf{z}^{-1}_{\{i_k\}}\mathbf{M}^{\mathrm{T}}_{\{i_k\}}\mathbf{Z}, \tag{6.24}$$

where

$$\mathbf{z}_{\{i_k\}} = diag(\mathbf{y}_{\{i_k\}})^{-1} - \mathbf{M}^{\mathrm{T}}_{\{i_k\}}\mathbf{Z}\mathbf{M}_{\{i_k\}}. \tag{6.25}$$

Since outages usually occur to very few branches at a time, $\mathbf{z}_{\{i_k\}}$ is very small and (6.24) has a complexity of $O(|\mathbf{V}|^2)$. However, if $\{i_k\}$ is a cut set, the update of $\mathbf{Z}$ has to be performed through singular value decomposition (SVD) of $\mathbf{Y}$ instead of (6.24), which has complexity $O(|\mathbf{V}|^3)$. Theoretically, the SVD is computationally expensive, and an alternative that searches islands and simulate events on each island, respectively, consumes less computational resources. However, tests on RTS-96 system show that the instances requiring SVD only account for less than 20% of the matrix update computations, so the influence of SVD on overall efficiency is not very significant. And the very optimized numerical computation libraries have well-supported high-performance SVD.

6.4.3 Backward Updating Risk Estimation Indices

After states on cascading outage paths are visited and recorded, reaching these states again in the future will not contribute to the increment of the risk. Therefore, after a cascading outage path is found, the risk estimation indices should also be updated. On the contrary to the searching direction from the root to terminals of the Markovian tree, the updating of risk estimation indices should go backward from terminals to the root. As shown in Fig. 6.15, assume that solid nodes are visited states and the node labelled as 3 in the bottom is the terminal of the path. Since searching to a visited terminal again will not make any contribution to the risk,

Fig. 6.15 Backward update of risk estimation indices on Markovian tree

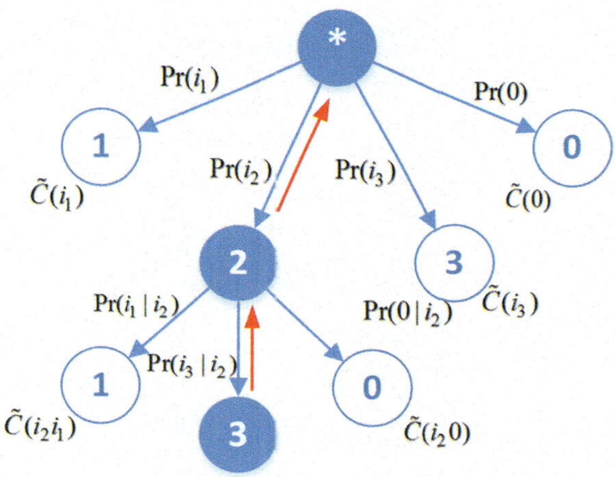

then for a terminal state i_{k_n} on a cascading outage path $i_{k_1} \cdots i_{k_n}$, assign its risk estimation index a sufficiently small value $\rho_{i_{k_n}} = \varepsilon_R$ to avoid visiting it again.

For a non-terminal state i_{k_r}, since it has been visited and the risk term on the state i_{k_n} itself will not contribute to the risk again, then its risk estimation index will only reflect risks of its subsequent states. Since risk estimation indices of all its next-level states $\rho_{i'_{k_{r+1}}|i_{k_1}\cdots i_{k_r}}$ must have been calculated or even updated, and the probabilities for searching are $\Pr^{calc}_{i'_{k_{r+1}}}$ according to (23), then the risk estimation index of i_{k_r} is

$$\rho_{i_{k_r}} = \sum_{i'_{k_{r+1}}} \rho_{i'_{k_{r+1}}|i_{k_1}\cdots i_{k_r}} \Pr^{calc}_{i'_{k_{r+1}}}. \tag{6.26}$$

Equation (6.26) represents recursive backward updating of risk estimation indices. In the risk assessment on Markovian tree, first do forward search along a path, and then reversely update risk estimation indices using (6.26).

6.4.4 Procedures of Risk Assessment with Markovian Tree Search

Figure 6.16 demonstrates the procedures of the proposed risk assessment method. To realize non-duplicated system state search, a search result table T_S is established to index and store the states that have been searched. During risk assessment, if a state to be simulated is found in T_S, then the state is directly retrieved and the simulation of this level of events is avoided. If the state is not found in T_S, then this level of outage is simulated.

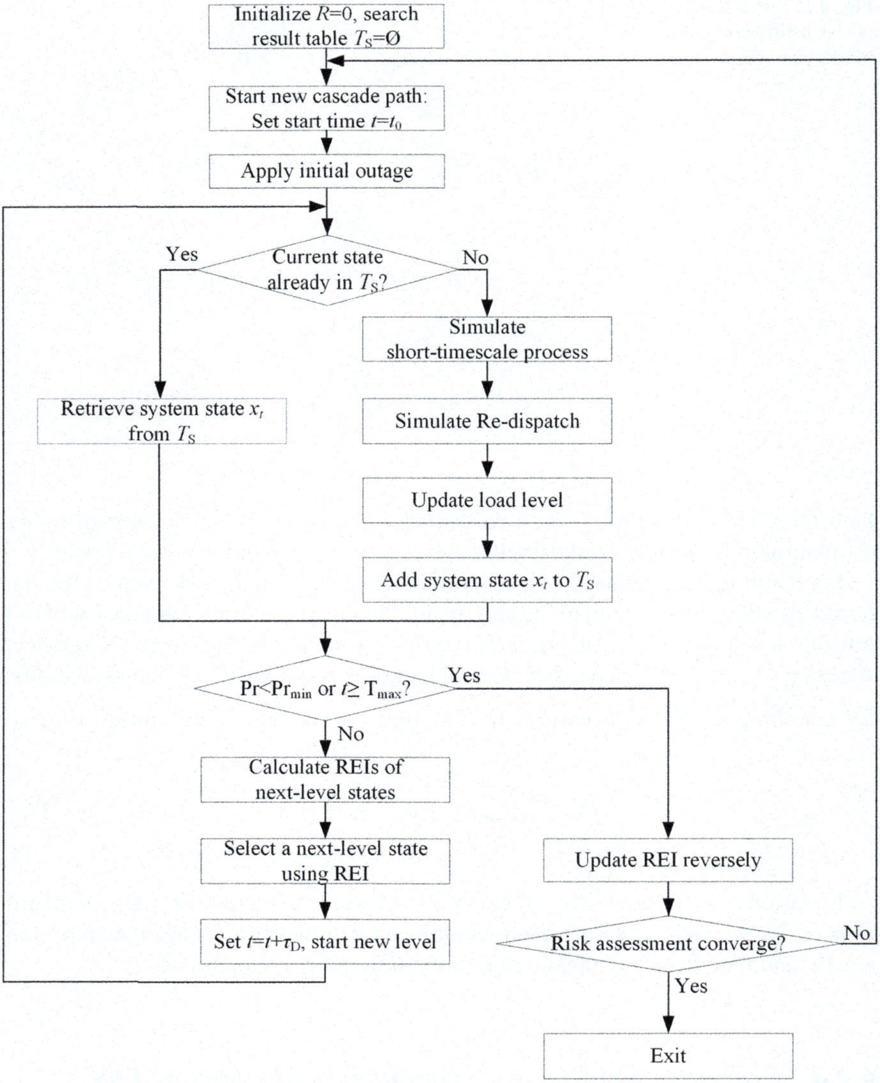

Fig. 6.16 Flowchart of risk assessment based on tree search

6.5 Risk Assessment Case Study on RTS-96 System

6.5.1 Performance of Risk Estimation Index

The approach is tested on a larger RTS-96 3-area system [29] that has 73 buses, 120 branches, 33 generator nodes, and 51 load nodes. Select outages of branches 22, 23,

Fig. 6.17 Correlation between risk and risk estimation index

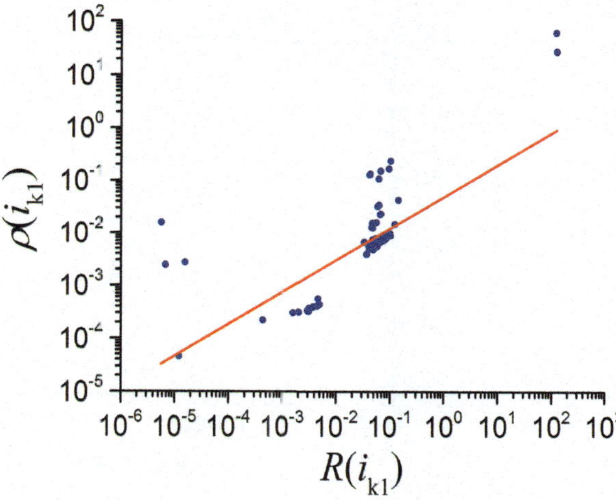

and 24 as initial outages, and use the tree-search-based approach to assess risk. In this case, set $\tau_D = 15$ min, maximum cascading outage duration $T_{\max} = 150$ min, and delay of re-dispatch $\Delta t_{Delay} = 30$ min.

The accuracy of risk estimation index will influence the accuracy and performance of risk assessment. So first we test the accuracy of the proposed risk estimation index in estimating the risk of subsequent cascading outage paths. Since the risk estimation index is proposed to facilitate the selection of next-level cascading outage path, we mainly compare the risk estimation indices of the states having the same previous state. For example, we study the risk estimation indices and the subsequent risk of the states on level 1. The subsequent risk of level-1 state i_{k_1} is

$$R(i_{k_1}) = \Pr(i_{k_1})C(i_{k_1}) + \Pr(i_{k_1})\sum_{k_2}\Pr(i_{k_2}|i_{k_1})C(i_{k_1}i_{k_2}) + \\ \Pr(i_{k_1})\sum_{k_2}\Pr(i_{k_2}|i_{k_1})\Pr(i_{k_3}|i_{k_1}i_{k_2})C(i_{k_1}i_{k_2}i_{k_3}) + \cdots \tag{6.27}$$

Figure 6.17 shows the relationship between subsequent risk of level-1 state $R(i_{k_1})$ and risk estimation index $\rho_{i_{k_1}}$ in a log–log plot. The plot contains 118 scattered data points. The results visually demonstrate positive relationship between $R(i_{k_1})$ and $\rho_{i_{k_1}}$. Linear regression of these points shows the approximated quantified relationship as

$$\log_{10}\rho_{i_{k_1}} = 0.6063\log_{10}R(i_{k_1}) - 1.2964,$$

and the Pearson's correlation coefficient is 0.712, indicating strong linear positive correlation between $R(i_{k_1})$ and $\rho_{i_{k_1}}$. The risks and risk estimation indices on other

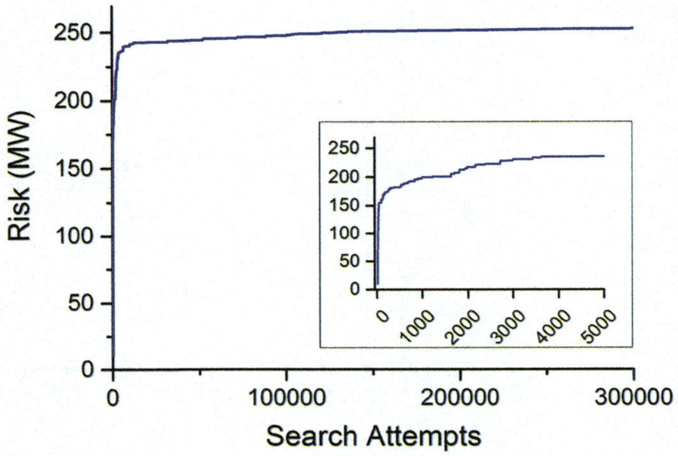

Fig. 6.18 Risk on RTS-96 test system

levels of Markovian tree generally show similar strong positive correlation, which verifies that the risk estimation index can effectively guide the tree search to cascading outage paths with higher risks.

6.5.2 Efficiency Test of Risk Assessment

According to (6.7), the number of possible cascading paths is about 3.5491×10^{20}, so it is unrealistic to enumerate them all and calculate theoretical risk value. Here use a relatively large number of search attempts N_S and regard the risk at N_S attempts R_{N_S} as the theoretical risk R. The test program is developed and tested in MATLAB on a workstation with 2.6 GHz processor and 32GB RAM.

Set $N_S = 300{,}000$ and get risk $R_{N_S} = 252.76$ MW through tree-search-based risk assessment. From Fig. 6.18, the risk rises sharply at the beginning and approaches quickly in the first several thousands of search attempts, and then its rising speed becomes much slower. As Table 6.1 shows, after 19 search attempts, the risk has reached $0.5R_{N_S}$, and then after 2709 attempts, the index reaches $0.9R_{N_S}$, with computation time less than 10 min. But reaching $0.99R_{N_S}$ takes a much larger amount of computation, consuming several hours. From the perspective of application, it is practically required to apply risk assessment with limited computation time. In this case, no more than 5000 attempts and 1000 seconds of computation time can account for more than 90% of the cascading risk.

Figure 6.19 shows the covered probability of simulated cascade paths along the process of risk assessment. At the beginning, the coverage of probability rises sharply over 0.9, and after the first 2000 search attempts, nearly 0.97 has been

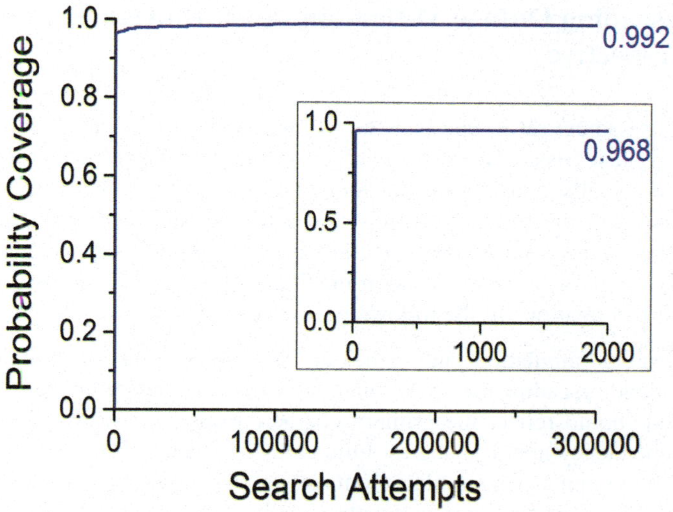

Fig. 6.19 Coverage of probability in risk assessment

Table 6.1 Performance of risk assessment on RTS-96 test system

$\geq R_{N_S}\%$	50%	90%	95%	99%	99.9%
Search attempts	19	2709	6259	129,134	259,856
Time (s)	3.87	552	1277	26,360	53,044
Probability coverage	0.961	0.969	0.973	0.991	0.992

covered. Therefore, most probable cascade paths have been simulated and assessed, and since the risk estimation index can effectively guide computation to cascade paths with major risks, the rest of the paths are expected to have minor contribution to the risk. With a high-performance computer and software-level optimization, the computation efficiency of this method is expected to meet the need for online applications. Moreover, using risk assessment results, all the simulated cascading outage paths and risks can be analyzed, and measures for lowering risk can be established [30] but are not covered in this chapter.

As for the memory usage, in this case where 300,000 cascading outages are simulated, about 1.8 million states are stored in the memory, and each state variable occupies 672 bytes, so the total memory usage for recording states is about 1 GB, which is easily satisfied on ordinary PCs. For larger systems, the requirement for memory space can also be satisfied on workstations, servers, or other high-performance platforms. Therefore, even though the memory usage of the Markovian tree search method is higher than some existing methods, the requirement for memory space is generally affordable for practical use.

6.6 Cascading Outage Mitigation: A Markovian Tree Perspective

The risk assessment result can be used to support decision-making for mitigating cascading outage risks. However, cascading outage mitigation is very complex problem, especially considering the numerous uncertain development paths that are coupled with the mitigation decisions. The uncertain and multi-stage nature of cascading outages poses challenges to the mitigation accuracy, and the huge problem space makes efficient decision-making challenging. To realize effective and efficient risk mitigation, all the following aspects are needed:

(1) Reasonable simulation and efficient risk assessment of multi-timescale-dependent cascading outages: First, a reasonable modeling and simulation method should reflect the primary characteristics of cascading outages, e.g., the dependency among outages, timescales of related processes, etc. [30, 31]; then, from numerous simulated outages, a risk assessment method needs to identify the most risky cascade paths; finally, the assessed risk is used in a risk mitigation method to reduce the risks of cascading outages.
(2) Risk metrics that directly evaluate the risk faced by the end users for effective risk mitigation, such as load or energy loss, and economic loss.
(3) A risk mitigation formulation allowing direct adjustment on reduction of risk, e.g., the amount of desired decrease in the load loss, energy loss, or economic loss, which will facilitate to determining the trade-off between risk reduction and control cost.
(4) An efficient risk mitigation algorithm to derive strategies for risk reduction on cascading outages.

We will discuss aspects 3 and 4 later. Regarding 1 and 2, the Markovian tree search formulation satisfies the needs. The Markovian tree formulation considers the multi-timescale modeling of cascading outages, and the risk metrics in the Markovian tree model can be load losses or economical losses, which provides straightforward insights of consequences. Take a look at the risk definition in Markovian tree again:

$$
\begin{aligned}
R = C_0 &+ \sum_{k_1} \Pr(i_{k_1})C(i_{k_1}) + \sum_{k_1} \Pr(i_{k_1}) \sum_{k_2} \Pr(i_{k_2}|i_{k_1})C(i_{k_1}i_{k_2}) \\
&+ \sum_{k_1} \Pr(i_{k_1}) \sum_{k_2} \Pr(i_{k_2}|i_{k_1}) \sum_{k_3} \Pr(i_{k_3}|i_{k_1}i_{k_2})C(i_{k_1}i_{k_2}i_{k_3}) + \cdots .
\end{aligned}
\tag{6.28}
$$

Suppose control actions are taken to reduce the cascading outage risk after the initial outages. Then the expansion (6.28) can be divided into two parts: C_0 is the cost of the control, and all the other terms are the risks of subsequent cascading outages, whose sum is denoted by R'. The risk is assessed by searching the MT and adding the risk terms into (6.28) corresponding to the newly visited states.

Cascading outages in the early stage usually develop slowly, so there is some time to adjust system states to reduce the risk of potential cascading outages after the initial outage. Control measures after the initial outages result in cost C_0 while reducing the subsequent risk R', so a compromise between the effect and cost of risk mitigation should be concerned. It is desirable that the risk of subsequent cascading outages is confined below a certain level R'_S, while the cost of control is minimized. Therefore, the basic formulation of risk mitigation can be written as

$$\begin{aligned} \min f &= C_0(\mathbf{x}^*) \\ s.t.\ R'(\mathbf{x}^*) &\le R'_S \\ \mathbf{g}(\mathbf{x}^*) &\le \mathbf{0}, \end{aligned} \tag{6.29}$$

where $\mathbf{x}^*$ is the target system state, $C_0(\mathbf{x}^*)$ is the cost of control, and $R'(\mathbf{x}^*)$ is the subsequent risk. The last constraint represents the constraints in operations, e.g., load, generation, and transmission capacity constraints. In real power systems, the commonly used control measures for overloading relief or power balance include re-dispatch of generation and load shedding initiated by operators or emergency control facilities. Moreover, to prevent voltage instability, reactive power compensation devices can also be adjusted. In emergency, the under-voltage load shedding, under-frequency load shedding, and controlled islanding can also be triggered to prevent system-wide collapse. In this chapter, cascading outages are modeled and simulated with DC power flow model, and the re-dispatch operations for overloading relief are considered, which include the re-dispatch of generators and curtailment of loads controlled by the operators. The re-dispatch is very commonly used in the operations of power systems, and it can be described with an optimal power flow model. The cost of re-dispatch $C_0(\mathbf{x}^*)$ can be described as follows:

$$C_0(\mathbf{x}^*) = -\mathbf{c}_D^{\mathrm{T}}(\mathbf{P}_d^* - \mathbf{P}_d) + \mathbf{c}_G^{\mathrm{T}}|\mathbf{P}_g^* - \mathbf{P}_g|. \tag{6.30}$$

Here $\mathbf{x}^* = [\mathbf{P}_d^{*\mathrm{T}}, \mathbf{P}_g^{*\mathrm{T}}]^{\mathrm{T}}$ is a vector of target re-dispatch state. $\mathbf{x} = [\mathbf{P}_d^{\mathrm{T}}, \mathbf{P}_g^{\mathrm{T}}]^{\mathrm{T}}$ is the pre-control system state. $\mathbf{c}_D$ and $\mathbf{c}_G$ are per unit costs of load shedding and generation adjustment, respectively. It should be noted that the proposed risk mitigation approach is not limited to the re-dispatch of active power of generators and loads. For example, since the simulation and risk assessment can also be implemented in an AC power flow model, in that case the control of reactive power compensators can also be realized.

However, quantifying the R' as a function of $\mathbf{x}^*$ in (6.29) is not straightforward. Since cascading outages involve complex dependent events, any change in the system state will affect all the following states, and thus the risk terms on all the levels of the MT are changed. Therefore, it is infeasible to analytically quantify R' as a function of $\mathbf{x}^*$, but the risk at the original target state $\mathbf{x}_0^*$ can be linearized to obtain the risk gradient:

$$\boldsymbol{\Gamma} = \left. \frac{\partial R'}{\partial \mathbf{x}^*} \right|_{\mathbf{x}^*=\mathbf{x}_0^*}, \tag{6.31}$$

and then the risk mitigation model (6.29) becomes

$$\begin{aligned} &\min f = C_0(\mathbf{x}^*) \\ &s.t.\ \boldsymbol{\Gamma} \cdot (\mathbf{x}^* - \mathbf{x}_0^*) \leq R_S' - R_0'. \\ &\mathbf{g}(\mathbf{x}^*) \leq \mathbf{0} \end{aligned} \tag{6.32}$$

6.7 Gradient Formulation on Markovian Tree

6.7.1 Derivative Chain of States on the MT

From (6.28) and (6.32), the risk gradient depends on the derivatives of conditional probability $\Pr(i_{k_r}|i_{k_1}\cdots i_{k_{r-1}})$ and cost $C(i_{k_1}\cdots i_{k_r})$ of states on each level of the MT. Such calculation requires the analysis of the chain of states on the cascading outage path. As shown in Fig. 6.20, denote the post-outage state on the rth level as $\mathbf{x}^{(r)\prime}$, and the state after re-dispatch as $\mathbf{x}^{(r)*}$. The costs of the short-timescale process and re-dispatch of all i_{k_r} $(r = 1, \cdots, n)$ are briefly denoted as vectors $\mathbf{C}_F^{(r)}$ and $\mathbf{C}_R^{(r)}$, respectively, and the total cost on level r is $\mathbf{C}^{(r)} = \mathbf{C}_F^{(r)} + \mathbf{C}_R^{(r)}$. Briefly denote outage probabilities as a vector $\mathbf{Pr}^{(r)}$.

In this chapter, assume that the probabilities and the costs are differentiable. If in reality these quantities are non-smooth or discontinuous functions of states, then they need to be treated in segments, or the sub-derivatives are used as an approximation. And if there are discrete variables, then they need to be treated as continuous ones temporarily. To calculate the gradient of risk, the terms $\frac{\partial \mathbf{Pr}^{(r)}}{\partial \mathbf{x}^{(0)}}$ and $\frac{\partial \mathbf{C}^{(r)}}{\partial \mathbf{x}^{(0)}}$ are necessary according to (6.28). In fact, according to Fig. 6.20, if the partial derivatives on levels up to r (i.e., $\frac{\partial \mathbf{C}^{(r)}}{\partial \mathbf{x}^{(r-1)}}$, $\frac{\partial \mathbf{Pr}^{(r)}}{\partial \mathbf{x}^{(r-1)}}$ and $\frac{\partial \mathbf{x}^{(r)}}{\partial \mathbf{x}^{(r-1)}}$) are obtained, then $\frac{\partial \mathbf{Pr}^{(r)}}{\partial \mathbf{x}^{(0)}}$

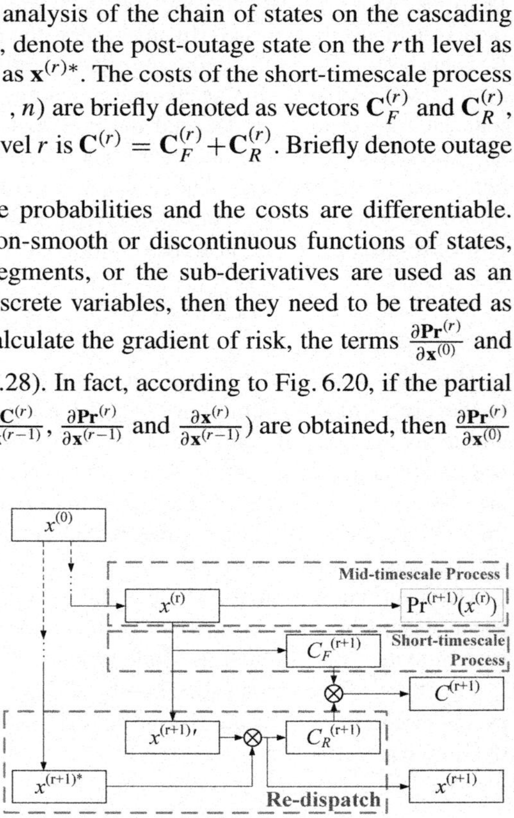

Fig. 6.20 State dependencies on one level of cascading outage

and $\frac{\partial \mathbf{C}^{(r)}}{\partial \mathbf{x}^{(0)}}$ can be calculated iteratively. The derivation of the partial derivatives on each level depends on the analysis of the cascading process, which will be elucidated below.

6.7.1.1 Mid-Term Random Outage

The probability of element i outage on the MT is [32]

$$\mathrm{Pr}_i^{\mathrm{MT}} = \frac{\lambda_i}{\sum_j \lambda_j}\left(1 - \mathrm{e}^{-\sum_j \lambda_j \tau_D}\right), \tag{6.33}$$

where λ_i is the failure rate of branch i, which is assumed to be a function of its branch flow F_i [18, 19, 33]. And F_i is a function of the system state $\mathbf{x}$. So the partial derivative of branch outage probability on level $r+1$ to $\mathbf{x}^{(r)}$ is

$$\frac{\partial \mathbf{Pr}^{(r+1)}}{\partial \mathbf{x}^{(r)}} = \frac{\partial \mathbf{Pr}^{(r+1)}}{\partial \boldsymbol{\lambda}} \cdot \frac{\partial \boldsymbol{\lambda}}{\partial \mathbf{F}} \cdot \left[-\mathbf{y}_D \mathbf{M} \mathbf{Y}^+, \mathbf{y}_D \mathbf{M} \mathbf{Y}^+\right], \tag{6.34}$$

where $\boldsymbol{\lambda}$ and $\mathbf{F}$ are vectors of λ_i and F_i, respectively. $\mathbf{y}_D$ is a diagonal matrix of branch admittances. $\mathbf{M}$ is a $|V| \times |E|$ matrix, and each of its column M_{i_k} corresponding to a branch $i_k = \{u, v\}$ satisfies $M_{i_k,u} = 1$, $M_{i_k,v} = -1$, and all the other entries are 0. $\mathbf{Y}^+$ is the Penrose–Moore pseudo-inverse of admittance matrix $\mathbf{Y}$.

6.7.1.2 Short-Timescale Process

As Fig. 6.21 shows, a short-timescale process may comprise several outages, and each outage may directly lead to cost on the loss of load due to load shedding [34]. The cost of short-timescale outages on level $r+1$ is the sum of costs caused by all outages.

$$\mathbf{C}_F^{(r+1)} = \sum_{k=1}^{n_{r+1}} \mathbf{C}_F^{(k|r+1)}, \tag{6.35}$$

where n_{r+1} is the number of short-timescale outages on level $r+1$, and the cost of the kth outage is $\mathbf{C}_F^{(k|r+1)}$. $\mathbf{x}^{(k|r+1)}$ is the state after the kth outage on level $r+1$

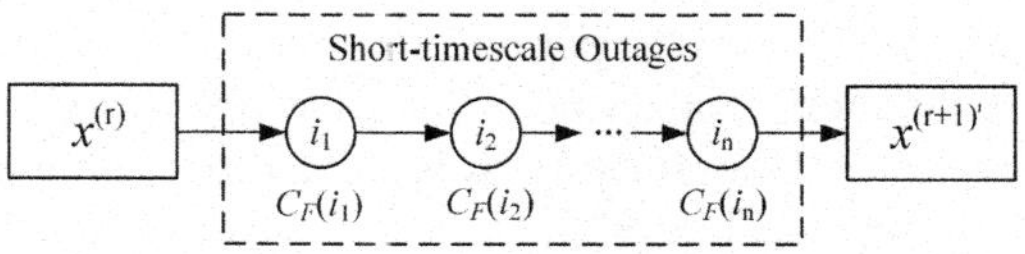

Fig. 6.21 Illustration of short-timescale outage events and costs

(note that $\mathbf{x}^{(0|r+1)} = \mathbf{x}^{(r)}$ and $\mathbf{x}^{(n_{r+1}|r+1)} = \mathbf{x}^{(r+1)\prime}$). Also $\frac{\partial \mathbf{x}^{(k|r+1)}}{\partial \mathbf{x}^{(k-1|r+1)}}$ and $\frac{\partial \mathbf{C}_F^{(k|r+1)}}{\partial \mathbf{x}^{(k-1|r+1)}}$ can be derived by sensitivity analysis of load shedding [35, 36]. So we can get:

$$\begin{aligned} \frac{\partial \mathbf{x}^{(k|r+1)}}{\partial \mathbf{x}^{(r)}} &= \frac{\partial \mathbf{x}^{(k|r+1)}}{\partial \mathbf{x}^{(k-1|r+1)}} \frac{\partial \mathbf{x}^{(k-1|r+1)}}{\partial \mathbf{x}^{(r)}} \\ \frac{\partial \mathbf{C}_F^{(k|r+1)}}{\partial \mathbf{x}^{(r)}} &= \frac{\partial \mathbf{C}_F^{(k|r+1)}}{\partial \mathbf{x}^{(k-1|r+1)}} \frac{\partial \mathbf{x}^{(k-1|r+1)}}{\partial \mathbf{x}^{(r)}}. \end{aligned} \tag{6.36}$$

From Fig. 6.21, the partial derivative of states in short-timescale outages $\frac{\partial \mathbf{x}^{(r+1)\prime}}{\partial \mathbf{x}^{(r)}}$ is obtained by applying (6.36) iteratively. And from (6.35) and (6.36), the partial derivative of the short-timescale outage cost is derived as

$$\frac{\partial \mathbf{C}_F^{(r+1)}}{\partial \mathbf{x}^{(r)}} = \sum_{k=1}^{n_{r+1}} \frac{\partial \mathbf{C}_F^{(k|r+1)}}{\partial \mathbf{x}^{(r)}}. \tag{6.37}$$

6.7.1.3 Re-dispatch

Re-dispatch is usually modeled as an optimization problem. Under the DC power flow assumption, the execution of re-dispatch can be modeled as a linear programming (LP) problem [32], which aims to minimize the distance between the actual post-dispatch state and the target state in a given time interval τ_D.

$$\begin{aligned} \min f &= \mathbf{c}_D^{\mathrm{T}} \left(\mathbf{P}_d - \mathbf{P}_d^*\right) + \mathbf{c}_G^{\mathrm{T}} \left|\mathbf{P}_g - \mathbf{P}_g^*\right| \\ s.t.\ & \mathbf{1}^{\mathrm{T}} \left(\mathbf{P}_g - \mathbf{P}_d\right) = 0 \\ & -\tau_D \mathbf{r}_g \le \mathbf{P}_g - \mathbf{P}_g' \le \tau_D \mathbf{r}_g \\ & \mathbf{P}_g^{\min} \le \mathbf{P}_g \le \mathbf{P}_g^{\max} \\ & \mathbf{P}_d^* \le \mathbf{P}_d \le \mathbf{P}_d', \end{aligned} \tag{6.38}$$

where $\mathbf{P}_d^*$ and $\mathbf{P}_g^*$ are the target load and generation given by dispatch center, and $\mathbf{P}_d'$ and $\mathbf{P}_g'$ are the load and generation before dispatch. $\mathbf{P}_d$ and $\mathbf{P}_g$ are the states to solve, i.e., the states after dispatch at time τ_D. $\mathbf{r}_g$ is the vector of ramping rates of all the generation buses. Equation (6.38) can be briefly denoted as follows:

$$\mathbf{x}^{(r+1)} = LP_e(\mathbf{p}^{(r+1)}, \mathbf{x}^{(r+1)\prime}, \mathbf{x}^{(r+1)*}, \tau_D), \tag{6.39}$$

where $\mathbf{p}^{(r+1)}$ is the parameter on the $r+1$th level, such as network topology, branch parameters, branch flow limit, etc. $\mathbf{x}^{(r+1)*}$ is the re-dispatch target state to fulfill, which is determined by solving another LP problem (the conventional DC-OPF in this chapter) [18, 32] as

$$\mathbf{x}^{(r+1)*} = LP_a(\mathbf{p}^{(r+1)}, \mathbf{x}^{(r+1)\prime}). \tag{6.40}$$

From (6.39), $\frac{\partial \mathbf{C}_R^{(r+1)}}{\partial \mathbf{x}^{(r+1)\prime}}$, $\frac{\partial \mathbf{C}_R^{(r+1)}}{\partial \mathbf{x}^{(r+1)*}}$, $\frac{\partial \mathbf{x}^{(r+1)}}{\partial \mathbf{x}^{(r+1)\prime}}$, and $\frac{\partial \mathbf{x}^{(r+1)}}{\partial \mathbf{x}^{(r+1)*}}$ can be obtained by means of Lagrange multiplier and sensitivity analysis [37]. Similarly, $\frac{\partial \mathbf{x}^{(r+1)*}}{\partial \mathbf{x}^{(r+1)\prime}}$ can be calculated from (6.40).

6.8 Risk Mitigation Based on Tree Search

6.8.1 Efficient Forward–Backward Algorithm for Risk Gradient

6.8.1.1 Iterative Calculation of Terms in Risk Gradient

With the analysis in III.A and the chain rule of derivatives, the terms $\frac{\partial \mathbf{Pr}^{(r)}}{\partial \mathbf{x}^{(0)}}$ and $\frac{\partial \mathbf{C}^{(r)}}{\partial \mathbf{x}^{(0)}}$ of each level r can be calculated. Assume for any level m $(1 \leq m \leq r)$, the terms $\frac{\partial \mathbf{x}^{(m)}}{\partial \mathbf{x}^{(0)}}$, $\frac{\partial \mathbf{x}^{(m)\prime}}{\partial \mathbf{x}^{(0)}}$, and $\frac{\partial \mathbf{x}^{(m)*}}{\partial \mathbf{x}^{(0)}}$ are obtained, then for level $r+1$ the terms $\frac{\partial \mathbf{x}^{(r+1)}}{\partial \mathbf{x}^{(0)}}$, $\frac{\partial \mathbf{x}^{(r+1)\prime}}{\partial \mathbf{x}^{(0)}}$, and $\frac{\partial \mathbf{x}^{(r+1)*}}{\partial \mathbf{x}^{(0)}}$ are obtained as follows:

$$\frac{\partial \mathbf{x}^{(r+1)\prime}}{\partial \mathbf{x}^{(0)}} = \frac{\partial \mathbf{x}^{(r+1)\prime}}{\partial \mathbf{x}^{(r)}} \frac{\partial \mathbf{x}^{(r)}}{\partial \mathbf{x}^{(0)}} \tag{6.41}$$

$$\frac{\partial \mathbf{x}^{(r+1)*}}{\partial \mathbf{x}^{(0)}} = \frac{\partial \mathbf{x}^{(r+1)*}}{\partial \mathbf{x}^{(r+1)\prime}} \frac{\partial \mathbf{x}^{(r+1)\prime}}{\partial \mathbf{x}^{(0)}} \tag{6.42}$$

$$\frac{\partial \mathbf{x}^{(r+1)}}{\partial \mathbf{x}^{(0)}} = \frac{\partial \mathbf{x}^{(r+1)}}{\partial \mathbf{x}^{(r+1)*}} \frac{\partial \mathbf{x}^{(r+1)*}}{\partial \mathbf{x}^{(0)}} + \frac{\partial \mathbf{x}^{(r+1)}}{\partial \mathbf{x}^{(r+1)\prime}} \frac{\partial \mathbf{x}^{(r+1)\prime}}{\partial \mathbf{x}^{(0)}}. \tag{6.43}$$

Since $\frac{\partial \mathbf{x}^{(1)*}}{\partial \mathbf{x}^{(0)}}$, $\frac{\partial \mathbf{x}^{(1)\prime}}{\partial \mathbf{x}^{(0)}}$, and $\frac{\partial \mathbf{x}^{(1)}}{\partial \mathbf{x}^{(0)}}$ can be obtained from III.A, so according to (6.41)–(6.43), for any r $(1 \leq r \leq n$, where n is the final level of cascading outage), $\frac{\partial \mathbf{x}^{(r)*}}{\partial \mathbf{x}^{(0)}}$, $\frac{\partial \mathbf{x}^{(r)\prime}}{\partial \mathbf{x}^{(0)}}$, and $\frac{\partial \mathbf{x}^{(r)}}{\partial \mathbf{x}^{(0)}}$ are obtained iteratively in the process of cascading outage path simulation. So $\frac{\partial \mathbf{Pr}^{(r)}}{\partial \mathbf{x}^{(0)}}$ and $\frac{\partial \mathbf{C}^{(r)}}{\partial \mathbf{x}^{(0)}}$ are calculated with

$$\frac{\partial \mathbf{Pr}^{(r)}}{\partial \mathbf{x}^{(0)}} = \frac{\partial \mathbf{Pr}^{(r)}}{\partial \mathbf{x}^{(r-1)}} \frac{\partial \mathbf{x}^{(r-1)}}{\partial \mathbf{x}^{(0)}} \tag{6.44}$$

$$\frac{\partial \mathbf{C}^{(r)}}{\partial \mathbf{x}^{(0)}} = \frac{\partial \mathbf{C}_F^{(r)}}{\partial \mathbf{x}^{(r-1)}} \frac{\partial \mathbf{x}^{(r-1)}}{\partial \mathbf{x}^{(0)}} + \frac{\partial \mathbf{C}_R^{(r)}}{\partial \mathbf{x}^{(r-1)\prime}} \frac{\partial \mathbf{x}^{(r-1)\prime}}{\partial \mathbf{x}^{(0)}} + \frac{\partial \mathbf{C}_R^{(r)}}{\partial \mathbf{x}^{(r-1)*}} \frac{\partial \mathbf{x}^{(r-1)*}}{\partial \mathbf{x}^{(0)}}. \tag{6.45}$$

6.8.1.2 Recursive Form of Risk Gradient

Define equivalent cascading outage cost $C'(i_{k_1}\cdots i_{k_r})$ as

$$C'(i_{k_1}\cdots i_{k_r}) \triangleq C(i_{k_1}\cdots i_{k_r}) + \sum_{i_{k_{r+1}}} \Pr(i_{k_{r+1}}|i_{k_1}\cdots i_{k_r})C(i_{k_1}\cdots i_{k_{r+1}}) + \cdots$$
$$= C(i_{k_1}\cdots i_{k_r}) + \sum_{i_{k_{r+1}}} \Pr(i_{k_{r+1}}|i_{k_1}\cdots i_{k_r})C'(i_{k_1}\cdots i_{k_{r+1}}). \tag{6.46}$$

Equation (6.46) shows a recursive relationship, so $C'(i_{k_1}\cdots i_{k_r})$ could be calculated and updated reversely from the terminal back to the root of the MT. Similarly, define

$$R'(i_{k_1}\cdots i_{k_{r+1}}) \triangleq \sum_{i_{k_{r+1}}} \Pr(i_{k_{r+1}}|i_{k_1}\cdots i_{k_r})C'(i_{k_1}\cdots i_{k_{r+1}}). \tag{6.47}$$

With given $i_{k_1}\cdots i_{k_r}$, abbreviate all $R'(i_{k_1}\cdots i_{k_{r+1}})$ as vector $\mathbf{R}^{(r)\prime}$, and $C'(i_{k_1}\cdots i_{k_r})$ as $\mathbf{C}^{(r)\prime}$, then

$$\frac{\partial R'(i_{k_1}\cdots i_{k_r})}{\partial \mathbf{x}^{(0)}} = \mathbf{C}^{(r)\prime,\mathrm{T}}\frac{\partial \mathbf{Pr}^{(r)}}{\partial \mathbf{x}^{(0)}} + \mathbf{Pr}^{(r)\mathrm{T}}\left(\frac{\partial \mathbf{C}^{(r)}}{\partial \mathbf{x}^{(0)}} + \frac{\partial \mathbf{R}^{(r)\prime}}{\partial \mathbf{x}^{(0)}}\right). \tag{6.48}$$

Equation (6.48) is also a recursive form. Note that $R^{(0)\prime} = R'$, so the gradient of risk can be computed with forward–backward scheme in the risk assessment based on MT search.

6.8.1.3 Forward–Backward Scheme of Risk Gradient Calculation

With (6.41)–(6.45), the partial derivatives are calculated in the process of forward searching with the risk assessment [32], as demonstrated in Algorithm 1.

Algorithm 1 Forward calculation of partial derivatives

Step 1. Initialize level on MT $r = 0$.
Step 2. Sample mid-timescale outages and calculate $\frac{\partial \mathbf{Pr}^{(r+1)}}{\partial \mathbf{x}^{(r)}}$ with (6.34).
Step 3. Simulate short-timescale outages [32], and calculate $\frac{\partial \mathbf{x}^{(r+1)\prime}}{\partial \mathbf{x}^{(r)}}$, $\frac{\partial \mathbf{C}_F^{(r+1)}}{\partial \mathbf{x}^{(r)}}$ with (6.35)–(6.37).
Step 4. Simulate re-dispatch. Calculate $\frac{\partial \mathbf{Pr}^{(r+1)}}{\partial \mathbf{x}^{(0)}}$, $\frac{\partial \mathbf{x}^{(r+1)}}{\partial \mathbf{x}^{(0)}}$, and $\frac{\partial \mathbf{C}^{(r+1)}}{\partial \mathbf{x}^{(0)}}$.
Step 5. If the cascading outage path ends, exit and start simulation of a new path. Otherwise, assign $r = r + 1$ and jump to Step 2.

After searching a cascading outage path on the MT, the risk gradient is updated reversely, as Algorithm 2 shows.

Algorithm 2 Backward update of risk gradient

Assume the cascade path is $i_{k_1} \cdots i_{k_n}$, the variables with superscript (r) correspond to $(i_{k_1} \cdots i_{k_r})$, and (0) corresponds to the root state on MT.

Step 1. Define $b^{(r)}, r = 1 \cdots n$. If state $(i_{k_1} \cdots i_{k_r})$ has been reached before, then $b^{(r)} = 0$; otherwise, $b^{(r)} = 1$.

Step 2. Assign $r = n$. Assign $\mathbf{S}^{(r)} = b^{(r)} \frac{\partial C^{(r)}}{\partial \mathbf{x}^{(0)}}$, $\Delta C^{(r)} = b^{(r)} C^{(r)}$.

Step 3. Reverse to the previous state on the cascade path. Assign $r = r - 1$.

Step 4. Let $\mathbf{S}^{(r)} = \mathrm{Pr}^{(r+1)} \mathbf{S}^{(r+1)} + \Delta C^{(r+1)} \frac{\partial \mathrm{Pr}^{(r+1)}}{\partial \mathbf{x}^{(0)}}$.

Step 5. Let $\Delta C^{(r)} = b^{(r)} C^{(r)} + b^{(r+1)} \mathrm{Pr}^{(r+1)} \Delta C^{(r+1)}$.

Step 6. If $r = 0$, then exit. Otherwise, jump to Step 3.

By applying Algorithms 1 and 2 repeatedly along with the forward searching and backward updating procedure of risk assessment [32], the $\mathbf{S}^{(0)}$ will converge to $\frac{\partial R^{(0)\prime}}{\partial \mathbf{x}^{(0)}}$. Note that in operations, $\mathbf{x}^{(0)}$ is indirectly changed by altering the dispatch target state $\mathbf{x}^*$, as Fig. 6.22 shows. The gradient of risk in the space of control variables is

$$\mathbf{\Gamma} = \mathbf{S}^{(0)} \frac{\partial \mathbf{x}^{(0)}}{\partial \mathbf{x}^*}. \tag{6.49}$$

6.8.2 *Implementation of Risk Management*

6.8.2.1 Full Optimization Model of Risk Mitigation (RM)

After obtaining the risk gradient in the space of control variables, the risk mitigation (RM) optimization model is established based on the generic form of (6.32) as

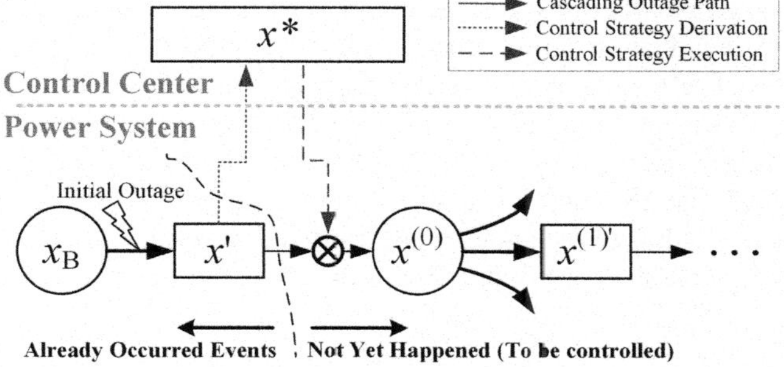

Fig. 6.22 Time series of the derivation and execution of control strategy

$$
\begin{aligned}
& \min f = -\mathbf{c}_D^{\mathrm{T}}\left(\mathbf{P}_d^* - \mathbf{P}_d\right) + \mathbf{c}_G^{\mathrm{T}}\left|\mathbf{P}_g^* - \mathbf{P}_g\right| \\
& s.t. \; -\boldsymbol{\Gamma} \cdot \begin{bmatrix} \mathbf{P}_d^* - \mathbf{P}_{d0}^* \\ \mathbf{P}_g^* - \mathbf{P}_{g0}^* \end{bmatrix} \le R_E - R'(\mathbf{x}_0^{(0)*}) \\
& \mathbf{1}^{\mathrm{T}}\left(\mathbf{P}_g^* - \mathbf{P}_d^*\right) = 0 \\
& -\mathbf{F}^{\max} \le \mathbf{y}_D \mathbf{M} \mathbf{Y}^{+}\left(\mathbf{P}_g^* - \mathbf{P}_d^*\right) \le \mathbf{F}^{\max} \\
& \mathbf{P}_g^{\min} \le \mathbf{P}_g^* \le \mathbf{P}_g^{\max} \\
& \mathbf{0} \le \mathbf{P}_d^* \le \mathbf{P}_d
\end{aligned}
\tag{6.50}
$$

,

where the first constraint is the risk constraint using the risk gradient. $\mathbf{x}_0^{(0)*} = [\mathbf{P}_{d0}^{*\,\mathrm{T}}, \mathbf{P}_{g0}^{*\,\mathrm{T}}]^{\mathrm{T}}$ is the target state from the original dispatch strategy, $R'(\mathbf{x}_0^{(0)*})$ is the risk of subsequent cascades of original dispatch strategy, and R_E is the expected risk after the RM. The other constraints are the limits of branches, generators, and loads. Here the variables are continuous, so the RM is an LP problem. If there are discrete variables, then RM will become a mixed-integer linear programming (MILP) problem.

The RM reduces risk by setting its solution as the target state of re-dispatch. The extent of reduction of cascading outage risk is adjusted by changing the expected subsequent cascade risk R_E. The constraint will force the solution to a less risky state, and the control cost is expected to be higher. Denote the expected risk decrease as $\Delta R = R'(\mathbf{x}_0^{(0)*}) - R_E \ge 0$; then the bigger ΔR, the more reduction of risk is expected.

6.8.2.2 Iterative Risk Mitigation (IRM)

The RM model is based on the linearization of risk at the original operating point. When ΔR goes outside an effective region of linearization, then there will be considerable linearization errors. To overcome such limitation, consider iterating the procedure of RM so as to accumulate the effect of linearization-based RM step by step. The procedure of iterative RM (IRM) is shown in Fig. 6.23.

The IRM first assesses risk on the original control strategy and solves the RM problem (6.50). The new dispatch strategy is then evaluated with risk assessment. If the risk is decreased, then the strategy is expected to be effective. Such a procedure is iterated until the risk does not decrease.

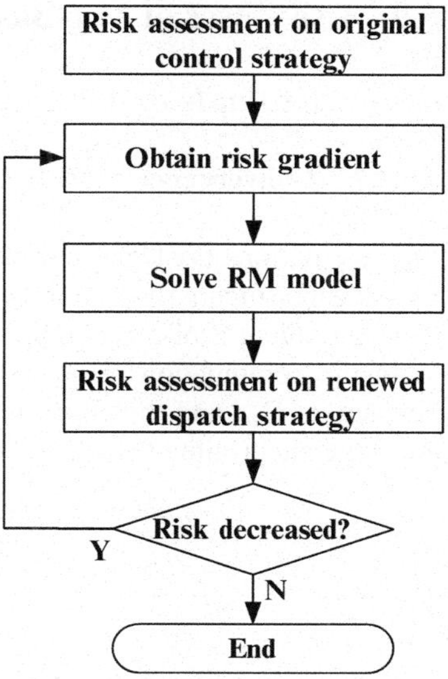

Fig. 6.23 The procedure of IRM

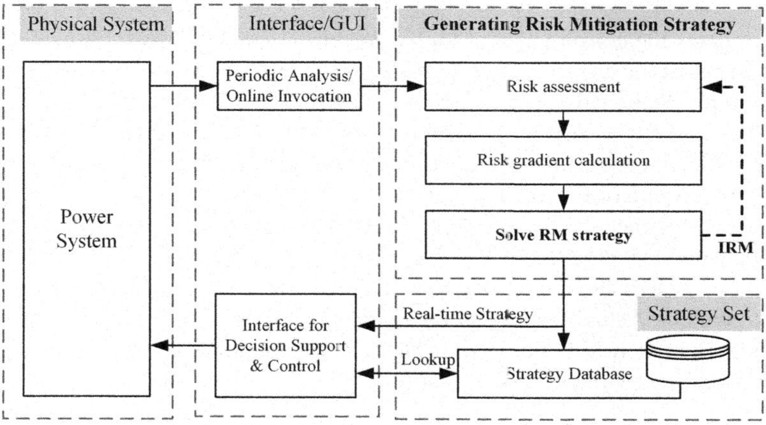

Fig. 6.24 The framework of RM/IRM application

6.8.2.3 Framework of RM/IRM Application

The RM and IRM can be used offline to generate control strategies on a given set of system working conditions. The generated strategies are stored in a database and can be extracted when corresponding events occur. Moreover, the RM and IRM also have potential of online assessment and decision support. The framework of RM/IRM application is demonstrated in Fig. 6.24.

6.9 Use Cases of MT-Based Risk Mitigation

6.9.1 Example of MT-Based Risk Gradient Calculation

6.9.1.1 Convergence of Risk Gradient

The calculation of risk gradient can be integrated into the forward–backward scheme of risk assessment. To efficiently derive risk management strategy, the convergence of risk gradient is of great significance.

In the computation process, the risk gradient after the mth search attempt is denoted as $\boldsymbol{\Gamma}_m$, and the converged risk gradient is denoted as $\boldsymbol{\Gamma}^*$. To evaluate the convergence profile of risk gradient, propose the following convergence indices:

$$\delta_m = \frac{\|\boldsymbol{\Gamma}_m - \boldsymbol{\Gamma}^*\|}{\|\boldsymbol{\Gamma}_1 - \boldsymbol{\Gamma}^*\|} \tag{6.51}$$

$$\delta_m^{\mathrm{dir}} = \left\| \frac{\boldsymbol{\Gamma}_m}{\|\boldsymbol{\Gamma}_m\|} - \frac{\boldsymbol{\Gamma}^*}{\|\boldsymbol{\Gamma}^*\|} \right\|. \tag{6.52}$$

δ_m reflects the convergence of the vector of risk gradient, and δ_m^{dir} evaluates the convergence of the direction of risk gradient. The test is conducted on the RTS-96 3-area system, in which the parameters are set as $T_{\max} = 150$ min, $\tau_D = 15$ min. After 10,000 times of tree search, the risk, δ_m, and δ_m^{dir} are all considered as converged. Figure 6.25 demonstrates the convergence profile of risk gradient, which is the same as the small figure of Fig. 6.18.

The convergence of risk gradient is slower than that of risk [32] (Fig. 6.26). However, the convergence of the direction of risk gradient only requires several hundreds of search attempts, which is much faster than the convergence of the vector of risk gradient. Actually, only obtaining the direction of risk gradient is enough for risk reduction, so in situations requiring fast computation, the tree search attempts can be significantly reduced.

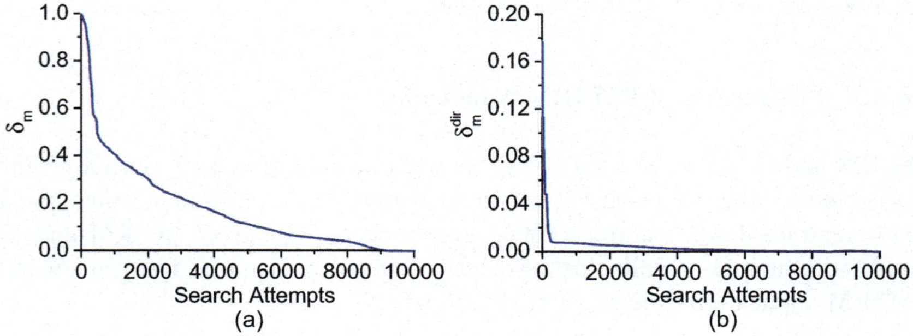

Fig. 6.25 Convergence of (**a**) risk gradient and (**b**) the direction of risk gradient

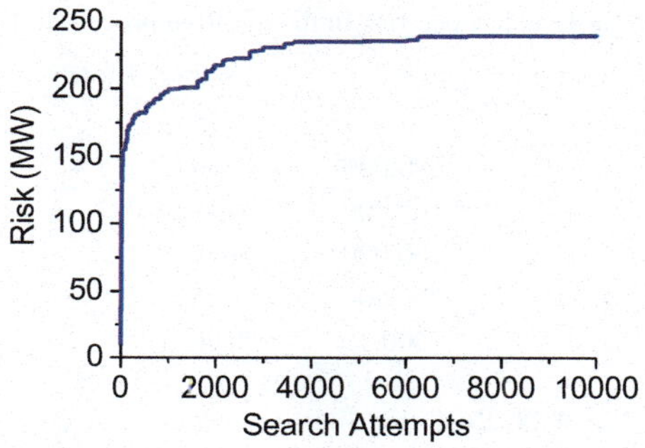

Fig. 6.26 Convergence profile of risk [32]

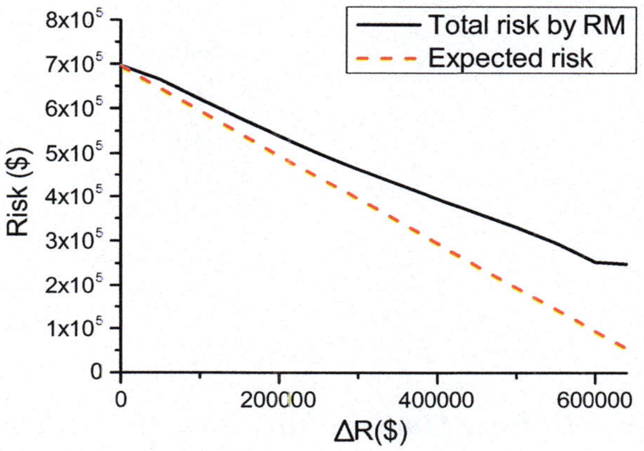

Fig. 6.27 Effect of single-step risk reduction based on risk gradient

6.9.1.2 Effectiveness of Risk Management Based on Risk Gradient

After validating the accuracy and the convergence of the calculation of risk gradient, the effectiveness of RM in the RTS-96 3-area system model is tested. Here all the cost and risk are converted into economic metrics. Assume that adjusting 1MW of generation in an interval τ_D costs \$100, and 1MW load loss in an interval corresponds to the loss of \$10,000. Set the initial failure on branches 22, 23, and 24. When utilizing conventional re-dispatch, the total risk (i.e., the cost of control plus the risk of subsequent cascading outages) is \$696,775. Then use the RM to reduce cascading outage risk and evaluate the performance with risk assessment. The effect of RM under different values of ΔR is shown in Fig. 6.27. The RM effectively decreases risk within a certain range, but the risk stops decreasing when ΔR reaches around \$600,000, which means the linearization is no longer effective and a new iterative step is needed.

Table 6.2 Cost–risk in the iteration process in RTS-96 test system

Round	ΔR ($)	Control cost ($)	Subsequent risk ($)	Total risk ($)
0	–	2475	694,300	696,775
1	600,000	3789.8	249,590	253,379.8
2	100,000	4141.6	174,840	178,981.6
3	100,000	5034.9	40,738	45,772.9
4	20,000	5456.4	12,434	17,890.4
5	1000	5590	13,132	18,722

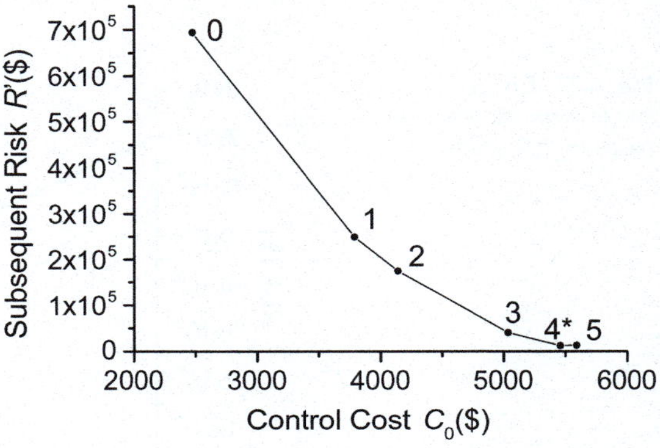

Fig. 6.28 Cost–risk trajectory of IRM in RTS-96 test system

6.9.2 Case Studies of Iterative Risk Mitigation

6.9.2.1 RTS-96 System

The performance of the RM is limited only within a range where linearization is effective. Figure 6.27 shows that with the RM, the risk stops decreasing at around $\$2.5 \times 10^5$. IRM keeps updating risk gradient at new operating points, and the risk is further reduced. Table 6.2 and Fig. 6.28 demonstrate the cost–risk relationship derived by IRM.

The results indicate that the IRM can effectively overcome the limitation of linearization with the RM and further reduce risk of cascading outages. After 4 rounds of iterations, the subsequent risk of cascading outages has reduced by 97.3%.

6.9.2.2 US–Canada Northeast System

Next, the IRM is tested on the US–Canada Northeast power grid model. The cost–risk characteristics derived by IRM are demonstrated in Fig. 6.29.

The results indicate that the total risk at the 6th iteration reaches the lowest among all iterations, where the total risk is expected to decrease by 54.5%, and the subsequent cascade risk drops significantly by 93.6%. The subsequent cascade risk is even lower in the 7th iteration, but the drop in subsequent cascade risk is offset

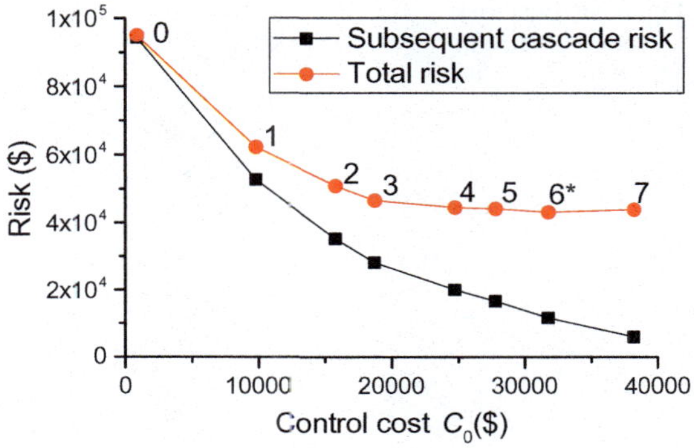

Fig. 6.29 Cost–risk trajectory of IRM in US–Canada Northeast system

Table 6.3 Time consumption of IRM in US–Canada Northeast system

Subprograms of IRM	Time consumption (s)
Cascading outage simulation and risk assessment	660.41
Computation of risk gradient	237.42
Solving RM model	3.48
Total	901.31

by a substantially high cost, which causes a higher total risk than that of the 6th iteration. In practice, the adopted strategy for risk management may vary depending on the risk preference. Therefore, the strategies in Fig. 6.29 can also be regarded as the results of multi-objective optimization of risk and cost, and the control strategy for risk management can be selected depending on the risk preference.

Regarding the computational efficiency, the average performance of a single IRM run in the US–Canada Northeast system model is shown in Table 6.3. The result shows that solving the RM models takes only several seconds, but much more time is consumed in cascading outage simulation, risk assessment, and calculating the risk gradient. The speed of computation can be further enhanced with parallel computing on high-performance computation platform, and this method also has potentials for online application on a period of 5–15 minutes for operators' decision support to prevent cascading outages.

As for the drawbacks of the proposed approach, it is observed that the accuracy of risk gradient calculation may decrease as the size of the power system grows. This is because as the system size grows, the number of nonlinear behaviors (e.g., switched active constraints in dispatch model, etc.) also grows. To maintain desired accuracy, more iterations in IRM may be necessary, and the computation speed will be adversely affected. The estimation of effective linearization region and the derivation of desirable step size in IRM will be studied in the future. Moreover, since the calculation of risk gradient requires to store all the sensitivity matrices $\frac{\partial \mathbf{x}^{(r)*}}{\partial \mathbf{x}^{(0)}}$, $\frac{\partial \mathbf{x}^{(r)\prime}}{\partial \mathbf{x}^{(0)}}$, and $\frac{\partial \mathbf{x}^{(r)}}{\partial \mathbf{x}^{(0)}}$, the memory usage in large-scale systems will be high. This problem can be alleviated by using compressed storage (CS) of the sensitivity matrices, since

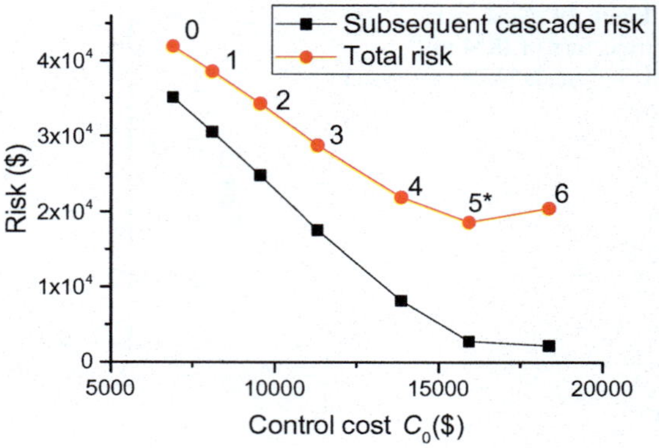

Fig. 6.30 Cost–risk trajectory of IRM in Mid-European system

Table 6.4 Time consumption of IRM in Mid-European system

Subprograms of IRM	Time consumption (s)
Cascading outage simulation and risk assessment	1806.17
Computation of risk gradient	1449.69
Solving RM model	6.079
Total	3261.94

most elements in the sensitivity matrices have very low absolute values (in large systems, generally less than 1% of the elements have absolute values larger than 10^{-3}, and less than 10% have absolute value larger than 10^{-5}). Compressing the matrices by dropping low-absolute-value elements practically does not affect the accuracy but can significantly decrease memory usage.

6.9.2.3 1354-Bus Mid-European System [38]

To further demonstrate the performance, the proposed approach is tested on a larger Mid-European backbone system model that appears in [38, 39]. The system has 1354 buses (including 260 generation buses and 688 load buses) and 1991 branches. The branch flow limits were modified to secure N-1. We maintain the same number of MT search attempts as the US–Canada Northeast System case and run the IRM. The effectiveness of IRM is shown in Fig. 6.30, and the performance is shown in Table 6.4. The computation can be finished within 1 hour on a desktop computer, which shows promising potential for online decision support. Also, the speed can be further enhanced with parallel computing on high-performance platforms (currently the approach is developed and tested on Matlab and has not been further optimized for performance).

For large-scale systems, the CS can be adopted to avoid memory overflow. Table 6.5 compares the performance with and without CS. The threshold of absolute sensitivity value is set as 10^{-5}. Although the CS causes some computational

Table 6.5 Time and memory usage of risk gradient computation

	Without CS		With CS	
Systems	Time (s)	Memory (GB)	Time (s)	Memory (GB)
US–Canada	237.42	4.87	670.10	0.76
Mid-European	2184.60	30.25	1449.69	3.32

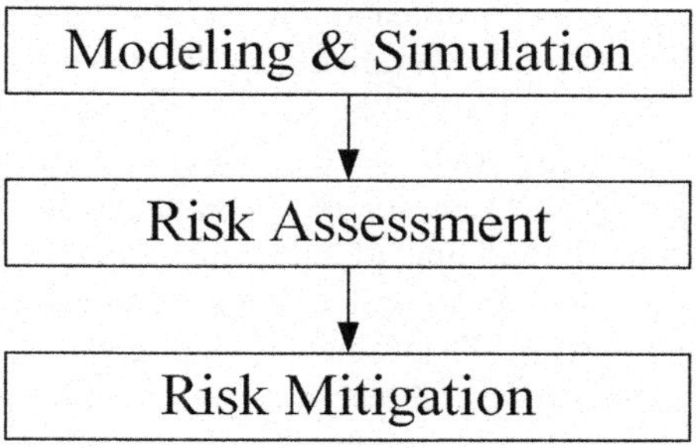

Fig. 6.31 Dependency of modeling, risk assessment, and mitigation of cascading outages

overhead in compressing and indexing matrices, the memory usage is significantly reduced. Per the observation from the test results on more systems, the computation time complexity when using the CS is approximately $O(N^2)$, where N is the number of buses. Moreover, it is observed that the number of remaining sensitivity elements after CS grows approximately linearly with system size, while the full sensitivity matrix grows quadratically with system size. Such a desirable spatial complexity of CS contributes to the less overall computation time in large-scale systems cases, as Table 6.5 shows. Therefore, the CS is recommended for large systems.

6.10 Summary and Discussions

Cascading outages as catastrophic events in the power grids have high values to be modeled, evaluated, and mitigated. The modeling, assessment, and mitigation of cascading outages have strong dependencies as Fig. 6.31 shows, which require proper modeling and simulation in assessment and accurate assessment in mitigation. And there are challenges of complex modeling of cascading outages, huge numbers of possible paths of outage development, and the long coupling chain of decision variables with subsequent outage patterns.

In this chapter, a multi-timescale framework based on quasi-dynamic method is established to realize more accurate simulation of interactions among dynamics in various timescales with explicit representation of time. The model overcomes the ambiguity of timescales in conventional quasi-static models, and time-consuming full-dynamic cascading outage simulation is avoided. Besides, this model also provides flexibility incorporation of dynamic simulation of short-term processes as needed. The proposed model improves simulation on reactive power-related dynamics as well as dispatcher actions. To better tackle the challenges of tractability in cascading outage risk assessment, this chapter also presents a Markovian tree formulation to model cascading outage paths and then derives a framework base on it for risk assessment and risk mitigation by using tree search techniques. Under the assumption that the probabilities of event occurrences during outage development can be well estimated, instead of using the sampling method that requires repetitive simulation and evaluation of same or similar outage paths, the risk metric can be directly calculated by simulating the path once (i.e., the path is reached/searched). This constitutes the foundation of the Markovian tree formulation efficient tree search methods for risk assessment and mitigation. Of course, the accurate and proper modeling of cascading outages is still important, and this chapter also presents how to weave the multi-timescale modeling of cascading outages into the framework. The framework for cascading outages risk assessment and mitigation is tested on several test systems and real transmission system models and the results show good potentials.

It is worth noting that the approaches presented still have lots of room to improve, including but not limit to even finer-grained modeling of multi-timescale cascading outages, the modeling and simulation of dynamics and transients, the consideration of inverter-based resources (IBRs), and variable renewable energy. Also on the computation domain, the tree formulation and tree search technique well associate with the recent advancements in machine learning, such as Monte-Carlo tree search (MCTS), decision tree, automatic gradients, etc., whose solutions also have potentials to be utilized to enhance this framework.

References

1. D. Burpee, H. Dabaghi, L. Jackson, F. Kwamena, J. Richter, T. Rusnov, K. Friedman, L. Mansueti, D. Meyer, US-Canada power system outage task force: final report on the implementation of task force recommendations, 2006
2. M. Vaiman, Y. Chen, B. Chowdhury, I. Dobson, P.D. Hines, M. Papic, P> Zhang, Risk assessment of cascading outages: Methodologies and challenges. IEEE Trans. Power Syst. **27**(2), 631–641 (2011)
3. P.M. Anderson, C.F. Henville, R. Rifaat, B. Johnson, S. Meliopoulos, *Power System Protection* (Wiley, 2022)
4. S. Imai, Undervoltage load shedding improving security as reasonable measure for extreme contingencies, in *IEEE Power Engineering Society General Meeting, 2005* (IEEE, 2005), pp. 1754–1759.
5. J. Qi, S. Mei, F. Liu, Blackout model considering slow process. IEEE Trans. Power Syst. **28**(3), 3274–3282 (2013)
6. P.P. Varaiya, F.F. Wu, J.W. Bialek, Smart operation of smart grid: Risk-limiting dispatch. Proc. IEEE **99**(1), 40–57 (2010)
7. J. Machowski, Z. Lubosny, J.W. Bialek, J.R. Bumby, *Power System Dynamics: Stability and Control* (Wiley, 2020)
8. Y. Sun, L. Cheng, H. Liu, S. He, Power system operational reliability evaluation based on real-time operating state, in *2005 International Power Engineering Conference* (IEEE, 2005), pp. 722–727

9. P. Simpson, R. Van Bossuyt, Tree-caused electric outages. J. Arboriculture **22**, 117–121 (1996)
10. I. Dobson, B.A. Carreras, D.E. Newman, How many occurrences of rare blackout events are needed to estimate event probability? IEEE Trans. Power Syst. **28**(3), 3509–3510 (2013)
11. Z. Wang, A. Scaglione, R.J. Thomas, A Markov-transition model for cascading failures in power grids, in *2012 45th Hawaii International Conference on System Sciences* (IEEE, 2012), pp. 2115–2124
12. M. Rahnamay-Naeini, Z. Wang, N. Ghani, A. Mammoli, M.M. Hayat, Stochastic analysis of cascading-failure dynamics in power grids. IEEE Trans. Power Syst. **29**(4), 1767–1779 (2014)
13. P. Linares, L. Rey, The costs of electricity interruptions in Spain. Are we sending the right signals? Energy Policy **61**, 751–760 (2013)
14. R. Yao, S. Huang, K. Sun, F. Liu, X. Zhang, S. Mei, A multi-timescale quasi-dynamic model for simulation of cascading outages. IEEE Trans. Power Syst. **31**(4), 3189–3201 (2016)
15. J. Yan, Y. Tang, H. He, Y. Sun, Cascading failure analysis with DC power flow model and transient stability analysis. IEEE Trans. Power Syst. **30**(1), 285–297 (2014)
16. P. Hines, K. Balasubramaniam, E.C. Sanchez, Cascading failures in power grids. IEEE Potentials **28**(5), 24–30 (2009)
17. IEEE standard for calculating the current-temperature relationship of bare overhead conductors, in *IEEE Std 738-2012 (Revision of IEEE Std 738-2006 - Incorporates IEEE Std 738-2012 Cor 1-2013)*, pp. 1–72, 2013
18. I. Dobson, B.A. Carreras, V.E. Lynch, D.E. Newman, An initial model for complex dynamics in electric power system blackouts, in *Proceedings of the Annual Hawaii International Conference on System Sciences* (Citeseer, 2001), pp. 51–51
19. J. Chen, J.S. Thorp, I. Dobson, Cascading dynamics and mitigation assessment in power system disturbances via a hidden failure model. Int. J. Electr. Power Energy Syst. **27**(4), 318–326 (2005)
20. D.P. Nedic, *Simulation of Large System Disturbances* (The University of Manchester (United Kingdom), 2003)
21. N. Bhatt, S. Sarawgi, R. O'keefe, P. Duggan, M. Koenig, M. Leschuk, S. Lee, K. Sun, V. Kolluri, S. Mandal et al., Assessing vulnerability to cascading outages, in *2009 IEEE/PES Power Systems Conference and Exposition* (IEEE, 2009), pp. 1–9
22. P. Henneaux, Probability of failure of overloaded lines in cascading failures. Int. J. Electr. Power Energy Syst. **73**, 141–148 (2015)
23. D. Lieber, A. Nemirovskii, R.Y. Rubinstein, A fast Monte Carlo method for evaluating reliability indexes. IEEE Trans. Reliab. **48**(3), 256–261 (1999)
24. Y. Wang, C. Guo, Q. Wu, A cross-entropy-based three-stage sequential importance sampling for composite power system short-term reliability evaluation. IEEE Trans. Power Syst. **28**(4), 4254–4263 (2013)
25. D.S. Kirschen, D. Jayaweera, D.P. Nedic, R.N. Allan, A probabilistic indicator of system stress. IEEE Trans. Power Syst. **19**(3), 1650–1657 (2004)
26. P. Rezaei, P.D. Hines, M.J. Eppstein, Estimating cascading failure risk with random chemistry. IEEE Trans. Power Syst. **30**(5), 2726–2735 (2014)
27. Y. Jia, P. Wang, X. Han, J. Tian, C. Singh, A fast contingency screening technique for generation system reliability evaluation. IEEE Trans. Power Syst. **28**(4), 4127–4133 (2013)
28. H. Liu, Y. Sun, L. Cheng, P. Wang, F. Xiao, Online short-term reliability evaluation using a fast sorting technique. IET Gener. Trans. Distrib. **2**(1), 139–148 (2008)
29. C. Grigg, P. Wong, P. Albrecht, R. Allan, M. Bhavaraju, R. Billinton, Q. Chen, C. Fong, S. Haddad, S. Kuruganty et al., The IEEE reliability test system-1996. A report prepared by the reliability test system task force of the application of probability methods subcommittee. IEEE Trans. Power Syst. **14**(3), 1010–1020 (1999)
30. R. Yao, X. Zhang, S. Huang, S. Mei, Z. Zhang, X. Li, Q. Zhu, Cascading outage preventive control for large-scale AC-DC interconnected power grid, in *IEEE PES General Meeting* (IEEE, 2014), pp. 1–5

31. M. Vaiman, K. Bell, Y. Chen, B. Chowdhury, I. Dobson, P. Hines, M. Papic, S. Miller, P. Zhang, Risk assessment of cascading outages: Methodologies and challenges. IEEE Trans. Power Syst. **27**(2), 631–641 (2012)
32. R. Yao, S. Huang, K. Sun, F. Liu, X. Zhang, S. Mei, W. Wei, L. Ding, Risk assessment of multi-timescale cascading outages based on Markovian tree search. IEEE Trans. Power Syst. **32**(4), 2887–2900 (2017)
33. M. Yang, J. Wang, H. Diao, J. Qi, X. Han, Interval estimation for conditional failure rates of transmission lines with limited samples. IEEE Trans. Smart Grid **9**(4), 2752–2763 (2016)
34. H. Ren, I. Dobson, B.A. Carreras, Long-term effect of the n-1 criterion on cascading line outages in an evolving power transmission grid. IEEE Trans. Power Syst. **23**(3), 1217–1225 (2008)
35. C. Reddy, S. Chakrabarti, S. Srivastava, A sensitivity-based method for under-frequency load-shedding. IEEE Trans. Power Syst. **29**(2), 984–985 (2014)
36. A.A. Girgis, S. Mathure, Application of active power sensitivity to frequency and voltage variations on load shedding. Electric Power Syst. Res. **80**(3), 306–310 (2010)
37. P.R. Gribik, D. Shirmohammadi, S. Hao, C.L. Thomas, Optimal power flow sensitivity analysis. IEEE Trans. Power Syst. **5**(3), 969–976 (1990)
38. C. Josz, S. Fliscounakis, J. Maeght, P. Panciatici, AC power flow data in matpower and QCQP format: iTesla, RTE Snapshots, and PEGASE. Preprint (2016). arXiv:1603.01533
39. R.D. Zimmerman, C.E. Murillo-Sánchez, R.J. Thomas, MATPOWER: Steady-state operations, planning, and analysis tools for power systems research and education. IEEE Trans. Power Syst. **26**(1), 12–19 (2011)

Chapter 7
Steady-State Simulation of Cascading Outages Considering Frequency

Wenyun Ju, Kai Sun, and Rui Yao

7.1 Introduction

Cascading outages within power transmission systems have led to widespread blackouts, arousing significant apprehensions regarding transmission system reliability [1–3]. While blackouts are infrequent, their occurrences could cause substantial economic losses and profound societal risks. Both academia and the power industry have dedicated substantial efforts to comprehending blackout mechanisms, evaluating cascading risks, and proposing strategies to mitigate such risks. The studies of blackouts triggered by cascading outages can be broadly categorized into three categories: cascading outage risk evaluation, mitigation of cascading outages, and modeling and simulation and cascading outages.

To evaluate the risk of cascading outages, a large number of samples of cascading outages are needed. They are either from simulation [4, 5] or from electric utility outage data [6]. However, since only partial cascading mechanisms are approximated in this process [7, 8], the accuracy of the cascading outages by simulation can be validated from the perspective of estimated blackout risk. Reference [9] proposes the concept of interaction graph (IG) for analysis of cascading outages using a dataset of simulated cascading outages. The IG describes the interactions between successive outaged components statistically, and it is not the actual power grid topology. The idea of a graph of interactions can be traced back to [10] in

W. Ju (✉)
Pacific Gas & Electric, Oakland, CA, USA

K. Sun
The University of Tennessee, Knoxville, TN, USA
e-mail: kaisun@utk.edu

R. Yao
Google LLC, Mountain View, CA, USA
e-mail: ruiyao@ieee.org

K. Sun (ed.), *Cascading Failures in Power Grids*, Power Electronics and Power Systems, https://doi.org/10.1007/978-3-031-48000-3_7

which a stochastic process at each graph node interacts with different strengths along the graph edges connecting that node to other nodes. A similar-graph-based model called the "influence graph" is studied in [11] and [12], in which nodes represent cascading events and edges measure influences between the cascading events. Reference [13] extends the single-layer IG to a multi-layer IG, where each layer focuses on one aspect of outage propagation, e.g., the number of line outages, the amount of load shed, and the electrical distance of the outage propagation. In [14], the authors extend the IG model in [9] to a dynamic IG and proposed an online mitigation strategy against cascading outages using an optimal power flow model. Reference [15] develops a Markovian influence graph from historical line outage data. Reference [16] includes material on both IGs and influence graphs. Reference [17] surveys various methods of constructing IGs and the reliability analysis based on IGs. There are a few different names for the graphs of interactions that depend on how they are formed, such as the correlation network, and the cascading faults graph. The IG can be developed from cascading events either from simulation or from electric utility outage data. More specially, the IG's nodes represent outages of single transmission lines, and its directed edges represent probabilistic interactions between successive line outages. The more probable edges correspond to the interactions between line outages that appear more frequently in the data. Cascades in the IG start with initial line outages at the nodes and spread probabilistically along the directed graph edges.

Based on the study of IG, the identification of critical components and upgrade of critical components to mitigate the propagation of cascading outages and reduce the blackout risk are becoming more and more popular research topics [9–22]. Reference [9] proposes its indices to identify the critical components based on the IG. Reference [12] forms its specific indices for identifying critical components based on the influence graph. Reference [18] applies a modified PageRank algorithm to identify critical lines. Reference [20] forms the influence graph using methods of both [12] and [19] and proposes a community-based measure to identify critical components. Reference [20] proposes its indices for identifying critical components, and the performance is better than other centrality indices based on network theory. Reference [13] identifies key components revealed by the multi-layer IG provide useful insights on the mechanism and mitigation of cascading outages, which cannot be obtained from any single layer. This multi-layer graph is suggested to mitigate cascades in system operation by providing the critical lines at different states of cascades. In [14], during the propagation of a specific cascade, the dynamic IG removes the interactions involving already outaged lines, and optimal power flow controls the power flow on the critical lines indicated by the dynamic IG. The dynamic IG model reduces the risk of large cascades more than the static IG. Reference [15] identifies critical components based on the asymptotic quasi-stationary distribution. The quasi-stationary distribution has a clear interpretation of specifying the probabilities that each of the lines is involved in large cascades. Reference [21] proposes an active islanding strategy based on IG to mitigate the propagation of cascading outages during the early stage of cascading outages. Reference [22] proposes a scheme to mitigate cascading outages by isolating the propagation of cascading outages in a local region by breaking inter-regional links of outaged components.

In the realm of modeling and simulating cascading outages, various models have been advanced, each offering distinct attributes and applications: (1) A number of steady-state models have been proposed for cascading outage simulations. These encompass the ***CASCADE*** model [23, 24], the ***DCSIMSEP*** model [11, 25], the ***Branching Process*** model [26, 27], the ***OPA*** model [28], the ***Improved OPA*** model [29], the ***AC OPA*** model [30], the ***TRELSS*** model [31], the ***Manchester*** model [32], the ***Hidden Failure*** model [33], and the ***Interaction Graph*** model [9, 13]. (2) Several quasi-dynamic models have been put forward such as the ***Multi-timescale*** model presented in [34]. It bridges the gap between purely steady-state and fully dynamic simulations, accounting for certain dynamics while maintaining computational efficiency. (3) A few dynamic models have been proposed to employ time-domain simulations to study cascading outages in depth. For example, the ***COSMIC*** model is detailed in [35], and the hybrid model is introduced in [36]. However, a key drawback of dynamic simulations is their time-intensive nature, particularly for large-scale power systems. The simulation of a realistic cascading outage scenario, from initial outages to system collapse, can demand significant time, often extending to tens of minutes. The implementation of full time-domain simulations for extensive time periods proves impractical due to the considerable time requirements. Additionally, the power system dynamic models utilized by electric utilities and reliability coordinators are primarily tailored for short-term transient stability assessments. These models may lack adequate validation for mid-term or long-term simulations, such as those involving cascading outages. Given these considerations, approaches based on steady-state power flow hold distinct advantages. They align well with the need for rapid simulation of cascading outages and can meet stringent time performance criteria. Furthermore, in most of the duration in cascading outages, the transients following each outage damp out quickly and the system often converges to a steady state prior to the occurrence of the subsequent outage. Therefore, power-flow-based steady-state or quasi-dynamic models are generally sufficient for the analysis of the overall loss and the mechanism in the propagation mechanism of cascading outages, as indicated in [34].

Frequency is a key indicator, reflecting the real-time balance between active power generation and load demand. Abnormal deviations of frequency can trigger under-frequency load shedding (UFLS) [37, 38] and generator frequency protection, incurring substantial loss in generation and load. Consequently, frequency emerges as a pivotal contributor to cascading outages and blackouts, corroborated by [39]. Conventional power flow models presume constant system frequency using swing buses to eliminate active power imbalances. Yet, the idealized swing buses with infinite capability of power balancing and frequency regulation do not exist in actual power systems. In practical scenarios, frequency control takes a distributed approach. Initially, generator governors adjust their speeds and active power outputs according to predefined strategies. Concurrently, frequency-sensitive loads vary their power consumption in response to frequency deviations. Consequently, it becomes crucial to incorporate frequency-related system behaviors and operations when simulating cascading outages.

Since the 1970s, endeavors have been directed toward integrating frequency deviation into power flow models [40]. In 1986, [41] proposed a "dynamic load flow" (DLF) algorithm where unbalanced active power is distributed among generators with speed controllers; however, this model does not obtain the frequency. An approach was presented in [42], introducing frequency as an unknown variable in DLF calculation. In the past few decades, researchers have introduced various power flow models [43–53] for fast simulation and analysis for complex power systems. These models primarily incorporate power-frequency characteristics into the power flow framework, taking into account power-frequency characteristics related to loads, speed governors of generators, and automatic generation control (AGC). References [44] and [45] integrate power-frequency and voltage-dependent attributes of loads and generator speed governors into the power flow model, specifically for dispatcher training simulators. Reference [46] expands this consideration to encompass power-frequency and voltage-dependent characteristics with loads, voltage-reactive power characteristics of generators, speed governors of generators, as part of power system security assessment. References [47–49] incorporate power-frequency characteristics of active loads and generator speed governors into the power flow model, enhancing the model's utility for risk assessment. Other fields to apply those similar power flow models include microgrid control [50, 51] and analyses involving wind generation [52, 53]. Analysis and simulation of cascading outages can also apply such models. References [54] and [55] incorporate frequency deviation into cascading outage simulation based on a DC power flow model, in which frequency deviation is calculated directly from power-frequency characteristics of generators and loads.

This chapter introduces a steady-state approach to simulate cascading outages while accounting for frequency-dependent system traits and operational actions. These actions encompass frequency deviation, power-frequency characteristics of generators and loads, under-frequency load shedding (UFLS) strategies, and generator frequency protection mechanisms. This approach is named as "SSCOF," and it has three key aspects. First, the SSCOF approach seamlessly integrates the computation of frequency deviation into a power flow model like [42], hereafter referred to as the "DLF model" to align with its development and inspiration from the DLF algorithm found in [41–49]. Consequently, the results of the power flow analysis can encompass active power imbalances and consider the power-frequency attributes of generators and loads. Second, an AC optimal power flow model considering frequency deviation (termed AC-OPFf) is employed. This model aims to identify remedial actions against potential system collapse, a situation indicated by divergent power flow computation. By factoring in frequency deviation, the DLF and AC-OPFf models enhance the credibility of steady-state simulations for power systems under cascading outages. Third, the SSCOF approach accommodates the modeling of UFLS strategies and generator frequency protection mechanisms. These factors, while pivotal, have not been adequately addressed by existing steady-state methodologies tailored for simulating cascading outages.

The remainder of this chapter is structured as follows. In Sect. 7.2, we detail the SSCOF approach designed for the simulation of cascading outages in [56]. Initially, this section describes the DLF model employed in the SSCOF approach, proceeds to unveil the AC-OPFf model, and subsequently presents the UFLS scheme along with generator frequency and line protection models within the SSCOF approach, and finally compares the steps of the SSCOF approach with those of a conventional approach utilized for cascading outage simulation. Section 7.3 is dedicated to comprehensive case studies. It first benchmarks the results of the DLF model with that of the time-domain simulation on a two-area system. Subsequently, this section evaluates the effectiveness of the SSCOF approach through numerous cascading outage scenarios simulated on the IEEE 39-bus system and the NPCC 48-machine, 140-bus systems. The simulated cascading outage scenarios against those generated through the conventional approach are compared and analyzed. Lastly, Sect. 7.4 encapsulates the findings in [56] and concludes the chapter.

7.2 Approach of SSCOF

In this section, a concise introduction for the DLF model is presented, and subsequently, the AC-OPFf model and a comparison with the conventional AC-OPF model are provided. Then, this section outlines the UFLS scheme and elaborates on the generator frequency protection and line protection models, all of which constitute integral components of the SSCOF approach.

7.2.1 Dynamic Load Flow Model

The power-frequency characteristics of a load at bus i, often referred to as static power-frequency characteristics (SPFCs), can be approximated by

$$P_{Di} = P_{D0i}(1 + D_i f_d), \quad f_d = f - f_n \tag{7.1}$$

where f represents the system frequency, f_n stands for the nominal frequency, f_d is the difference between the system frequency and the nominal frequency, P_{D0i} represents the power load when the frequency is at its nominal value, and constant D_i quantifies frequency–sensitivity of the load, showing how active the load changes with frequency deviation.

A frequency deviation arises when the system's active power balance is not satisfied at the nominal frequency. The speed governor of a generator at bus i can adjust its steady-state output P_{Gi} based on its regulation factor R_i automatically:

$$P_{Gi} = P_{G0i} - \frac{f_d}{R_i}, \quad P_{Gi,min} \le P_{Gi} \le P_{Gi,max} \tag{7.2}$$

where P_{G0i} is the active power output of generator i at f_n, and $P_{Gi,min}$ and $P_{Gi,max}$ are the lower and upper limits of P_{G0i}.

For an n-bus power system, there are m buses known as PQ buses, which are numbered from 1 to m. Then, we have $n-m-1$ PV buses, numbered from $m+1$ to $n-1$. Finally, there is one swing bus labeled as bus No. n. The calculation of DLF model is to eliminate the active power mismatches of all n buses and reactive power mismatches among the m PQ buses.

$$\Delta P_i = P_{Gi} - P_{Di} - V_i \sum_j V_j (G_{ij} \cos\theta_{ij} + B_{ij} \sin\theta_{ij}) \quad i = 1, \ldots, n \tag{7.3}$$

$$\Delta Q_i = Q_{Gi} - Q_{Di} - V_i \sum_j V_j (G_{ij} \sin\theta_{ij} - B_{ij} \cos\theta_{ij}) \quad i = 1, \ldots, m \tag{7.4}$$

where P_{Gi} and P_{Di} are calculated using Eqs. (7.1)–(7.2), which are frequency-dependent. Q_{Gi} represents the reactive power output of generator at bus i and Q_{Di} corresponds to the reactive power of load at bus i, which are frequency-independent. V_i denotes the voltage magnitude at bus i. $\theta_{ij} = \theta_i - \theta_j$ represents the phase angle difference between buses i and j. G_{ij} and B_{ij} are real and imaginary elements in the bus admittance matrix.

The mismatches ΔP_i and ΔQ_i collectively form two vectors: $\Delta \mathbf{P}_{n\times1}$ for all buses and $\Delta \mathbf{Q}_{m\times1}$ for PQ buses. There are a total of n+m unknown variables. These include n-1 voltage angles, m voltage magnitudes, and the frequency deviation f_d. The N-R (Newton–Raphson) method can be used to solve (7.5).

$$\begin{bmatrix} \Delta \mathbf{P}_{n\times1} \\ \Delta \mathbf{Q}_{m\times1} \end{bmatrix} = \begin{bmatrix} \Delta \mathbf{P}_{(n-1)\times1} \\ \Delta P_n \\ \Delta \mathbf{Q}_{m\times1} \end{bmatrix} = \mathbf{J} \begin{bmatrix} \Delta \boldsymbol{\theta}_{(n-1)\times1} \\ \Delta f_d \\ \mathbf{V}_{m\times m}^{-1} \Delta \mathbf{V}_{m\times1} \end{bmatrix} \tag{7.5}$$

$$\text{where} \quad \mathbf{J} = \begin{bmatrix} \mathbf{J}1_{(n-1)\times(n-1)} & \frac{\partial \Delta P}{\partial f_d}{}_{(n-1)\times1} & \mathbf{J}2_{(n-1)\times m} \\ \frac{\partial \Delta P_n}{\partial \boldsymbol{\theta}}{}_{1\times(n-1)} & \frac{\partial \Delta P_n}{\partial f_d} & \mathbf{N}_{1\times m} \\ \mathbf{J}3_{m\times(n-1)} & \mathbf{0} & \mathbf{J}4_{m\times m} \end{bmatrix}.$$

ΔP_n represents the mismatch of active power at the swing bus. Unlike in conventional power flow calculations, this mismatch will be eliminated. $\Delta \mathbf{P}_{(n-1)\times1}$ encompasses the mismatches of active power for all other buses. $\Delta \boldsymbol{\theta}_{(n-1)\times1}$ is a vector that holds angle corrections for all buses except the swing bus. Δf_d denotes the correction for system frequency deviation. $\mathbf{V}_{m\times m}^{-1}$ is a diagonal matrix composed of the reciprocals of V_i's of the m PQ buses. $\Delta \mathbf{V}_{m\times1}$ is a vector containing corrections for V_i's for PQ buses.

The Jacobian matrix $\mathbf{J}$ is an $(n+m)$-dimensional square matrix containing partial derivatives of the active and reactive power injections with respect to voltage angles, magnitudes, and f_d. The elements of $\mathbf{J_1}$, $\mathbf{J_2}$, $\mathbf{J_3}$, and $\mathbf{J_4}$ in the i-th row and j-th column are $\frac{\partial P_i}{\partial \theta_j}$, $\frac{\partial P_i}{\partial V_j} V_j$, $\frac{\partial Q_i}{\partial \theta_j}$, and $\frac{\partial Q_i}{\partial V_j} V_j$, i.e., the same as the corresponding elements in the Jacobian matrix of conventional power flow model. Let the bus angle θ_n of the swing bus be zero. The other elements of **J** are:

The Jacobian matrix **J** is a square matrix with dimensions $(n + m)$. It contains the partial derivatives of active and reactive power injections with respect to voltage angles, magnitudes, and the frequency deviation. The i-th row and j-th column in submatrices $\mathbf{J_1}$, $\mathbf{J_2}$, $\mathbf{J_3}$, and $\mathbf{J_4}$ is $\frac{\partial P_i}{\partial \theta_j}$, $\frac{\partial P_i}{\partial V_j}V_j$, $\frac{\partial Q_i}{\partial \theta_j}$, and $\frac{\partial Q_i}{\partial V_j}V_j$, respectively. They are analogous to those corresponding elements in the Jacobian matrix of the conventional power flow model. Additionally, it is assumed that the bus angle θ_n of the swing bus is 0. Other elements within **J** can be described as follows:

$$\left[\frac{\partial \Delta \mathbf{P}}{\partial f_d}\right]_i = \frac{\partial \Delta P_i}{\partial f_d} = \begin{cases} -P_{D0i}D_i & \text{for PQ buses} \\ -(P_{D0i}D_i + \frac{1}{R_i}) & \text{for PV and swing buses} \end{cases} \quad i=1,\ldots,n \quad (7.6)$$

$$\left[\frac{\partial \Delta P_n}{\partial \boldsymbol{\theta}}\right]_j = \frac{\partial \Delta P_n}{\partial \theta_j} = -V_n V_j (G_{nj} \sin\theta_j + B_{nj}\cos\theta_j) \quad j = 1,\ldots,n-1 \quad (7.7)$$

$$[\mathbf{N}]_j = \frac{\partial \Delta P_n}{\partial V_j} V_j = -V_n V_j (G_{nj}\cos\theta_j - B_{nj}\sin\theta_j) \quad j = 1,\ldots,m \quad (7.8)$$

The DLF model described by (7.1)–(7.8) can be solved by the N-R method. The computational load does not significantly increase compared to solving the conventional power flow model. This is because only one additional unknown variable and one more equation are introduced. For solving the DLF model, the limit of active power generation is considered, and the constraint $P_{Gi,min} \le P_{G0i} - \frac{f_d}{R_i} \le P_{Gi,max}$ would be checked during each iteration of the N-R method. If this constraint is violated, the value of P_{Gi} would be fixed at the limit to ensure it does not go beyond its allowed range.

In (7.5), the Jacobian matrix **J** in the DLF model shares a similar sparsity characteristic with that of the conventional power flow model. It has more non-zero elements than that of the conventional power flow model due to the inclusion of frequency deviation f_d. But it has at most $2n + m - 1$ more non-zero elements. The total number of elements in the **J** with the DLF model is $(n + m)^2$, and for the conventional power flow, it is $(n+m-1)^2$. The ratio of $2n+m-1$ to $(n+m-1)^2$ or $(n + m)^2$ is very small for a large system. Take the NPCC 140-bus system as the example, and the ratio $(2n+m-1)/(n+m-1)^2$ is equal to 0.007. It indicates that the **J** with the DLF model is quite sparse.

A few remarks are summarized here:

(1) For the steady-state simulation of cascading outages, AGC is typically disabled. Therefore, AGC or secondary frequency regulation is not taken into account as part of the SSCOF approach.
(2) Reactive power loads tend to be less affected by variations in frequency compared to active power loads. It is commonly assumed that reactive power

loads are frequency-independent [47–49]. This same assumption holds true in the SSCOF approach as well.

(3) In a conventional power flow model, buses are categorized into PQ buses, PV buses, and swing buses. The DLF model may inherit those bus types [40], which indicate the quantities that are basically unchanged. For instance, P_{D0i} in (7.1) and P_{G0i} in (7.2), respectively, correspond to "P" components in PQ and PV buses. In fact, P_{D0i} and P_{G0i} slightly vary with frequency deviation around certain constant values. Strictly speaking, PV and PQ buses in a DLF model only maintain constant voltage magnitudes and reactive power injections. Finally, only one swing bus is required for the DLF model, primarily serving as a reference bus for voltage angles.

7.2.2 AC Optimal Power Flow Model Considering Frequency

The calculation with the DLF model using (7.1)–(7.8) could be divergent under the simulation of cascading outages. This indicates a highly stressed system condition or even collapse. The remedial control actions including load shedding and generation redispatch can be employed to tackle these conditions. An AC optimal power flow model considering frequency deviation (AC-OPFf) including the objective function and constraints is presented by (7.9)–(7.10) to model a centralized remedial control action. As a comparison, a conventional AC-OPF model described by (7.11)–(7.12) is also presented.

In the AC-OPFf model, the main objective is to preserve the largest remaining active power load after remedial control measures are applied. To determine how crucial the load at bus i is, a weighting factor λ_i is used. The control variables include P_{Gi}, P_{Di}, Q_{Gi}, Q_{Di}, V_i, θ_i, and f_d. Among them, P_{Gi} and P_{Di} correspond to Eqs. (7.1) and (7.2). The first two constraints involve power flow equations (7.10a) and (7.10b). The other constraints are related to power generation (7.10c) and (7.10d), bus voltage magnitudes (7.10e) and (7.10f), loads (7.10g) and (7.10h), branch flows (7.10i), maintaining a constant power factor (1.10j), and frequency deviation (7.10k). It is important to note that in the AC-OPFf model, one equality constraint (7.10a) and three inequality constraints (7.10c), (7.10g), and (7.10k) involve f_d. As a result, the final calculated f_d may not necessarily reach to its upper or lower limit specified in constraint (7.10k) if any limit set in constraints (7.10c) or (7.10g) is reached.

In the AC-OPF model, the primary goal (7.11) is the same as the AC-OPFf model. The control variables encompass P_{Gi}, P_{Di}, Q_{Gi}, Q_{Di}, V_i, and θ_i. The constraints described in (7.12a)–(7.12j) are quite similar to those seen in the AC-OPFf model, with one notable difference: the AC-OPF model does not take into account frequency deviation.

From what has been described above, the AC-OPFf model is more general by taking frequency deviation f_d into account when compared to the AC-OPF model. In essence, the AC-OPF model can be seen as a specific instance of the AC-OPFf model when $f_d = 0$.

$$\min \quad -\sum_i \lambda_i P_{Di} \tag{7.9}$$

Subject to

$$\Delta P_i = P_{Gi} - P_{Di} - V_i \sum_j V_j (G_{ij} \cos\theta_{ij} + B_{ij} \sin\theta_{ij}) = 0 \tag{7.10a}$$

$$\Delta Q_i = Q_{Gi} - Q_{Di} - V_i \sum_j V_j (G_{ij} \sin\theta_{ij} - B_{ij} \cos\theta_{ij}) = 0 \tag{7.10b}$$

$$P_{Gi,min} \le P_{Gi} \le P_{Gi,max} \tag{7.10c}$$

$$Q_{Gi,min} \le Q_{Gi} \le Q_{Gi,max} \tag{7.10d}$$

$$V_{i,min} \le V_i \le V_{i,max} \tag{7.10e}$$

$$-\frac{\pi}{2} \le \theta_i \le \frac{\pi}{2} \tag{7.10f}$$

$$0 \le P_{Di} \le P_{D0i} \tag{7.10g}$$

$$0 \le Q_{Di} \le Q_{D0i} \tag{7.10h}$$

$$|S_{ij}| \le S_{ij,max} \tag{7.10i}$$

$$Q_{Di} P_{D0i} = \frac{P_{Di} Q_{D0i}}{(1 + D_i f_d)} \tag{7.10j}$$

$$f_{dmin} \le f_d \le f_{dmax} \tag{7.10k}$$

$$\min \quad -\sum_i \lambda_i P_{Di} \tag{7.11}$$

Subject to

$$\Delta P_i = P_{Gi} - P_{Di} - V_i \sum_j V_j (G_{ij} \cos\theta_{ij} + B_{ij} \sin\theta_{ij}) = 0 \tag{7.12a}$$

$$\Delta Q_i = Q_{Gi} - Q_{Di} - V_i \sum_j V_j (G_{ij} \sin\theta_{ij} - B_{ij} \cos\theta_{ij}) = 0 \tag{7.12b}$$

$$P_{Gi,min} \le P_{Gi} \le P_{Gi,max} \tag{7.12c}$$

$$Q_{Gi,min} \le Q_{Gi} \le Q_{Gi,max} \tag{7.12d}$$

$$V_{i,min} \leq V_i \leq V_{i,max} \tag{7.12e}$$

$$-\frac{\pi}{2} \leq \theta_i \leq \frac{\pi}{2} \tag{7.12f}$$

$$0 \leq P_{Di} \leq P_{D0i} \tag{7.12g}$$

$$0 \leq Q_{Di} \leq Q_{D0i} \tag{7.12h}$$

$$|S_{ij}| \leq S_{ij,max} \tag{7.12i}$$

$$Q_{Di} P_{D0i} = P_{Di} Q_{D0i} \tag{7.12j}$$

.

The optimality of the final solution in the AC-OPF or AC-OPFf depends on the choice of optimization algorithm, and the level of deviation from the true global optimum is considered acceptable. However, when it comes to simulating cascading outages, the AC-OPF or AC-OPFf model serves a different purpose. It is not about finding the absolute best control strategy. Instead, the primary goal is to replicate the remedial actions taken by a central control room, akin to the OPF module in OPA models [28–30]. In practical terms, the AC-OPF or AC-OPFf model is used to identify a new feasible power flow solution when a system collapse occurs. These models can be easily replaced by the central control room's own remedial control strategies to address critical situations if the SSCOF approach is applied in electric utilities or Independent System Operators (ISOs).

7.2.3 *Under-Frequency Load Shedding Scheme in the SSCOF Approach*

Since frequency deviation is considered in the DLF model, it enables the UFLS scheme to be modeled under cascading outages when there is a significant and unacceptable drop in frequency. Typically, such a scheme is set to shed around 25–30% of the total system load in predefined steps within each reliability coordinator region [1]. Furthermore, if the frequency decline persists, more load is progressively shed.

In the SSCOF approach, an UFLS scheme as outlined in Table 7.1 is adopted. This scheme is based on the North American Electric Reliability Corporation (NERC) UFLS reliability standard "PRC-006 NPCC" for the NPCC region, as specified in [57].

The percentage of active power load to be shed is determined based on the frequency thresholds listed in Table 7.1 when the UFLS scheme is activated. Typically, the amount of active power load shed $\Delta P_{Di,\text{UFLS}}$ also leads to a reduction in reactive power load $\Delta Q_{Di,\text{UFLS}}$. The corresponding change in reactive power load is determined by (7.13). This equation assumes that a constant power factor is maintained throughout the shedding process.

Table 7.1 UFLS scheme for different load sizes [56]

100 MW or more load		50 MW or more and less than 100 MW		25 MW or more and less than 50 MW	
f_t(Hz)	L_p(%)	f_t(Hz)	L_p(%)	f_t(Hz)	L_p(%)
59.5	6.5	59.5	14	59.5	28
59.3	6.5	59.1	14	–	–
59.1	6.5	–	–	–	–
58.9	6.5	–	–	–	–

Note: f_t and L_p represent frequency threshold and percentage of load shed

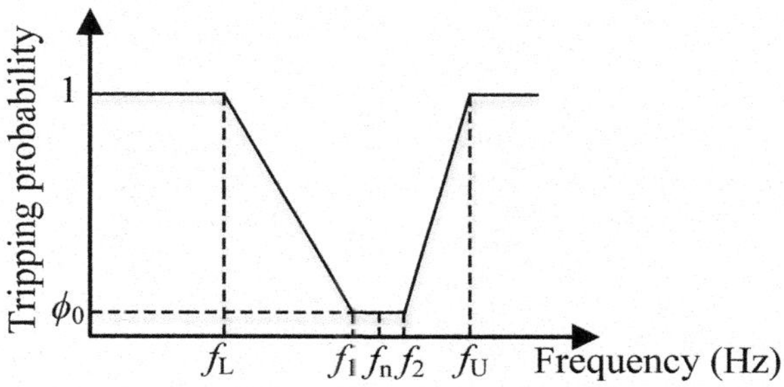

Fig. 7.1 Relationship between generator tripping probability and frequency [56]

$$\frac{\Delta Q_{Di,\text{UFLS}}}{\Delta P_{Di,\text{UFLS}}} = \frac{Q_{Di}}{P_{Di}} \tag{7.13}$$

7.2.4 *Protections of Generator Frequency and Transmission Line*

When protective actions related to generators and transmission lines are considered, they may bring more uncertainties and challenges to the system operation, as mentioned in [58, 59]. Thus it is valuable to model them in the simulation of cascading outages. Thanks to the DLF model, it is able to model certain protective actions related to frequency. Specifically, generator frequency protection and transmission line protection could be embedded into the SSCOF approach. This allows the study for the influence of these protections on cascading outages.

As shown in Fig. 7.1, the frequency range of a generator can be categorized into three types based on the characteristics of its turbine and power plant auxiliaries [59]: (1) The normal operation range is bounded by f_1 and f_2; (2) Two restricted time operation ranges are outside the normal range, and they are bounded by a lower limit f_L and an upper limit f_U, i.e., intervals $[f_L, f_1]$ and $[f_2, f_U]$; (3) There are two prohibited ranges, one is lower than f_L and the other is higher than f_U.

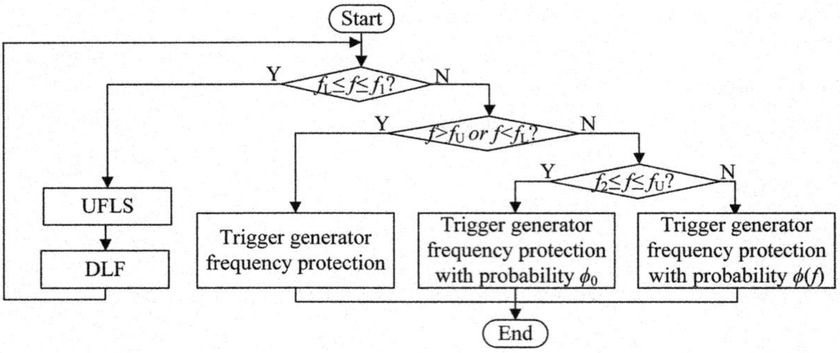

Fig. 7.2 Module of UFLS and generator frequency protection [56]

The probability of a generator i being tripped, denoted as $\phi(f)$, is determined by its frequency according to (7.14). In this equation, ϕ_0 represents the base probability of an unexpected action occurring due to generator frequency protection.

$$\phi(f)\begin{cases} 1 & f < f_L \text{or} f > f_U \\ \frac{(\phi_0-1)f+f_1-\phi_0 f_1}{f-f_L} & f_L \leq f \leq f_1 \\ \phi_0 & f_1 < f < f_2 \\ \frac{(1-\phi_0)f+\phi_0 f_U-f_2}{f_U-f_2} & f_2 \leq f \leq f_U \end{cases}. \tag{7.14}$$

Both the UFLS scheme and generator frequency protection are modeled in the SSCOF approach. Their relay actions have different response times. For instance, the UFLS scheme typically acts with a time delay of 0.1 second in the Eastern Interconnection, as noted in [57]. In contrast, the time delays for generator frequency protection can vary widely, ranging from 0.1 seconds to several hundred seconds, depending on the severity of the frequency deviation. To manage these different timeframes effectively, the SSCOF approach employs a specialized module, as depicted in Fig. 7.2. This module coordinates the actions of the UFLS scheme and generator frequency protection. It prioritizes UFLS, activating it whenever the conditions for triggering it are met. Only if the UFLS is not triggered, then the generator frequency protection could be activated, and this occurs with a probability denoted as $\phi(f)$ defined by (7.14). This coordination module is integrated into the SSCOF approach, as shown in Fig. 7.3, and is represented by the block labeled "UFLS and generator frequency protection module."

As cascading outage propagates, a transmission line i–j is overloaded when its apparent power, denoted as S_{ij}, exceeds its transmission capacity, represented as $S_{ij,max}$. When a line becomes overloaded, it would be tripped with a probability denoted as β. And all the other lines are assumed to be tripped with a probability of $\epsilon \times |\frac{S_{ij}}{S_{ij,max}}|^{\tau}$. ϵ serves as a base probability for any unwanted protection operation. Importantly, it should increase as the line's loading ratio grows, as explained in [29].

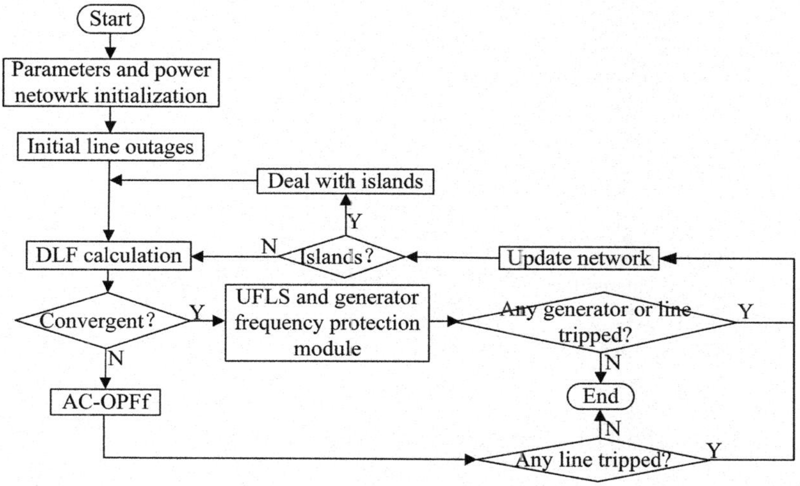

Fig. 7.3 Simulation procedure for the SSCOF approach [56]

Note that the SSCOF approach relies on a steady-state power flow model, which means it does not provide explicit information about time evolution. Unlike time-domain simulations that track the tripping sequence and the dynamic behavior of generators in detail, the described approach simplifies this aspect. In the SSCOF approach, if a generator is tripped, it would not be brought back online until the simulation is completed.

7.2.5 Simulation Procedure of the SSCOF Approach

Two different approaches for simulating cascading outages are illustrated here. The SSCOF approach can be seen in Fig. 7.3. The conventional approach is shown in Fig. 7.4. The SSCOF approach uses the DLF and AC-OPFf models and takes into account the UFLS scheme and generator frequency protection. The conventional approach does not consider the UFLS scheme and generator frequency protection, and it replaces the DLF model and the AC-OPFf model with the conventional power flow and AC-OPF models. In both Figs. 7.3 and 7.4, there is a step "parameters and power network initialization," and this step performs conventional power flow calculation to establish a base operating condition where the system operates at the nominal frequency of 60Hz. This initial condition serves as a starting point before any line outages are introduced in the subsequent steps.

A practical method suggested in [60] is to designate the bus with the highest active power generation limit in each island as the swing bus for that island in cases where the system separates into islands during the propagation of cascading outages.

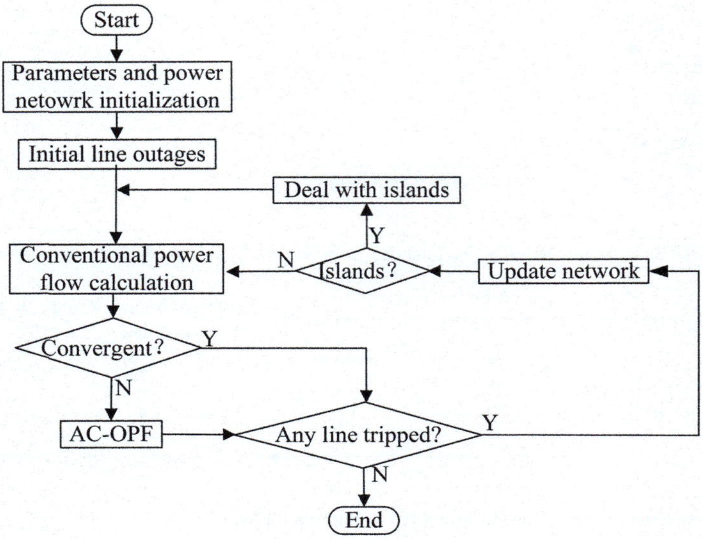

Fig. 7.4 Simulation procedure for a conventional approach [56]

In the context of the DLF model, it is essential to clarify the role of the swing bus within each island. Unlike a conventional power flow model that a swing bus is used to compensate active power imbalance, in the DLF model, the swing bus does not aim to eliminate these imbalances by itself. Instead, both the swing bus and other PV buses work together to eliminate active power imbalances based on their power-frequency characteristics. Essentially, the swing bus serves as a reference bus for voltage angles within each island. In addition, the AC-OPFf or AC-OPF model would be used to search for a new operating point if a significant imbalance in active power occurs and leads to power flow divergence.

7.3 Case Studies and Analysis

In this section, the frequency calculated by the DLF model is first benchmarked with the steady-state frequency extracted from the time-domain simulation on Kundur's two-area, 4-machine power system [61]. Subsequently, the verification is extended to two larger power systems: the IEEE 39-bus system and the NPCC 48-machine, 140-bus system [62, 63]. The comparison of the cascading outages simulated by the SSCOF approach and the conventional approach, all implemented in MATLAB, is performed to verify the effectiveness of the SSCOF approach. These time-domain simulations are carried out using TSAT, a software tool developed by Powertech Labs.

Table 7.2 Turbine–Governor model "TGOV1" [56]

Parameter	Value	Unit
Speed regulation factor R	0.05	pu
Turbine damping coefficient D_t	0	pu
Main steam control valve max limit V_{max}	1	pu
Main steam control valve min limit V_{min}	0.3	pu
Governor time constant T_1	0.5	s
Steam chest time constant T_2	1.0	s
Reheater time constant T_3	1.0	s

In the DLF model, R_i in (7.2) is set as 0.0056 pu for all generators. This value is based on the system base of 100 MVA, following the conversion from the value of R listed in Table 7.2, which is based on the generator base. For all loads in (7.1), D_i is set as 1 pu. In the AC-OPFf and AC-OPF models, λ_i is set as 1 for all loads. The maximum frequency deviation in the AC-OPFf model is set as 0.5 Hz in constraint (7.10k). This means f_{dmin} is -0.5 Hz, and f_{dmax} is 0.5 Hz. The threshold that triggers the UFLS scheme is set at 59.5 Hz.

For the generator frequency protection, the following parameters in (7.14) are set: $\phi_0 = 0.002$; $f_L = 57$ Hz; $f_1 = 59.5$ Hz; $f_n = 60$ Hz; $f_2 = 60.5$ Hz; $f_U = 61.7$ Hz.

For the protection of transmission line, $\beta = 0.999$, $\epsilon = 0.001$, and $\tau = 10$, which are the same as the values used in [29].

For the time-domain simulation on the two-area system, the generators are modeled with the 2nd order classic model and the steam turbine–governor model "TGOV1" [64].

For the NPCC power system, 24 generators are modeled with the classic model. The other 24 generators are equipped with a detailed round rotor model "GENROU" and an exciter model "ESDC1A" with PSS/E v32 [65]. Additionally, the "TGOV1" governor model is modeled for each generator with the same parameters listed in Table 7.2. For the loads in the system, they are represented with frequency-dependent loads, specifically using the "IEELBL" in PSS/E v32. It is assumed the reactive powers of the loads to be constant throughout the simulation if the AC-OPFf model and the UFLS scheme are not triggered.

In the IEEE 39-bus system, the MATPOWER 6.0 toolbox provides the transmission capacity limit $S_{ij,max}$ for each transmission line. The transmission capacity limits for the lines in the NPCC system are set using the $N-1$ criterion to have the initial limits and then increase the limits by 20% to provide additional margins.

7.3.1 Case Study on the Two-Area System

In the two-area system, there are loads at buses 7 and 9. The steady-state frequencies from the DLF model and the time-domain simulation are compared for the following three scenarios of load variation:

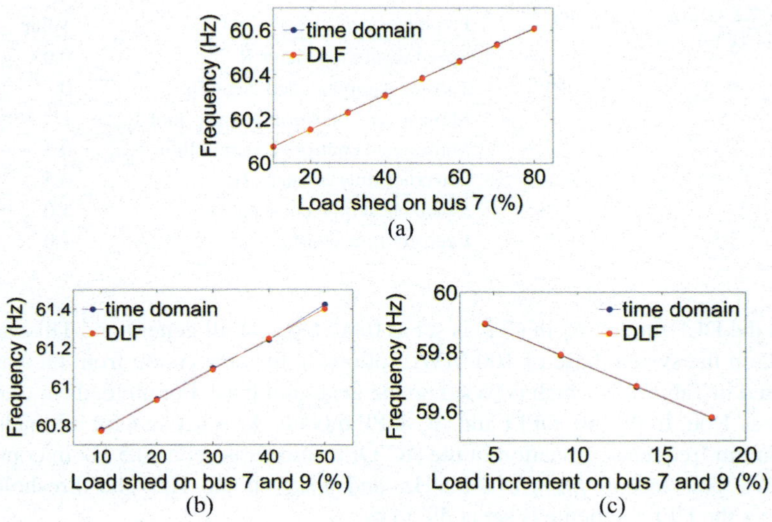

Fig. 7.5 Frequency variations from the DLF model and the time-domain simulation [56]

(a) Load shed at bus 7.
(b) Load shed at both bus 7 and bus 9.
(c) Load increase at both buses.

Figure 7.5 shows that the steady-state frequencies calculated from the DLF model match well with those from the time-domain simulation in all three scenarios. This verifies the accuracy for capturing the steady-state frequency by the DLF model.

7.3.2 Case Study on the IEEE 39-Bus System

Four sets of tests are conducted using the IEEE 39-bus system, each with a specific purpose:

(1) The first set of tests is to verify the accuracy of the steady-state frequency calculated by the DLF model and the convergence characteristic with the DLF model. Scenarios 1 and 2 are used.
(2) The second set of tests is to study the UFLS scheme and generator frequency protection module. The influence of SPFCs of loads on frequency is also investigated. Scenario 3 is used.
(3) The third set of tests is to investigate the influence of active power generation limits on frequency. Scenario 4 is intentionally designed to make some generators hit their generation limits after line outages.

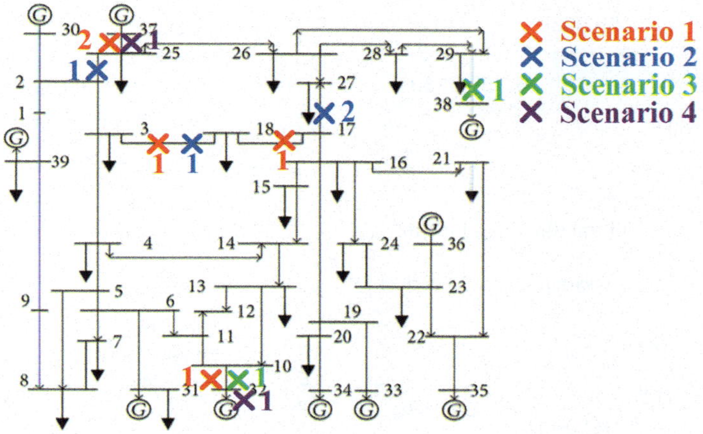

Fig. 7.6 Scenarios 1, 2, 3, and 4 on the IEEE 39-bus system (outages are marked with crosses and labeled with stages) [56]

(4) The last set of tests is to statistically compare the scenarios of cascading outages generated from the SSCOF approach as shown in Fig. 7.3 and the conventional approach as shown in Fig. 7.4. The scenarios of cascading outages that start from $N-2$ initial outages are considered.

Scenarios 1, 2, 3, and 4 are marked in different colors on the IEEE 39-bus system shown in Fig. 7.6.

7.3.2.1 Verification of Steady-State Frequency with the DLF Model

Two typical scenarios are selected. Scenario 1 contains line outages that do not split the system into separate parts, and Scenario 2 does lead to a separation for the system. Both of these scenarios bring large disruptions by tripping lines, which results in significant imbalance in active power and large variation in frequency. The steady-state frequencies after line outages are captured in both time-domain and DLF models first and then compared.

Scenario 1: Lines 10–32, 17–18, and 3–18 are tripped in stage 1 and line 25–37 is tripped in stage 2.

In the time-domain simulation, two stages of outages are designed to have 100 seconds waiting period to ensure that the frequency settles into a steady state. From Fig. 7.7, the frequencies from both the time-domain simulation and the DLF model for all stages can be seen. Note that the comparison is only for steady-state frequencies. At stage 2, generator 32 is tripped. At stage 3, generator 37 is tripped. The frequencies calculated by the DLF model in stages 2 and 3 closely match those obtained from the time-domain simulation. Any slight difference between them are

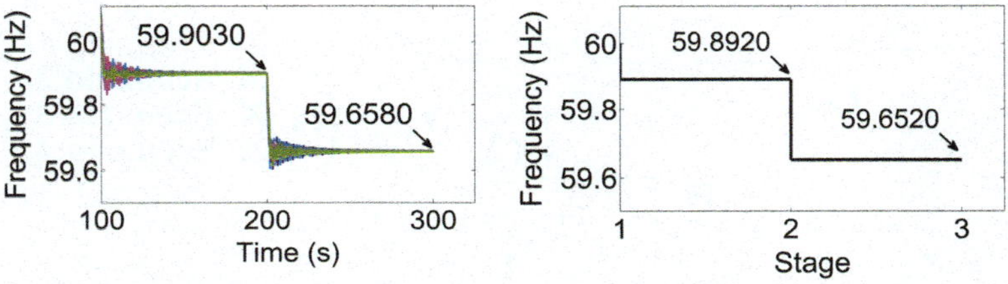

(a) From the time-domain (a) From the DLF model

Fig. 7.7 Frequency variations on Scenario 1 [56]

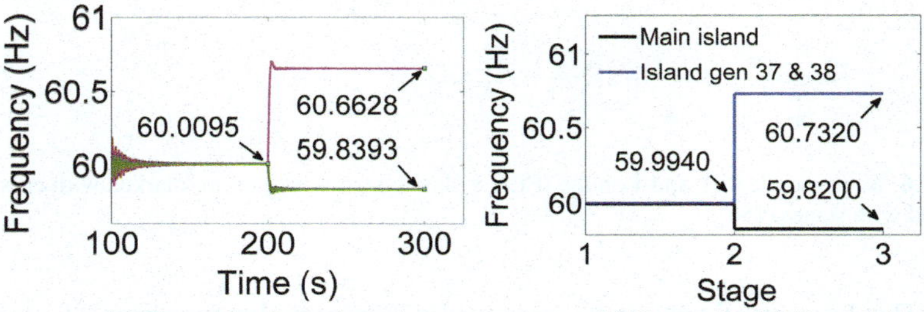

(a) From the time-domain (b) From the DLF model

Fig. 7.8 Frequency variations on Scenario 2 [56]

because the power flow results are not exactly identical. This verifies the accuracy of frequency calculated by the DLF model.

Scenario 2: Lines 2–25, 3–18 are tripped in stage 1 and line 17–27 is tripped in stage 2.

After the trip of lines in stage 2, the system separates into two parts. One is with 8 generators as the main island, and the other contains generators 37 and 38. The calculated steady-state frequencies by the DLF model in stages 1 and 2 are compared to those observed in simulations. They are very close as shown in Fig. 7.8.

In the DLF model, the convergence of the N-R method for Scenarios 1 and 2 is investigated. From Fig. 7.9, the variations of mismatches of equations with iterations for the outages in two stages can be seen. The mismatches drop below a small tolerance of 10^{-9} (pu), and the power flows converge after just 3 or 4 iterations. The largest value among all ΔP_is and ΔQ_is at each iteration is used here and plotted on the y-axis in Fig. 7.9.

If the initial guess is sufficiently close to the solution, the N-R method could make the iteration converge with a quadratic rate. For a series x_k converging to x^* with a quadratic rate, Eq. (7.15) holds true. This is discussed in [66, 67].

$$\frac{|x_{k+1} - x^*|}{|x_k - x^*|^2} \leq M \quad if \quad M > \frac{|h^{''}(x^*)|}{2|h^{'}(x^*)|}. \tag{7.15}$$

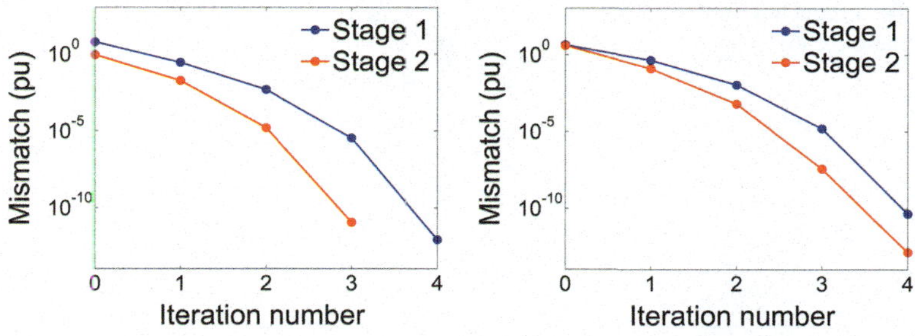

(a) From Scenario 1 (b) From Scenario 2

Fig. 7.9 Convergence of the N-R method with the DLF model [56]

Table 7.3 Convergence rate estimation of the N-R method with the DLF model [56]

Scenario	Stage	ρ_k		
		$k = 0$	$k = 1$	$k = 2$
Scenario 1	Stage 1	0.040	0.325	0.380
	Stage 2	0.829	1.353	0.000
Scenario 2	Stage 1	0.027	0.468	0.394
	Stage 2	0.460	0.341	0.413

For the DLF model or a conventional power flow model, the function h in (7.15) represents (7.3) and (7.4), and $h^{'}$ and $h^{''}$ represent the first and second derivatives of h. The power flow solution, denoted as x^*, is selected from the last iteration of the DLF model. The value of $\rho_k = \frac{|x_{k+1}-x^*|}{|x_k-x^*|^2}$ at each iteration step k of the N-R method on the DLF model is listed in Table 7.3. The results show that the values of ρ_k for various k values are all roughly in the same scale. This tells that the N-R method to solve the DLF model has a quadratic convergence rate.

7.3.2.2 UFLS Scheme and Generator Frequency Protection Module

In the SSCOF approach to simulate cascading outages, the UFLS scheme and generator frequency protection module is activated in some scenarios. Scenario 3 with the trip of lines 32–10 and 38–29 is investigated.

The frequency f drops to 59.39 Hz after the two lines are tripped. The frequency is within the range of 57 Hz and 59.5 Hz, and the UFLS is triggered to bring the frequency up to 59.55 Hz after 384.07 MW load shed. Since this new frequency falls within the range of 59.5 Hz to 60.5 Hz, the generator frequency protections are activated with a low probability of 0.002. Ultimately, in this scenario, none of the generators is tripped.

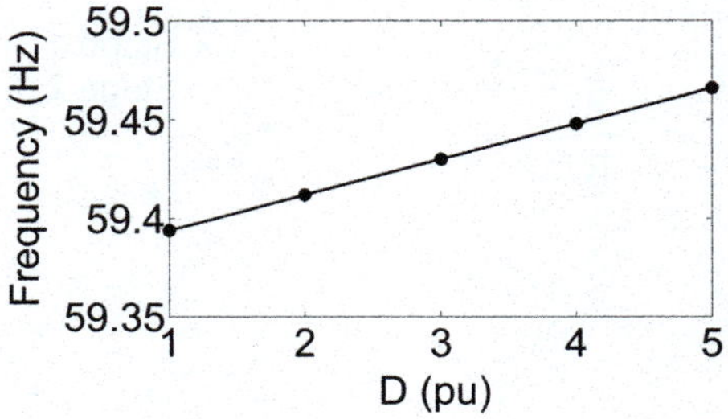

Fig. 7.10 Frequency vs. D on Scenario 3 [56]

Table 7.4 Active power outputs of generators with and without limits [56]

		Without limits		With limits	
Generator	P_{Gmax}(MW)	P_G(MW)	f_d(Hz)	P_G(MW)	f_d(Hz)
30	350.00	395.66	−0.49	350.00	−1.29
31	1145.60	507.88		749.74	
33	732.00	777.66		732.00	
34	608.00	653.66		608.00	
35	750.00	795.66		750.00	
36	660.00	705.66		660.00	
38	930.00	975.66		930.00	
39	1100.00	1145.66		1100.00	
32	750.00	–	–	–	–
37	640.00	–	–	–	–

The impact of SPFCs of loads on frequency is also examined in Scenario 3. The result in Fig. 7.10 demonstrates that when the *D* parameter is bigger, the frequency deviation becomes smaller.

7.3.2.3 Study of Active Power Generation Limits on Frequency Deviation

In the DLF model, the active power generation limits are considered. How these limits affect the frequency is investigated in Scenario 4. Table 7.4 shows a comparison of the power outputs of generators and the system's frequency after lines 10–32 and 25–37 are tripped. The results with and without considering the limits of active power generation are listed.

It is found that the frequency drops by 1.29 Hz in Table 7.4 when active power generation limits are considered, which is larger than the drop of 0.49 Hz when active power generation limits are not considered. When some generators hit their generation limits, they are not able to provide more power outputs, while others, like generator 31, which still has room to produce more output, continue to do so. If it does not account for active power generation limits, the frequency deviation might be underestimated.

7.3.2.4 Statistical Comparison Between the SSCOF and the Conventional Approaches

Some measures can be used to quantify the severity of cascading outage for those scenarios simulated by the two approaches as shown in Figs. 7.3 and 7.4. Here the number of line outages and the amount of load shed are used. The scenarios of cascading outages start from all $N-2$ initial outages since the system is $N-1$ secure. There are a total of 1,035 pairs of cascading outage scenarios. The ones that do not spread beyond the initial outages are excluded. A total of $K = 1{,}028$ scenarios for each approach are used here for comparison.

Two indices shown as below are used to compare the scenarios of cascading outages generated by the SSCOF approach and the conventional approach.

$$R_{i,\text{path}} = \frac{|A_i|}{|B_i|}, \quad R_{i,\text{load}} = \frac{\text{Load}_{A_i}}{\text{Load}_{B_i}}, \quad i = 1, 2, \ldots K \tag{7.16}$$

where A_i and Load_{A_i} represent the set of line outages and the amount of load shed in a scenario i of cascading outages, using the conventional approach; B_i and Load_{B_i} represent the set of line outages and the amount of load shed in the same scenario i, but using the SSCOF approach; $|\cdot|$ counts the number of elements in a set.

In Fig. 7.11a, we can see that in most scenarios, $R_{i,\text{path}} < 1$. This means that line outages tend to spread more for the SSCOF approach compared to the conventional approach. Figure 7.11b and Table 7.5 show that the SSCOF approach results in more load shed than the conventional approach. This happens because the SSCOF approach takes into account the UFLS scheme and generator frequency protection. It is found that the conventional approach underestimates the spread of outage propagation because it does not consider frequency variation and frequency-related actions. The SSCOF approach does a better job of capturing how outages propagate and load loss due to frequency-related characteristics.

In addition, the similarity of the sets of line outages between the two approaches is quantified to see how closely their simulated outage paths match. A measure

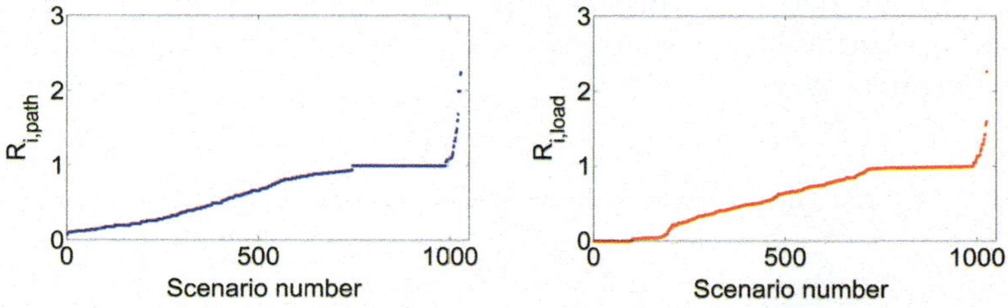

(a) Length of cascading outage (b) Load shed of cascading outage

Fig. 7.11 Ratio between the SSCOF and the conventional approaches (scenario numbers in **(a)** and **(b)** are, respectively, ordered by values of the ratio from small to large) [56]

called "average overlap ratio," denoted as R_{avg} [35], is used to compare the sets of line outages in cascading outages simulated by two approaches:

$$R_{\text{avg}} = \frac{1}{K}\sum_{i=1}^{K} R_{i,\text{overlap}}, \quad R_{i,\text{overlap}} = \frac{|A_i \cap B_i|}{|A_i \cup B_i|}. \tag{7.17}$$

The value of R_{avg} is 0.61, which suggests that the two approaches have noticeably different characteristics when it comes to outage propagation.

Out of 1028 scenarios, the UFLS is activated in 255 of them. On average, the amount of load shed by the UFLS scheme is 282.23MW during these activations. The highest load shed is 599.60MW, while the lowest is 40.85MW. The study highlights an important point: if the effect of frequency variation and the UFLS scheme are not considered in the simulation, it might underestimate the risk of cascading outages.

7.3.3 Case Study on the NPCC 140-Bus System

7.3.3.1 Verification of Steady-State Frequency by the DLF Model

The accuracy of the steady-state frequency calculated by the DLF model is tested on the NPCC 48-machine, 140-bus power system.

Two scenarios (Scenario 5 and Scenario 6) that cause significant imbalances of active power are selected. These two scenarios could cause over- and under-frequency conditions. Note that the AC-OPFf model, the UFLS scheme, and generator frequency protection module are disabled in these two scenarios since the main purpose is to verify the calculated steady-state frequency.

In Scenario 5, there are two stages of outages as listed in Table 7.6. The steady-state frequencies calculated by the DLF model with those obtained from time-domain simulations are listed in Table 7.7. Scenario 6 has three stages of outages listed in Table 7.8, and the frequencies are compared in a similar way (Table 7.9). It is found the frequencies calculated by the DLF model closely match the benchmarking results, showing that the DLF model could accurately capture the steady-state frequency.

Table 7.5 Statistical comparison between the SSCOF and the conventional approaches for 1028 scenarios [56]

Approach	Average no. of line outages	Average amount of load shed (MW)
Conventional	7.76	1848.9
SSCOF	12.89	3376.5

Table 7.6 Propagation path of cascading outages in Scenario 5 [56]

Stage	Line outages
1	130–131, 131–133, 131–135, 131–139
2	124–128, 125–128, 126–128, 127–128, 128–130

Table 7.7 Comparison of steady-state frequencies for Scenario 5 [56]

	Frequency (Hz)	
Approach	Stage 1	Stage 2
DLF	60.137	60.244
Time domain	60.147	60.268

Table 7.8 Propagation path of cascading outages in Scenario 6 [56]

Stage	Line outages
1	85–86, 85–105
2	78–79
3	131–133, 132–133, 133–135

Table 7.9 Comparison of steady-state frequencies for Scenario 6 [56]

	Frequency (Hz)		
Approach	Stage 1	Stage 2	Stage 3
DLF	59.779	59.622	59.498
Time domain	59.802	59.649	59.532

Note: Three generators are tripped one by one after stages 1, 2, and 3

7.3.3.2 Detailed Comparison Between the SSCOF and the Conventional Approaches

In this section, detailed comparisons are performed between the SSCOF approach and the conventional one in two more scenarios, numbered 7 and 8. From Figs. 7.12 and 7.13, the path of outages, the amount of load shed, and frequency variation can be seen.

In Scenario 7, both approaches show the same outage propagation paths following the same initial outages seen from Fig. 7.12. In the early stage, the frequency deviation is not large. By stage 2, the frequency only drops by 0.033 Hz. After the line outages in stage 3, power flow calculations from the DLF model and the conventional power flow model diverge, and then AC-OPFf and AC-OPF are triggered to find a new feasible operating condition. The value of system frequency reaches to 60.498Hz after AC-OPFf, and the UFLS scheme is not triggered since the frequency is within the normal range. After these calculations, no more overloaded lines and the outages stop under both approaches. The comparison for Scenario 7 shows that the SSCOF approach and the conventional approach behave similarly when there is only a small frequency variation. However, it is more practical for system operators to monitor the frequency variation using the SSCOF approach.

In Scenario 8, the propagation paths of cascading outages from two approaches triggered by the same initial outages are compared. At the first stage, both approaches show the same outage path, but they start to differ at stage 2 (Fig. 7.13).

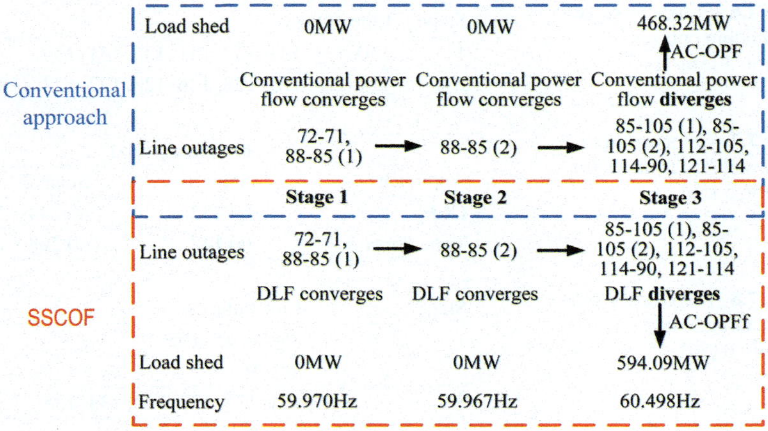

Fig. 7.12 Comparison between the SSCOF and the conventional approaches on Scenario 7 [56]

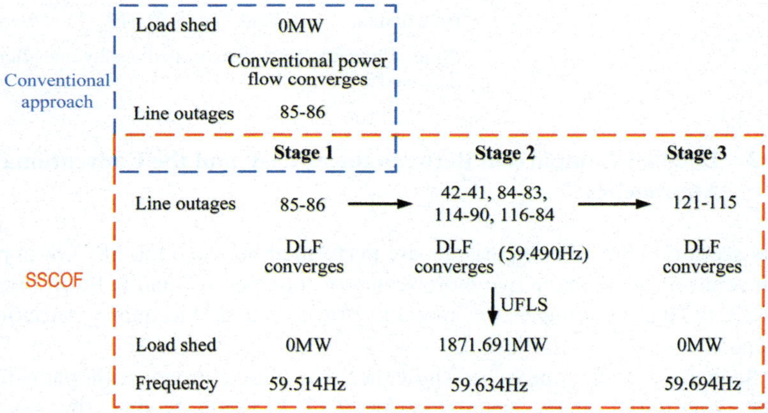

Fig. 7.13 Comparison between the SSCOF and the conventional approaches on Scenario 8 [56]

With the SSCOF approach, the outages at stage 2 drop the system frequency by 0.51 Hz. This triggers the UFLS scheme to shed some load and bring the frequency back to 59.634 Hz. After stage 3, the frequency remains at 59.694 Hz, and there are no more UFLS activations. The power flow profiles resulting from the outage propagation for these two approaches are quite close in the first stage and then become quite different as the outage propagates. This shows that the variation of system frequency during the propagation of cascading outages cannot be ignored, especially in the later stages. Ignoring the variation of frequency could lead to the underestimation of the impact of cascading outages.

Table 7.10 Statistical comparison between the SSCOF and the conventional approaches for 10,000 scenarios [56]

Approach	Average no. of line outages	Average load shed (MW)	Average load shed by UFLS (MW)
Conventional	8.79	222.45	0
SSCOF	14.27	985.02	146.97

Table 7.11 Comparison of time performance [56]

Number of scenarios	Conventional	SSCOF
10,000	14.50 hours	16.78 hours

7.3.3.3 Statistical Comparison Between the SSCOF and the Conventional Approaches

A large amount of simulations of cascading outages are conducted on the NPCC system using the SSCOF approach and the conventional one. Here "$N - 2$" contingency is considered as the initial outages.

In Table 7.10, the average results from 10,000 different scenarios for both approaches are listed. The average number of line outages, the average amount of load shed by remedial actions, and the average amount of load shed by the UFLS scheme are compared. The scenarios of cascading outages simulated from the SSCOF approach tend to spread more and are more severe than those generated from the conventional approach. This shows that taking into account frequency variations and frequency-related actions is very critical when simulating cascading outages.

The performance in terms of time cost for two different approaches on a regular desktop computer with an Intel Core i7-3770K 3.40GHz processor and 4GB of RAM. The total time costs for both approaches to handle the same number of scenarios are listed in Table 7.11. The SSCOF approach takes about 16% more time compared to the conventional approach. This happens because, in some scenarios with significant frequency deviations, the SSCOF approach has to simulate the cascading outages over more stages, which required more computational work for N-R computations.

7.4 Conclusion

In this chapter, a steady-state approach for simulating cascading outages in power systems called SSCOF is described. The SSCOF approach combines a DLF model and an AC-OPFf model and enables frequency deviation to be considered in the steady-state simulation of cascading outages. The SSCOF approach could accurately capture the variation of steady-state frequency and model load loss due to voltage collapse and frequency violations. From this perspective, the SSCOF approach works better for simulating real-world grid operations compared to the conventional steady-state approach that does not take into account the frequency deviation. Also, it highlights the importance for considering frequency variations in the simulation of cascading outages. Also, the SSCOF approach allows remedial actions and protections related to frequency to be modeled, like the UFLS scheme and generator frequency protection. The calculated steady-state frequency from the DLF model has been benchmarked with the results of time-domain simulation on both small and large systems. The comparison between the SSCOF approach and the conventional one has been performed and validated using detailed and statistical analyses. This helps prove the advantages of the SSCOF approach. In a short summary, the SSCOF approach can calculate the steady-state frequency accurately in the propagation of cascading outages. However, it might not fully account for the risk of frequency violations since the dynamic system behaviors in frequency are not captured. The future work will integrate the SSCOF approach and the time-domain simulation for a more efficient hybrid simulation approach for cascading outages.

References

1. U.S.-Canada Power System Outage Task Force, Final report on the August 14, 2003 blackout in the United States and Canada: Causes and Recommendations, Apr. 2004
2. A. Atputharajah, T.K. Saha, Power system blackouts–Literature review, in *Proc. 4th Int. Conf. Industrial and Information Syst.*, Sri Lanka, 2009
3. IEEE CAMS Task Force on Understanding, Prediction, Mitigation, Restoration of Cascading Failures, "Risk assessment of cascading outages: Methodologies and challenges. IEEE Trans. Power Syst. **27**(2), 631–641 (2011)
4. R. Baldick, B. Chowdhury, I. Dobson et al., Initial review of methods for cascading failure analysis in electric power transmission systems, in *IEEE PES General Meeting*, Pittsburgh, PA, USA, Jul. 2008
5. M. Papic, K. Bell, Y. Chen et al., Survey of tools for risk assessment of cascading outages, in *IEEE PES General Meeting*, Detroit, MI, USA, Jul. 2011.
6. B.A. Carreras, D.E. Newman, I. Dobson, North American blackout time series statistics and implications for blackout risk. IEEE Trans. Power Syst. **31**(6), 4406–4414 (2016)
7. J. Bialek et al., Benchmarking and validation of cascading failure analysis tools. IEEE Trans. Power Syst. **31**(6), 4887–4900 (2016)
8. E. Ciapessoni et al., Benchmarking quasi-steady state cascading outage analysis methodologies, in *Prob. Methods Applied to Power Syst.*, Boise, ID, USA, Jun. 2018
9. J. Qi, K. Sun, S. Mei, An interaction model for simulation and mitigation of cascading failures. IEEE Trans. Power Syst. **30**(2), 804–819 (2015)
10. C. Asavathiratham, S. Roy, B. Lesieutre, G. Verghese, The influence model. IEEE Control Syst. Mag. **21**(6), 52–64 (2001)
11. P. Hines, I. Dobson, E. Cotilla-Sanchez et al., "Dual graph" and "random chemistry" methods for cascading failure analysis, in *Proc. 46th Hawaii Intl. Conf. System Sciences*, Maui, HI, USA, Jan. 2013
12. P. Hines, I. Dobson, P. Rezaei, Cascading power outages propagate locally in an influence graph that is not the actual grid topology. IEEE Trans. Power Syst. **32**(2), 958–967 (2017)
13. W. Ju, K. Sun, J. Qi, Multi-layer interaction graph for analysis and mitigation of cascading outages. IEEE Trans. Emerg. Sel. Topics Circuits Syst. **7**(2), 239–249 (2017)

14. C. Chen, W. Ju, K. Sun, S. Ma, Mitigation of cascading outages using a dynamic interaction graph-based optimal power flow model. IEEE Access **7**, 168,637–168,648 (2019)
15. K. Zhou, I. Dboson, A Markovian influence graph formed from utility line outage data to mitigate large cascades. IEEE Trans. Power Syst. **35**(4), 3224–3235 (2020)
16. K. Sun, Y. Hou, W. Sun, J. Qi, *Power System Control Under Cascading Failures: Understanding, Mitigation, and System Restoration* (Wiley-IEEE Press, 2019)
17. U. Nakarmi, M. Rahnamay-Naeini, M.J. Hossain, M.A. Hasnat, Interaction graphs for reliability analysis of power grids: A survey. Preprint (2019). arXiv:1911.00475 [physics.soc-ph]
18. Z. Ma, C. Shen, F. Liu, S. Mei, Fast screening of vulnerable transmission lines in power grids: A PageRank-based approach. IEEE Trans. Smart Grid **10**(2), 1982–1991 (2019)
19. Y. Yang, T. Nishikawa, A.E. Motter, Vulnerability and cosusceptibility determine the size of network cascades. Phys. Rev. Lett. **118**(4), 048301 (2017)
20. U. Nakarmi, M. Rahnamay-Naeini, H. Khamfroush, Critical component analysis in cascading failures for power grids using community structures in interaction graphs. IEEE Trans. Netw. Sci. Eng. **7**(3), 1079–1093 (2019)
21. W. Ju, R. Yao, K. Sun, Interaction graph-based active islanding to mitigate cascading outage, in *IEEE Power Energy Society General Meeting*, Atlanta GA, Aug. 2019
22. C. Chen, S. Ma, K. Sun, Mitigation of cascading outages by breaking inter-regional linkages in the interaction graph. IEEE Trans. Power Syst. **38**(2), 1501–1511 (2022)
23. B.A. Carreras, V.E. Lynch, I. Dobson, D.E. Newman, Dynamical and probabilistic approaches to the study of blackout vulnerability of the power transmission grid, in *37th HICSS*, Hawaii, 2004
24. I. Dobson, B.A. Carreras, D.E. Newman, A loading-dependent model of probabilistic cascading failure. Probab. Eng. Inf. Sci. **19**(1), 15–32 (2005)
25. P. Rezaei, P.D.H. Hines, M.J. Eppstein, Estimating cascading failure risk with random chemistry. IEEE Trans. Power Syst. **30**(5), 2726–2735 (2015)
26. I. Dobson, J. Kim, et al., Testing branching process estimators of cascading failure with data from a simulation of transmission line outages. Risk Anal. **30**, 650–662 (2010)
27. J. Qi, W. Ju, K. Sun, Estimating the propagation of interdependent cascading outages with multi-type branching processes. IEEE Trans. Power Syst. **32**(2), 1212–1223 (2017)
28. I. Dobson, B.A. Carreras, et al., An initial model for complex dynamics in electric power system blackouts, in *34th HICSS*, Hawaii, 2001
29. S. Mei, F. He, X. Zhang, et al., An improved OPA model and blackout risk assessment. IEEE Trans. Power Syst. **24**(2), 814–823 (2009)
30. S. Mei, Y. Ni, G. Wang, et al., A study of self-organized criticality of power system under cascading failures based on AC-OPA with voltage stability margin. IEEE Trans. Power Syst. **23**(4), 1719–1726 (2008)
31. M. Bhavaraju, N. Nour, TRELSS: A computer program for transmission reliability evaluation of large-scale systems, in *Electr. Power Res. Inst.*, Palo Alto, CA, USA, Tech. Rep. EPRI-TR-100566, 1992
32. D.P. Nedic, I. Dobson, D.S. Kirschen, B.A. Carreras, V.E. Lynch, Criticality in a cascading failure blackout model. Electr. Power Energy Syst. **28**(9), 627–633 (2006)
33. J. Chen, J.S. Thorp, I. Dobson, Cascading dynamics and mitigation assessment in power system disturbances via a hidden failure model. Int. J. Elect. Power Energy Syst. **27**(4), 318–326 (2005)
34. R. Yao, S. Huang, K. Sun, et al., A multi-timescale quasi-dynamic model for simulation of cascading outages. IEEE Trans. Power Syst. **31**(4), 3189–3201 (2016)
35. J. Song, E. Cotilla-Sanchez, G. Ghanavati, P.D.H. Hines, Dynamic modeling of cascading failure in power systems. IEEE Trans. Power Syst. **31**(2), 1360–1368 (2016)
36. P. Henneaux, P.-E. Labeau, J.-C. Maun, L. Haarla, A two-level probabilistic risk assessment of cascading outages. IEEE Trans. Power Syst. **31**(3), 2393–2403 (2016)

37. Q. Shi, F. Li, Q. Hu, Z. Wang, Dynamic demand control for system frequency regulation: concept review, algorithm comparison, and future vision. Electr. Power Syst. Res. **154**, 75–87 (2018)
38. H. Pulgar-Painemal, Y. Wang, H. Silva-Saravia, On inertia distribution, inter-area oscillations and location of electronically-interfaced resources. IEEE Trans. Power Syst. **33**(1), 995–1003 (2018)
39. G. Andersson, P. Donalek, R. Farmer, et al., Causes of the 2003 major grid blackouts in North America and Europe, and recommended means to improve system dynamic performance. IEEE Trans. Power Syst. **20**(4), 1922–1928 (2005)
40. M. Okamura, S. Hayashi, K. Uemura, et al., A new power flow model and solution method including load and generator characteristics and effects of system control devices. IEEE Trans. Power Apparatus Syst. **94**, 1042–1050 (1975)
41. R. Ramanathan, Dynamic load flow technique for power system simulators. IEEE Trans. Power Syst. **1**(3), 25–30 (1986)
42. I. Roytelman, S.M. Shahidehpour, A comprehensive long term dynamic simulation for power system recovery. IEEE Trans. Power Syst. **9**(3), 1427–1433 (1994)
43. M.S. Ćalović, V.C. Strezoski, Calculation of steady-state load flows incorporating system control effects and consumer self-regulation characteristics. Int. J. Elect. Power Energy Syst. **3**, 65–74 (1981)
44. Y. Ping, A fast load flow model for a dispatcher training simulator considering frequency deviation effects. Electr. Power Energy Syst. **20**(3), 177–182 (1998)
45. Y.Q. Hai, X. Wei, W.X. Fen, The improvement of dynamic power flow calculation in dispatcher training simulator. Autom. Elect. Power Syst. **23**(23), 20–22 (1999)
46. D.P. Popović, An efficient methodology for steady-state security assessment of power systems. Int. J. Elect. Power Energy Syst. **10**, 110–116 (1988)
47. Y. Duan, B. Zhang, Security risk assessment using fast probabilistic power flow considering static power-frequency characteristics of power systems. Electr. Power Syst. Res. **60**, 53–58 (2014)
48. X. Ye, W. Zhong, X. Song, et al., Power system risk assessment method based on dynamic power flow, in *International Conference on Probabilistic Methods Applied to Power Systems*, 2016
49. P. Bei, B. Zhang, H. Li, et al., Probabilistic dynamic load flow algorithm considering static security risk of the power system, in *International Conference on Electric Utility Deregulation and Restructuring and Power Technologies*, Changsha, China, 2015
50. Y.H. Liu, Z.Q. Wu, S.J. Lin, et al., Application of the power flow calculation method to islanding microgrids, in *International Conference on Sustainable Power Generation and Supply*, Nanjing China, 2009
51. L. Rese, A.S. Costa, A.S. e Silva, A modified load flow algorithm for microgrids operating in islanded mode, in *IEEE PES Conference on Innovative Smart Grid Technologies*, DC Washington, 2013
52. Y. Duan, B. Zhang, An improved fast decoupled power flow model considering static power–frequency characteristic of power systems with large-scale wind power. IEEE Trans. Electr. Electron. Eng. **9**(2), 151–157 (2014)
53. S. Li, W. Zhang, Z. Wang, Improved dynamic power flow model with DFIGs participating in frequency regulation. IEEE Trans. Electr. Energy Syst. **27**, 1–13 (2017)
54. O.A. Mousavi, et al., Blackouts risk evaluation by Monte Carlo Simulation regarding cascading outages and system frequency deviation. Electr. Power Syst. Res. **89**, 157–164 (2012)
55. O.A. Mousavi, et al., Inter-area frequency control reserve assessment regarding dynamics of cascading outages and blackouts. Electr. Power Syst. Res. **107**, 144–152 (2014)
56. W. Ju, K. Sun, R. Yao, Simulation of cascading outages considering frequency using a dynamic power flow model. IEEE Access **6**(1), 37784–37795 (2018)
57. Standard PRC-006-NPCC-1 Automatic Underfrequency Load Shedding, February 9, 2012. [online] available: http://www.nerc.com/files/PRC-006-NPCC-1.pdf

58. S. Imai, T. Yasuda, UFLS program to ensure stable island operation, in *IEEE PES Power Systems Conference and Exposition*, 2004
59. IEEE Standard C37.106–2003, in *IEEE Guide for Abnormal Frequency Protection for Power Generating Plants*, 2004
60. A.G. Exposito, J.L.M. Ramos, J.R. Santos, Slack bus selection to minimize the system power imbalance in load-flow studies. IEEE Trans. Power Syst. **19**(2), 987–995 (2004)
61. P. Kundur, *Power System Stability and Control* (McGraw-Hill Education, New York, 1994)
62. W. Ju, J. Qi, K. Sun, Simulation and analysis of cascading failures on an NPCC power system test bed, in *IEEE Power and Energy Society General Meeting*, Denver CO, Jul. 2015
63. M. Variani, S. Wang, K. Tomsovic, Study of flatness-based Automatic Generation Control Approach on an NPCC system model, in *IEEE Power and Energy Society General Meeting*, Denver CO, Jul. 2015
64. P.M. Anderson, M.A. Mirheydar, A low-order system frequency response model. IEEE Trans. Power Syst. **5**(3), 720–729 (1990)
65. PSS/E V32 User Manual, Siemens Power Transmission & Distribution, Inc., Dec 2007
66. A. Melman, Geometry and convergence of Euler's and Halley's methods. SIAM Rev. **39**(4), 728–735 (1997)
67. I. Shames, F. Farokhi, M. Cantoni, Guaranteed maximum power point tracking by scalar iterations with quadratic convergence rate, in *IEEE 55th Conference on Decision and Control*, Las Vegas NV, 2016

Chapter 8
Industrial Practices and Criteria Against Cascading Failures

Milorad Papic

8.1 Introduction

Cascading failures in power systems occur when an initial outage, or combination of outages, triggers the successive outage of multiple system elements [1]. Initial outage may include component failure due to aging, natural disasters, poor component design or operating settings, and transmission line or generator outages [2, 3]. As a result of component failure, the redistribution of power flow happened across other components in the system. The effects of a component failure may be either contained locally or it may propagate to components further away, potentially causing widespread damage to the power system, or a blackout [2]. Although infrequent, large blackouts are expensive, and its impact can propagate into other sectors [4]. Current efforts to understand and mitigate cascading failures can be demonstrated by papers published by the IEEE Cascading Failure Working Group (CFWG) [5–10].

Papers [8–10] indicate that as a result of component failure, the reactions typically occur in rapid succession such that human intervention is unable to stop the process. The reactions include sequential tripping of transmission lines and generators. The initiating event or events can include a line sagging into vegetation, or a short on a transmission line caused by natural causes such as high wind or lightning. Other initiating events can include human actions (or inaction), the particular network topology at the time, and/or imbalances between load and generation. A common part of cascading outages is the tripping of lines and generators by the equipment's protective equipment to protect the equipment from damage. This protective equipment, built to sense faults that threaten the equipment

M. Papic (✉)
Independent Consultant, Boise, ID, USA
e-mail: mpapic@ieee.org

K. Sun (ed.), *Cascading Failures in Power Grids*, Power Electronics and Power Systems, https://doi.org/10.1007/978-3-031-48000-3_8

they protect, may trip due to what it senses is a fault, but is in fact high currents or low voltages as a result of earlier line or generator tripping.

A considerable amount of published research work in the recent past has been devoted to various aspects of cascading calculations in power systems [2–25]. However, practical applications of cascading methodologies have been lagging these efforts. Cascading failure is common in power grids when one of the elements fails (completely or partially) and shifts its load to nearby elements in the system. Those nearby elements are then pushed beyond their capacity, so they become overloaded and shift their load onto other elements. Large-scale blackouts result from cascading failures, which occur when equipment failure or other disruption cause further failures or disruptions in large portions of the power grid. These events are relatively rare but potentially very costly. In recent years, much effort has been undertaken by researchers and utility engineers to comprehensively evaluate the vulnerability of the power system and analyze its ability to withstand extreme contingencies. Two programs often used by utilities to perform cascading studies are presented in [24, 25]. The methodologies implemented in these two tools are presented in Sect. 8.5.

Maintaining adequacy and operational reliability becomes more challenging than ever before, especially as the frequency of extreme environmental conditions increases, together with their impacts on variable energy resources (VER), which count on the environment for the energy they need to operate. A cascading effect is an unforeseen chain of events that occurs when an event in a system has a negative impact on other related systems. As have been pointed out, a cascading outage is a sequence of events in which an initial event triggers a sequence of one or more dependent component outages. In general, cascading outage is influenced by the system state, such as components out for maintenance and the patterns of power transfers, and the automatic and manual system procedures [12–19].

Major blackouts are usually the outcome of cascading outages, which North American Electric Reliability Corporation (NERC) defines "as the uncontrolled loss of any system facilities or load, whether because of thermal overload, voltage collapse, or loss of synchronism, except those occurring as a result of fault isolation" [26]. The complexity of cascading outages makes enumeration of all possibilities impossible. Some degree of approximation is therefore necessary, whether in terms of the individual events by which cascades might be propagated or the modeling of the physical phenomena they involve.

8.2 Operating States

The simplest description of a cascading outage is given in Wenyuan Li's book [27] "as the event in which a failure of the first component causes a series of successive outages 2, . . . N caused by many factors such as weather, incorrect operation of the power system equipment." The state-space diagram of a cascading outage as a looped transition process is presented in Fig. 8.1. State 0 corresponds to the normal state, state 1 to the failure of the trigger component, state 2 to

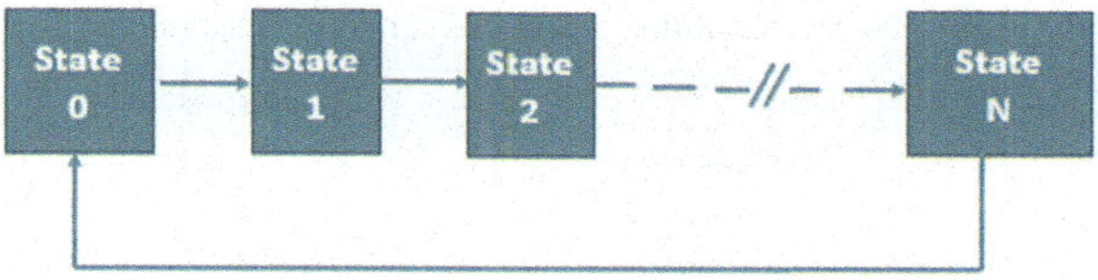

Fig. 8.1 State-space diagram for a cascading outage

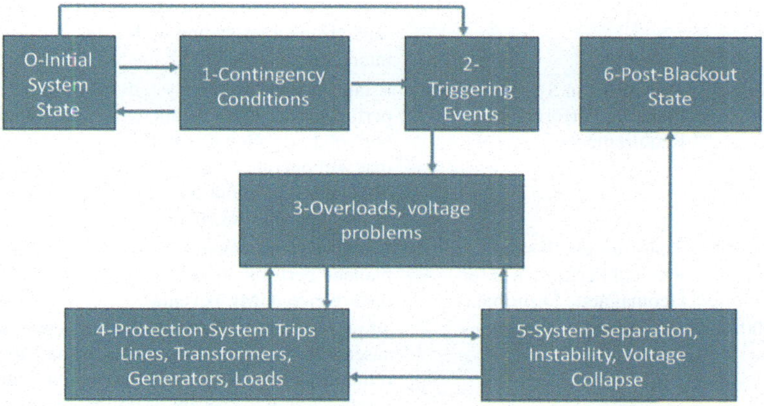

Fig. 8.2 Generic scenario of cascading failures

the second component's failure, and so on. In this diagram, it's assumed that all failed components will return to the normal state at the same time which might not happened in real life.

More detailed operating process of a power system during cascading that may result in the blackout event can generally be characterized by diagram with six states presented in Fig. 8.2 [28].

State 0: Initial Pre-outage Operating State A power system in state 0 operates within pre-specified operating constraints and doesn't cause any unsolved violations by transitioning from this state to state 1 under a single (n-1) outage. In this state, the system is supposed to operate reliably. According to NERC, the term "reliable operation" means operating the elements of the bulk-power system within equipment and electric system thermal, voltage, and stability limits so that instability, uncontrolled separation, or cascading failures of such system will not occur as a result of a sudden disturbance or unanticipated failure of system elements [29].

State 1: Contingency Operating State

A power system in state 1 operates in contingency state under single n-1, double n-2 such as two lines on the same tower or two lines on a common corridor, or n-k outage such as common breaker failure. These different categories of contingencies are defined by NERC planning standard TPL-001-5 [26]. Table 8.1 in this standard lists initial system conditions and contingency categories P1–P8 and extreme category contingencies. Categories P1–P2 are "events resulting in the loss of a single

Table 8.1 NERC reliability standards related to cascading [26]

Standard	Title	Purpose
CIP-014-2	Physical Security	To identify and protect transmission stations and transmission substations, and their associated primary control centers, that if rendered inoperable or damaged as a result of a physical attack could result in instability, uncontrolled separation, or cascading within an interconnection.
TPL-001-5	Transmission System Planning Performance Requirements	Establish transmission system planning performance requirements within the planning horizon to develop a bulk electric system (BES) that will operate reliably over a broad spectrum of system conditions and following a wide range of probable contingencies
TPL-007-4	Transmission System Planned Performance for Geomagnetic Disturbance	Establish requirements for transmission system planned performance during geomagnetic disturbance (GMD) events
TOP-001-5	Transmission Operations	To prevent instability, uncontrolled separation, or cascading outages that adversely impact the reliability of the interconnection by ensuring prompt action to prevent or mitigate such occurrences
TOP-002-4	Operations Planning	To ensure that transmission operators and balancing authorities have plans for operating within specified limits
IRO-008-2	Reliability Coordinator Operational Analyses and Real-Time Assessments	Perform analyses and assessments to prevent instability, uncontrolled separation, or cascading
IRO-009-2	Reliability Coordinator Actions to Operate Within IROLs	To prevent instability, uncontrolled separation, or cascading outages that adversely impact the reliability of the interconnection by ensuring prompt action to prevent or mitigate instances of exceeding IROL

element." Categories P3–P5 are "events resulting in the loss of multiple system elements with initial conditions for P3 loss of generating unit with adjustments, and for P4–P5 normal state. Category P6 is an event resulting in the loss of two overlapping single elements. Initial condition for category P6 event is the loss of transmission circuit, transformer, shunt device, and single pole of a DC line. Category P7 is an event resulting in the loss of multiple elements associated with common structure. Initial condition for category P7 event is normal state. Category P8 is an event resulting in the loss of multiple elements as a result of "fault plus non-redundant component of a protection system failure to operate." Initial condition for category P8 event is normal state. Extreme category contingency is an "event resulting in the loss of multiple elements," and as a result, cascading as an outcome might happened. In planning, this category is simulated by the removal of all elements that protection systems and automatic controls are expected to disconnect for each contingency.

State 2: Triggering Events

A variety of contingencies (single *n*-1, double *n*-2 such as two lines on the same tower or two lines on a common corridor, or *n*-*k* outage such as common breaker failure) that bring system in state 2 may develop as a triggering event of the cascading outage. There are many causes of triggering events that lead to a cascading outage such as forced outages of system elements (lines, transformers, buses, circuit breakers, etc.), protection system failures or misoperation, natural incidents, reduced or lack of system awareness, cyberattacks, failures of information communication technology systems used in protection and control, human errors, terrorist threats, unpredictable fluctuating characteristics of renewable energy sources, etc. The failures of transmission equipment (generators, transmission lines, transformers, etc.) can be grouped as [30] independent outages, dependent outages, common-cause outages, and station originated outages. Another cause of a triggering event that results in abnormal operating conditions in the system is incorrect relay operations caused by an undetected defect of the relay. This type of failure is often referred to as hidden failure [31]. A detailed analysis of historical multiple outage events in North American power grid, collected in NERC Transmission Availability Data System (TADS), is performed in [32–35].

State 3: Power Flow Surges, Overloads, and Voltage Problems In this state, the triggering event, as well as the subsequent events, causes power flow surges, overloads, and voltage problems. These problems in their turn are causes of the subsequent events in the sequence. The result of the triggering events in terms of system limits and impacts can be defined as:

- Thermal and voltage limits are within acceptable rating and the system is stable.
- Thermal and voltage limits are violated.
- Curtailed firm transfers and loss of load.
- Cascading outages.

State 4: Protection System Trips Lines, Transformers, and Generators The power system relay protection plays a very important role in a development of cascading event. Its action could be caused directly by the system problems when the protective relays reacted as if the high flows or low voltages were due to a short circuit, or indirectly, when the system problem causes real short circuits or instability, e.g., when the overheated conductors contact a tree. The protection system isolates the equipment or a group of equipment from the rest of the network. Some load loss may accompany this process. This can result in more power flow surges, overloads, and voltage problems, and so on. Many of cascading failures were initiated by a single fault. During the fault, protection devices may malfunction due to various reasons such as overcurrent, voltage drop, or protection device failures.

State 5: System Separation, Instability, and Voltage Collapse On the advanced stages of a cascading event, system separation, instability, and voltage collapse can occur. As a result, a significant load loss may be inflicted. Load loss due to islanding could potentially help to balance generation and load and relieve system problems in the remaining part of the interconnection as well as in some isolated parts within the separated grid.

State 6: Blackout State After a number of subsequent phases of the developing cascading process, the blackout state may happen. This state is the starting point to begin the system restoration process. The following events may turn into a blackout:

- System and equipment faults (e.g., line contact with trees)
- Overloaded equipment
- Voltage, transient, and/or small signal instability
- Protection equipment hidden failures triggered by events, such as outdated settings and HW failures
- Inadequate or faulty alarm and monitoring equipment
- Human error or slow operator response
- Intertie separations
- Lack or inadequate remedial action scheme to prevent spreading of the disturbance
- Voltage instability: inability to maintain voltage so that both power and voltage are controllable

The corrective actions such as generation rescheduling, load shedding, etc. are used to alleviate problems associated with triggering events in operating states 3, 4, and 5. The operators of the bulk transmission system must be constantly vigilant of these events that could lead to violations of system operating limits (SOLs) and might bring the system into an insecure state.

8.3 NERC Standards Related to Cascading

Power system performance in the planning environment is measured by performance indicators for planning events (P1–P8) and extreme events defined in Table 8.1 of the NERC Standard TPL-001-5 [26]. Power system performance in the operating environment is measured by interconnection reliability operating limits (IROL) and SOL. An IROL exceedance as a result of a contingency can lead to system instability, uncontrolled separation, or cascading. The NERC Standard CIP-014-2 has been adopted to identify and protect transmission stations and substations and their associated primary control centers that, if rendered inoperable or damaged as a result of a physical attack, could result in widespread instability, uncontrolled separation, or cascading within an interconnection [26]. Historically, the system is designed to perform service reliably without interruption for all *n*-1 and credible *n*-

2 contingencies. However, widespread interruption and cascading remain possible, especially if these contingencies are accompanied by other system vulnerabilities that result in complex *n*-*k* contingencies.

Electric utilities maintain sufficient generation, transmission, and distribution capacity to ensure continuity of electric service to their customers under normal and various abnormal conditions, including uncertainties associated with load variations and variability of renewable energy sources. There have been extensive efforts to prevent cascading failures. For example, the NERC requires the power grids to operate under the *n*-1 security criterion: the system remains safe after any single failure. This preventive security criterion improves grid reliability against failures; however, multiple failures may happen simultaneously. To make things worse, hidden failures, such as human operating errors, may enlarge the impact of failures. Improving grid reliability becomes crucial as the power grid becomes increasingly stressed by more volatile supply and demand fluctuations. The fact that power system operation is subject to an enormous number of random events makes cascading analysis a rather complex issue. Increased penetration of VER generation such as wind and solar photovoltaic (PV) creates uncertainties an order of magnitude greater than those in legacy power systems and increased the variability in power flow patterns, volatility of system stress, reserve capacity requirements, cycling of thermal units, and ramping capacity requirements. Since uncontrolled, cascading outages may have such a widespread effect and take extensive time and difficulty to recover from, NERC, under its transmission planning standards [26], requires analysis of the planning events P1–P8 and extreme outage events. NERC reliability standards define the requirements for both planning process and operations of the bulk power systems in the following areas: modeling, planning, operation, protection, emergency, and schedule and dispatch.

- Modeling standards such as NERC MOD-032 and MOD-033 [26] specify the requirements, assessments, and validation of modeling data and reporting procedures for development of cases necessary to support analysis of the reliability of the interconnected transmission system.
- Planning standards specify technical and design criteria and procedures in the planning and development of transmission systems, such as NERC TPL-001-5 (Transmission Planning) standards.
- Operation standards [26] specify the operations to protect the reliability and security of power supply and operation under normal and abnormal operating conditions, such as NERC TOP-001-5 (Transmission Operations) standards, NERC IRO-008-2 (Interconnection Reliability Operations and Coordination) standards, and NERC VAR-001-5 (Voltage and Reactive) standards.
- Protection and CIP standards [26] specify the coordination and responsibilities of the protection, such as NERC PRC-023-4 (Protection and Control) standards, and NERC CIP (Critical Infrastructure Protection) standards such as CIP-014-2.
- Emergency standards [26] specify the procedures, implementing plans, and responsibilities relating to operating emergencies, such as NERC EOP-008-3 (Emergency Preparedness and Operations) standards.

Table 8.1 lists an extract of the representative NERC reliability standards that directly or indirectly deal with cascading in power systems.

Transmission planning studies can determine system planning and operational requirements, generator interconnection requirements, and protective relay coordination and can help ensure reliable system operation and quick system restoration. Transmission planning studies should identify any potential SOL and IROL violations. Transmission planning studies can identify potential IROL violations, propose system preventive actions to avoid IROL violations, and take corrective actions to mitigate an IROL violation if it occurs. IROL violations may be caused by single or multiple contingencies occurring at the same time, or successive tripping or contingencies related to the prior contingencies. They can also be performed for purposes of identifying potential blackouts, replicating actual system disturbances, and validating system models. SOL is the value (such as MW, MVAr, amperes, frequency, or volts) that satisfies the most limiting of the prescribed operating criteria for a specified system configuration to ensure operation within acceptable reliability criteria. IROL is the limit at which instability, uncontrolled separation, or cascading occurs [1].

8.4 Planning and Operating Cases and Study Assumptions

Developing adequate planning and operating cases is an essential first step in performing cascading analysis. Results of the cascading depend not only on methods and computing tools but also on the data (equipment data, generation data, interconnection data, new planned facilities, scheduled outages, load forecast data, and system operating limits). Operating real-time state estimator (SE) cases are used by RTCA (real-time contingency analysis). After every system estimate with valid solution, the base case is monitored for violations such as over-voltages, under-voltages, angle pairs, interfaces, and branch flows. Data in planning and operation cases need to be reviewed and updated regularly. Planning cases should be developed from data consistent with what is defined by NERC standards MOD-031 and MOD-032 [26]. Planning and operating cases can operate under normal or stressed conditions

(a) Normal conditions

- All facilities are modeled to reflect normal operating conditions and limits.
- The loading of lines and equipment shall be within normal rating limits.
- Voltage levels shall be maintained within plus or minus 5% of nominal voltage.
- Electrical demand shall be supplied, and all contracted firm (non-recallable reserved) transfers shall be maintained.
- Stability of the studied system shall be maintained.
- Cascading outages shall not occur.

(b) Stressed conditions/sensitivity cases

- Studied cases are stressed to identify potential future transmission system weaknesses and limiting facilities.
- Deferral of expected in-service dates of planned transmission facilities.
- Changes in generation dispatch scenario.
- Changes in forecasted load.
- Changes in reactive resource capability.
- Changes in area interchange.
- Changes in reactive resource capability.
- Dynamic load model changes.
- Changes in forecasted load.
- Changes in area interchange.

For the operating and planning cases used for performing cascading analysis by utilities, the following steps are usually performed:

- Cases are checked for thermal (flow) violations on all lines and transformers modeled in the base case and for voltage violations on all buses modeled in the case under normal (n-0) conditions.
- Cases are also tested by studied engineers to ensure no violations exist under n-1 contingencies.
- In addition, cases are stressed by load or transfer to determine if there is any thermal/voltage violation under n-1 outages. Each violation was mitigated either by existing remedial action schemes (RAS) or by projects identified to alleviate those problems.

In normal state, the operating parameters of the BES should lie within acceptable limits. A power system in this state operates within pre-specified operating constraints and will not cause any unsolved violations. However, noticeable deviations of parameters might be observed during extreme conditions (summer and winter peak), changes in topology, natural disasters, extreme weather, etc. which can be taken into account by developing stressed cases. Stressed cases are used to identify transmission system weaknesses and limiting facilities in order to minimize the negative impact of cascading process. For monitoring system performance, the following equipment trip settings need to be defined:

- Relay loadability limits for transmission facilities, defined by transmission system providers (TSP)
- Generator over- and under-protection voltage relays (defined by generator owners)
- Buses with under-voltage load shedding (UVLS) protection schemes where voltages go below the under-voltage triggering level

Criteria for monitoring potential system cascading are as follows:

- System-wide voltage collapse occurs upon applying initiating contingency or as the result of cascading outages.

- Power flow fails to converge following the event.
- Islanding with the total MW of load in island greater than pre-defined threshold.
- Interface MW flow exceeds "stability" interface limit by pre-defined % level.
- Total MW loss of load exceeds pre-defined threshold.
- Total MW loss of generation exceeds pre-defined threshold.
- Cascade propagates beyond balancing area footprint.
- The number of cascade levels exceeds user-defined tiers.

8.5 Cascading Methodologies

Power system reliability evaluation in general is concerned with the ability of the system to supply adequate and suitable electrical energy to customers and to withstand the extreme cascading outages. Many published references have been devoted to various methodologies for assessing of cascading outages in power systems [2–23]. The presented methodologies have a common goal to focus on the identification of modeling characteristics needed for adequate representation of cascading failure in power system simulations.

Simulation methodologies for analysis of cascading outages to identify the potential cascading modes [2–23] are normally applied via the following steps:

- Creating a list of initiating events
- Identifying the list of critical outages (events that cause SOL or IROL violations)
- Applying the existing RAS and other resources in the system to mitigate some of the violations
- Monitoring thermal and voltage tripping thresholds.
- Applying outages consecutively until system fails to solve or tripping thresholds are violated
- Summarizing the results and listing the outages that lead to cascading instability

Many achievements have been made on methodologies for cascading assessment of power systems. The Electric Power Research Institute (EPRI) report [2] presents a range of views and recommendations by representatives of government, universities, private industries, utilities, and EPRI on cascading failures at the jointly organized workshop by the National Science Foundation (NSF) and EPRI. The main objective of the workshop was to exchange ideas and approaches to improve understanding, prediction, and prevention of cascading failures, as well as restoring from such failures, and to document research needs in this area, for the purpose of coordinating research in the industry.

IEEE Power and Energy Society (PES) Computing and Analytical Methods Subcommittee (CAMS) Task Force on understanding, prediction, mitigation and restoration of cascading failures has published a paper on "an initial review of methods for cascading failure analysis in electric power transmission systems" [3]. This paper defines cascading failure for blackouts and gives an initial review of the current understanding, industrial tools, and the challenges and emerging methods of analysis and simulation.

Bhatt et al. published a paper that addresses the testing and implementation of a fast process for sequential contingency simulation in order to identify potential cascading modes due to thermal overloads [4]. It also presents computation of the vulnerability index of cascading, based on the estimated likelihood and consequences of cascading outages. The approach described in this paper offers a unique capability to automatically identify initiating events that may lead to cascading outages. The results of the study indicate that initiating events and possible cascading chains may be quickly identified, ranked, and visualized in online and offline environments.

The cascading cluster-based methodology is presented in paper [5]. A power system network may be represented as a number of clusters (groups) connected to the rest of the network via critical lines. A cluster approach is based on system representation via three types of clusters: load clusters, generation clusters, and a connecting cluster. This representation of the system helps to quickly identify possible initiating events that may lead to cascading outages and automatically determine possible cascading chains. The network is divided into three types of clusters:

- Generator clusters
- Load clusters
- Connecting cluster

An initial bus in each generator cluster is a generator bus. If a bus connected to a cluster already belongs to another generator cluster, these clusters are merged. Cut set of lines that connect the cluster with other clusters is identified; usually these are high-voltage lines. The sum of flows on the lines comprising the cut set and the sum of ratings of these lines are computed. Those clusters that have large flows on the cut set are of interest within this analysis framework.

Papers published by CFWG [6–10] present various methodologies and tools to analyze the cascading in power systems. Paper [6] presents the state of the art in cascading failure modeling tools for analysis and prediction of cascading failure events. Paper [7] discusses the challenges of cascading failure and summarizes a variety of state-of-the-art analysis and simulation methods, including analyzing observed data, and simulations relying on various probabilistic, deterministic, approximate, and heuristic approaches. Paper [8] presents the basic methodologies for mitigation, summarizes currently deployed special protection schemes, and lists cases of successful and unsuccessful mitigation of cascading outages and lessons learned. Paper [9] reviews and synthesizes how benchmarking and validation of tools for cascading failure analysis can be done properly. Paper [10] benchmarks the performance of several widely used quasi-steady-state (QSS) cascading outage methodologies based on the reliability test system (RTS) and the results are compared.

Papers [11–13] define and quantify a measure that relates to the vulnerability of the power grid to cascading outages. The presented approach simulates and identifies potential cascading modes (PCM) and computes the probability and impact of each stage of the cascade and displays the risk of each potential cascade.

Paper [14] presents the event tree approach to determine possible sequences of cascading failures with severe impact on a given power system. The approach takes into account protection systems' hidden failures and transmission system equipment overload.

Paper [15] explores the suitability of using steady-state, *N-k* analysis to identify a smaller set of contingencies that require detailed cascading analysis, including protection system operation and dynamic simulation. The method is demonstrated on the IEEE 300-bus test system.

Paper [16] presents results on the dynamic modeling and analysis of cascading failures in large-scale electric grids. It shows that simulations utilizing full dynamic modeling of power system elements have the ability to demonstrate cascading failures.

Paper [17] proposes an AC power flow-based cascading failure model that explicitly considers external weather conditions and extreme temperatures in particular and evaluates the impact of extreme temperature on the initiation and propagation of cascading blackouts. The effectiveness of the proposed model is verified by simulation results on the RTS-96 3-area system.

Paper [18] describes a stochastic "random chemistry" (RC) algorithm to identify large collections of multiple contingencies that initiate large cascading failures in a simulated power system. The paper discusses various ways that RC-generated collections of dangerous contingencies could be used in power system planning and operations.

Paper [19] compares random chemistry algorithm with Monte Carlo simulation. It shows that the random chemistry method is at least an order of magnitude faster than Monte Carlo simulation.

Papers [20–22] estimate the amount by which line outages propagate from standard utility data that is reported annually to NERC by transmission owners for the TADS. The papers also estimate the distribution of the total number of outages after cascading from the amount of propagation and a probabilistic branching process model of the cascading.

Paper [23] presents the event tree methodology for modeling cascading outages in power systems. The presented methodology is a special simulation approach (SA) that has been implemented into EPRI reliability analysis tool Transmission Reliability Evaluation of Large Scale Systems (TRELSS) to simulate system vulnerability to cascading failures [24].

The cluster-based cascading methodology is implemented in a PCM program [25]. The PCM program capabilities include:

- Quickly identifying initiating events
- Automatically identifying cascading chains
- Ranking cascading outages
- Visualization of cascading outages

The following steps in planning are taken in performing the cascading analysis:

- Studied cases have no violations under normal conditions (*n*-0) and *n*-1 contingencies.
- *n*-1 and *n*-2 contingencies are generated by combining outages of lines, transformers, and generators including the operation of the existing RAS.
- The thresholds for thermal overloads and voltages at generation and load buses to simulate the operation of protection systems are specified.
- Initiating events are consecutively applied until the following occurs:
  - The system fails to solve due to voltage instability.
  - Islanding with an imbalance of load and/or generation within an island.
  - Thermal/voltage violations are alleviated or no longer exceed the thresholds.
- The loss of load and generation is simultaneously monitored and reported at each tier.

Prevention of cascading outages starts with system planning and continues through system operations. In the system planning area, there is a need for a long-term vision, rather than the driving of system design as a reaction to system events. In the system operations area, entities that operate the power system need a workable organizational structure, an outage coordination process that addresses both short-term and long-term issues, and a resulting operational plan that ensures the ability to respond to a contingency and alert operator to potential operational issues. The successful operation requires quality training, reliable and accurate operator tools, and questioning operators who do not blindly follow energy management system (EMS) output. The operators need to think through the contingency response beyond the *N*-1 contingency to the 30 minutes in advance when redispatch or other operational procedure occurs to bring the system back within *N*-1 criteria after a contingency. The measures to prevent or minimize the impact of cascading outages can be summarized as follows [8]:

- MW redispatch
- MVAr re-dispatch
- Circuits' (lines and transformers) switching in and out
- Load curtailment
- PAR transformer phase angle change
- Capacitor and reactor switching
- ULTC transformer tap change
- Reactive devices' placement
- Employed operating procedures
- Installation, modification, retirement, or removal of transmission and generation facilities and any associated equipment
- Installation, modification, or removal of protection systems or RAS
- Installation or modification of automatic generation tripping as a response to a single or multiple contingencies to mitigate stability performance violations

8.6 Industry Practices in the Analysis of Cascading Outages

This section covers results of the survey conducted by IEEE CAMS CFWG in 2019 to assess the present state of industry practice in assessment of cascading outages [36] and best practices by utilities to apply RAS and analyze cascading outages in planning and operation of the power grid. Through various case studies and results referenced in this section, power industry practitioners have demonstrated the benefits of performing cascading analysis in planning and operation [37–72].

8.6.1 IEEE CAMS CFWG Cascading Survey

The CAMS CFWG conducted a survey in 2019 to assess the state of industry practice in assessment of cascading outages [36]. The survey's questions addressed the types of analyses being done, the contexts and purposes of these analyses, data needed for analysis, and the software tools in use.

The survey was administered by the strategic research staff of IEEE headquarters. A total of 200 responses were received from North America, Europe and Asia. Of these, 54, or 27%, indicated that that their organizations are using cascading analysis as part of NERC compliance studies (TPL-001-4, CIP-014-2) and IROL computation.

Performing cascading analysis in the modern power grid is a challenging problem that utilities face today due to many causes such as frequent extreme events (e.g., failure of multiple physical components, extreme weather events, and other natural disasters), high penetration of intermittent renewable sources, and the increasing complexity of energy system infrastructure.

Results of the survey illustrate the present state of practical applications used by utility industry. The responses on two survey questions are presented in graphical form in Figs. 8.3 and 8.4.

- How often is the analysis of cascading outages performed in your organization?
- In which domain does your organization study cascading events?

The reader can find more details about other responses from this survey in reference [36].

Another paper [37] presents the survey that highlights the root causes of different blackouts around the globe including blackout and cascading analysis methods and the consequences of blackouts. Research directions and issues to be considered in future power system blackout studies are also proposed.

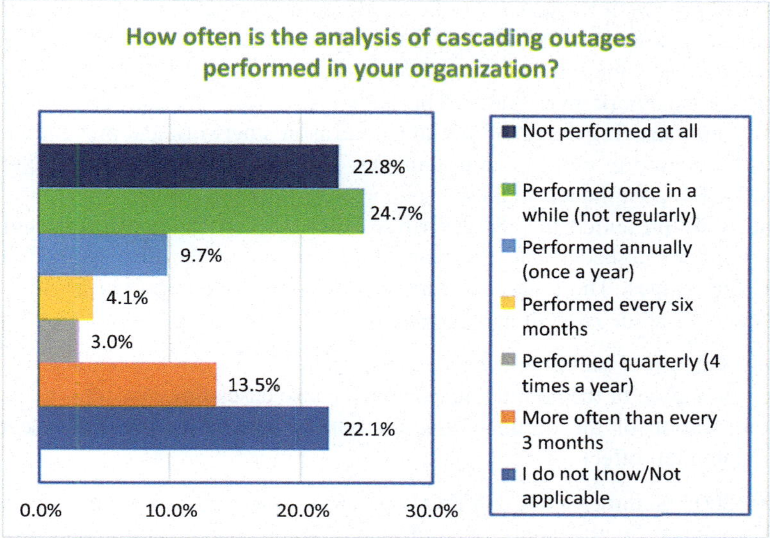

Fig. 8.3 How often is the analysis of cascading outages performed in your organization?

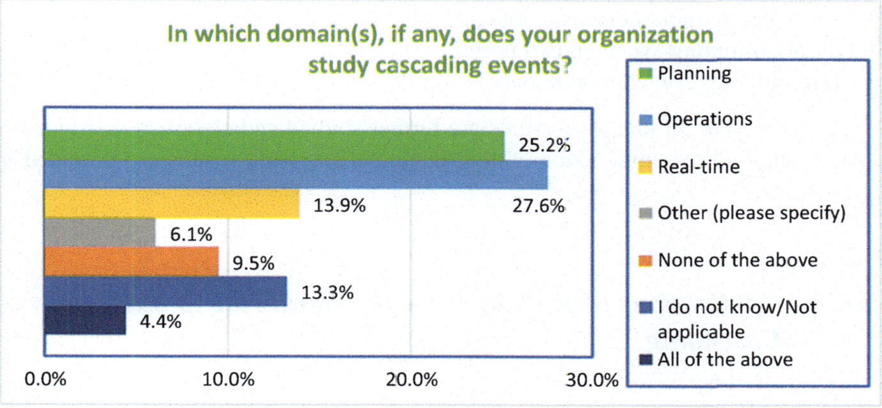

Fig. 8.4 In which domain does your organization study cascading events?

8.6.2 *Prevention of Cascading Outages in Con Edison's Network*

Con Edison has been actively involved in a number of research and proof-of-concept studies devoted to prediction and prevention of cascading outages caused by thermal overloads. Paper [38] presents the framework for prevention of cascading outages caused by thermal overloads. The approach presented here is a fast, flexible, and automated process for assessment of cascading outages and their prevention. It was implemented using the 2007 NYISO Summer Peak planning case, and an extensive analysis of cascading outages was performed and their effect on the Con Edison's transmission network investigated. The study results show that this approach is very effective in improving the reliability of Con Edison's network and preventing major blackouts, caused by the cascading outages. The analysis may also be included as part of NERC compliance studies.

The study presented in [38] describes Con Edison's experience in identifying and predicting cascades and the framework for comprehensive analysis of the cascading outages. The proposed framework covers three important issues related to assessment of the cascading outages:

- Identification of initiating events/contingencies
- Determination of the contingencies which cause cascading
- Identification of the optimal remedial actions needed to prevent cascades or mitigate their effect

The 2007 Summer Peak NYISO load flow case with approximately 50,000 buses was used for this study. A total of 250 single and 31,000 double formed a list of initiating events. Thermal overload constraints are monitored on all system elements. Cascading results are summarized by tiers:

- Seventeen initiating events in one tier
- Eleven initiating events in two tiers
- Ten initiating events in three tiers

These total of 38 initiating events are further studied and classified in groups by type of applied remedial actions. More details about study results are provided in [38].

8.6.3 Applications of RAS by Industry Worldwide to Mitigate Cascading

Considerable effort over the last several decades has been devoted to the research, various applications, and operation issues of RAS [39–55]. In the past, several committees of the International Council on Large Electric Systems (CIGRE) and IEEE have conducted surveys on the operation performance and reliability of remedial action schemes installed across the globe [39–42]. CIGRE report [39] provides a roadmap for the development of RAS to mitigate extreme contingencies and CIGRE report [40] focuses on corrective measures (protection and emergency control) against the system instability or breakdown. IEEE-CIGRE reports [41, 42] present the results of the survey on various RASs implemented by the industry with a global participation from members of IEEE and CIGRE.

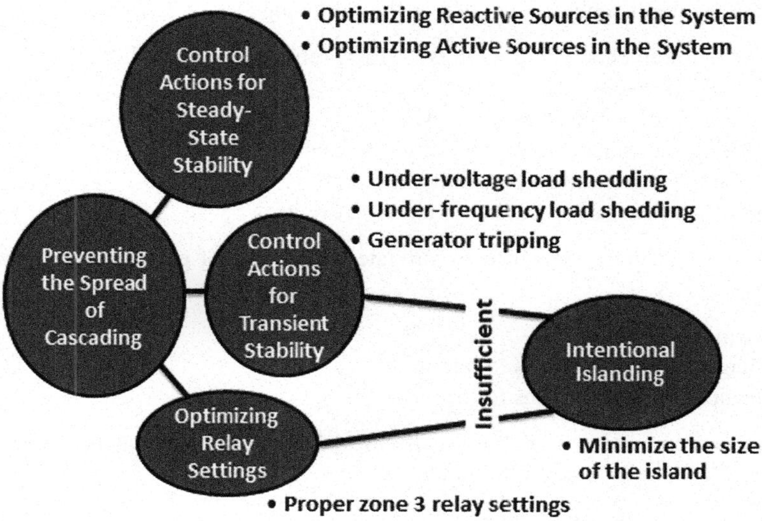

Fig. 8.5 Preventive measures/islanding for different types of cascading events

CIGRE defines RAS as a set of coordinated automated schemes that can minimize the risk of impending disturbances cascading to widespread blackouts. The NERC glossary defines a RAS as an automatic protection system designed to detect abnormal or predetermined system conditions and take corrective actions other than and/or in addition to the isolation of faulted components to maintain system reliability [1]. Current industry standards that deal with RAS are given in [26]. NERC standards PRC-012–PRC-017 address issues related to RAS and ensure that RAS are properly designed and coordinated with other protection systems. As a result of the major blackout events that happened across the globe in the past, many utilities have developed RAS to mitigate or minimize the impact of extreme contingencies that usually lead to the cascading. A variety of remedial action schemes as part of a defense plan have been introduced by utilities throughout the world to alleviate the start of cascading or to reduce the impact or likelihood of blackouts. Measures for mitigating and/or preventing cascading outages depend on the type of an event. The process of determining preventive measures [8] is given in Fig. 8.5.

A RAS is normally designed to detect critical system conditions such as:

- Generation patterns
- Transmission line loadings
- Load patterns
- Reactive power reserves
- System response as determined from the data provided by wide area measurement systems
- Other unsustainable conditions identified by studies of system characteristics

RAS accomplish objectives such as the following [1]:

- Meet requirements identified in the NERC reliability standards.
- Maintain BES stability.
- Maintain acceptable BES voltages.
- Maintain acceptable BES power flows.
- Increase system transfer capability.
- Limit the impact of cascading or extreme events.

RAS are generally designed to mitigate three types of power system problems and time scales:

- Thermal (minutes)
- Voltage stability (seconds to minutes)
- Transient stability (cycles to seconds)

Various remedial actions are usually available to improve system performance. These may include but are not limited to:

- Islanding/configuration changes
- Generator tripping
- Generator runback
- Load tripping (direct, underfrequency, undervoltage)
- Braking resistors
- HVDC ramping
- Static VAR control units
- Shunt capacitor/reactor switching
- Series capacitor/reactor switching

The following subsections briefly describe RASs installed in the USA, Canada, Brazil, and Italy.

8.6.3.1 Remedial Action Schemes at Western Electricity Coordinating Council (WECC)

The development and practical applications of the RAS across WECC are presented in [8, 43–47]. WECC member utilities Bonneville Power Administration (BPA), California Independent System Operator (CAISO), etc. use RAS extensively in planning and operation to ensure adequate system reliability, maintain or increase the transmission system capability, mitigate certain low probability/high consequence system events, and prevent events spreading out across large regions.

The most common RAS in WECC by percentage use are given in Fig. 8.6 [8].

There are over 270 RASs in WECC transmission system, and their number has grown in the recent past [43].

WECC identifies three types of RASs, depending on their potential impact:

- Local Area Protection Scheme (LAPS)

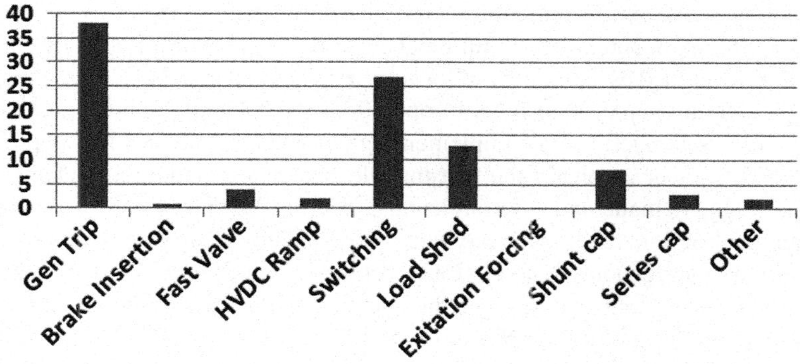

Fig. 8.6 Percentages of typical RAS actions

- Wide Area Protection Scheme (WAPS)
- Safety Net (SN)

LAPS are used to meet an owner's performance requirements within their system. WAPS is needed to meet WECC performance requirements and operating standards. SN scheme provides defense against extensive cascading or complete system collapse. Typically, planning studies conducted to comply with NERC standard TPL-001-5 are used to detect system problems and determine the need for a RAS and the type and characteristics needed to mitigate the problems. These actions are achieved via simple control principles, typically one or more of the following [43]:

- Event-based
- Parameter-based
- Response-based
- Combination of the above

Event-based schemes directly detect outages and/or fault events and initiate actions such as generator/load tripping to fully or partially mitigate the event impact. Parameter-based schemes measure variables for which a significant change confirms the occurrence of a critical event. Response-based schemes monitor system response during disturbances and incorporate a closed-loop process to react to actual system conditions.

8.6.3.2 Remedial Action Schemes at BPA and CAISO

BPA, a large company in WECC, has implemented several RASs to prevent or minimize the impact of cascading events that lead to blackout. For example, BPA implemented online the response-based Wide-Area stability and voltage Control System (WACS). The control system comprises phasor measurements at many substations, fiber-optic communications, real-time deterministic computers, and transfer trip output signals to circuit breakers at many substations and power plants. WACS is developed as a flexible platform to prevent blackouts [44]. BPA experience of implementing some other RAS in operations and planning studies is presented in [45]. CAISO experience of implementing automated RAS in EMS is presented in [46]. This paper emphasis the fact that electric grid is undergoing changes as a result of rising demand and environmental concerns. Among investigated solutions, the remedial action schemes are typically applied to solve violations as a result of extreme contingencies. The CAISO operation group is responsible for reliable operation of its system as portion of the WECC grid. Details of the remedial action schemes implemented in the EMS of the California ISO can be found in [47].

8.6.3.3 Remedial Action Schemes at Electric Reliability Council of Texas (ERCOT)

The ERCOT Operating Guides [48] describe RASs in ERCOT as "protective relay systems designed to detect abnormal ERCOT System conditions and take pre-planned corrective action (other than the isolation of faulted elements) to provide acceptable ERCOT System performance." RAS actions include changes in demand, generation, or system configuration. A RAS does not include underfrequency or under-voltage load shedding. ERCOT classifies RAS into two groups: Type 1 (wide-area impact) and Type 2 (local-area impact). Type 1 RAS is designed:

(a) To change generation output or constrain generation or imports over DC ties
(b) To open 345 kV transmission lines or other lines that interconnect transmission and/or distribution service providers and impact transfer limits

ERCOT's Type 2 RAS has only local-area impact and involves only the facilities of the owner.

8.6.3.4 Remedial Action Schemes at Hydro-Quebec and BC Hydro

Hydro-Quebec has undertaken many actions to enhance the reliability of its transmission system in the past. Paper [49] indicates that much effort of Hydro-Quebec has focused on increasing the system's ability to withstand extreme contingencies. As a result, Hydro-Quebec has adopted a number of RAS called defense plan against extreme contingencies. Also, the paper presents the specific automatic actions of RAS including:

- 735 kV shunt reactor switching system (called MAIS)
- Generation rejection and/or remote load-shedding system (called RPTC)
- Underfrequency load-shedding system

BC Transmission Corporation (BCTC) experience with the EMS-based RAS (EMS-RAS) is presented in [50]. This paper specifically describes three RASs to illustrate the EMS-RAS capability and flexibility with specific details on improving the reliability and transfer limits.

8.6.3.5 Remedial Action Schemes in Italy

The Italian defense plan [51] includes remedial actions aimed at:

(a) Preventing cascade tripping and consequent uncontrolled network separations
(b) Limiting the impact of network separation in case measures identified in (a) do not meet their target

To avoid the cascading and potential network separation, RAS includes system for automatic load shedding. The amount and location of load shedding depend on lines where out of service in the pre-fault conditions, lines where threshold has been exceeded, and which line has been tripped. Remedial actions are defined by offline steady-state and transient studies on different grid configurations and load flow conditions.

8.6.3.6 Remedial Action Schemes at ENTSOE

The CIGRE report [52] presents technical recommendations and rules for automatic actions to manage critical system conditions to prevent the Continental Europe (CE) synchronous area or parts of it from the loss of stability and cascading effects that could potentially lead to a system blackout. The report presents the analysis of defense plan procedures and provides the technical recommendations for defense plan of the CE synchronous area.

8.6.3.7 Remedial Action Schemes in Brazil

The Brazilian defense plan against extreme contingencies was developed after two major disturbances took place in March 1999 and January 2002 and initially published in [53]. The defense plan has been further refined by practically implementing the following actions by the National Operator [54]:

- Actions for minimizing the probability of the disturbance occurrence
- Actions for minimizing the propagation of the cascading disturbance
- Actions for optimizing the restoration times

8.6.3.8 Remedial Action Schemes in China

Paper [55] presents ways how to maintain the system integrity to prevent the cascading blackout. The paper indicates that 210 major disturbances, including out-of-step oscillations, frequency collapse, and voltage collapse, that happened in the period from 1970 to 1980 in China were prevented to lead to the blackout cascading event.

8.6.4 Cascading Analysis at Idaho Power Company (IPC)

IPC belongs to a small group of US utilities that implemented a practical approach to study cascading as part of NERC compliance planning studies. Paper [28] presents a comprehensive, practical approach to identify and analyze the multiple contingencies that lead to cascading outages in IPC's network. Paper [56] presents a comprehensive, practical approach for identifying and analyzing the multiple contingencies that could lead to voltage stability violations, widespread disturbance, or cascading in Idaho Power's network. Paper [57] presents a practical approach for performing the steady-state cascading analysis of a power system represented as a node/breaker (NB) model using real-time state estimator (SE) cases. This paper discusses advantages of using the NB model in cascading analyses compared to the bus/branch (BB) model. Paper [58] addresses the development of a risk-based contingency analysis for planning and operation of a power grid. The primary focus of the paper is on assessing power system performance following the consecutive loss of two bulk transmission elements (n-1-1 contingency analysis). The developed approach can be extended to perform the risk-based analysis of other complex types of contingencies.

8.6.4.1 Prediction and Prevention of Cascading Outages in Idaho Power Network

The primary focus of paper [28] is to identify the system's most vulnerable places and double outages that lead to widespread power disruptions and cascading, evaluate their consequences, and identify possible remedial actions to prevent cascading or mitigate its effect. Understanding the effects of cascading on vulnerabilities of Idaho Power's system is needed to determine when a disruption of service is likely to occur and to take appropriate steps to reduce the associated risk. The analysis of multiple outages that lead to cascading chains and cascading instability is performed on five WECC cases—three summers (2015, 2019, and 2023) and two winters (2014) and (2015). IPC's system is part of the interconnected transmission network of the WECC which is described in [28]. Two WECC paths (Idaho to Northwest—Path 14 and West of Borah—Path 17) play an important role in serving load and transferring power across the Idaho Power system.

The cases used for this study are approved WECC cases. The following assumptions were made for this study:

- Cases are checked for thermal (flow) violations on all lines and transformers modeled in the case and for voltage violations on all buses modeled in the case under normal (*n*-0) conditions.
- Cases are checked for voltages greater than or equal to 1.05 per unit (pu) or less than or equal to 0.95 pu for buses with a nominal voltage rating of 345 kV and below. A voltage analysis is also checked for voltages greater than or equal to 1.1 pu or less than or equal to 0.95 pu for buses with a nominal voltage rating of 500 kV. All studied cases show no thermal or voltage violations under normal (*n*-0) conditions.
- Cases are further stressed: 2014HW, 2019HS, and 2023HS by load increase and 2014LW and 2015HS by increasing MW flow across WECC paths 14 and 17.

These stressed cases are used to determine if there is any thermal/voltage violation under *n*-1 outages. Each violation was mitigated either by existing remedial action schemes (RAS) or by projects identified to alleviate those problems.

The principal schematic diagram for an analysis of cascading outages is shown in Fig. 8.7. The following steps are taken in performing the analysis:

- Step 1: All cases have been tested to ensure no violations exist under normal conditions (*n*-0) and *n*-1 contingencies.
- Step 2: *n*-1 contingencies are defined based on actual operation of breakers and include the operation of the existing RAS.
- Step 3: The initiating *n*-2 events in all five cases have been generated automatically from the list of *n*-1 contingencies by combining outages of lines, transformers, and generators.
- Step 4: Define the thresholds for thermal overloads and voltages at generation and load buses to simulate the operation of protection systems.
- Step 5: Initiating events are consecutively applied with associated threshold logic as described in step 4 until the following occurs:
  - The system fails to solve due to voltage instability.
  - Islanding with an imbalance of load and/or generation within an island.
  - Thermal/voltage violations are alleviated or no longer exceed the thresholds.
- Step 6: The loss of load and generation is simultaneously monitored and reported at each tier.

The key results of the cascading analysis and identification of preventive actions to mitigate cascading impacts are summarized in Table 8.2. Table 8.2 shows that violations from five initiating events in 2014HW, six in 2014LW, one in 2015HS, and two in the 2023HS case have resulted in cascading instability and were therefore strong candidates for further evaluation.

The type of identified violations as a result of performed cascading analysis is shown in Fig. 8.8.

Total results of cascading analysis in this example are as follows:

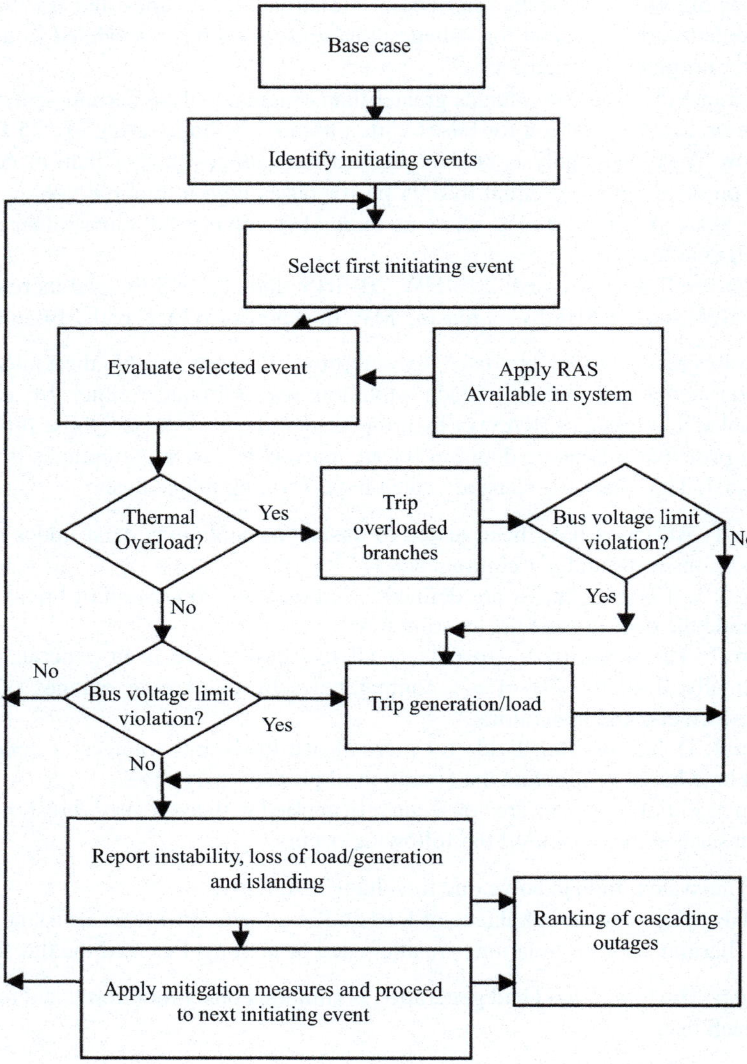

Fig. 8.7 Schematic diagram of cascading methodology

Table 8.2 Critical and mitigated contingencies in five studied cases

Case	Initiating events	Stability[a]	Cascading instability	Cascading instability after OPM[b]
14hw	20,910	2	5	3
14lw	20,503	4	6	2
15hs	20,910	1	1	0
19hs	21,528	0	0	0
23hs	23,436	0	2	0

[a]Failure to obtain a solution under the initiating event
[b]Optimal mitigation measures (OPM)

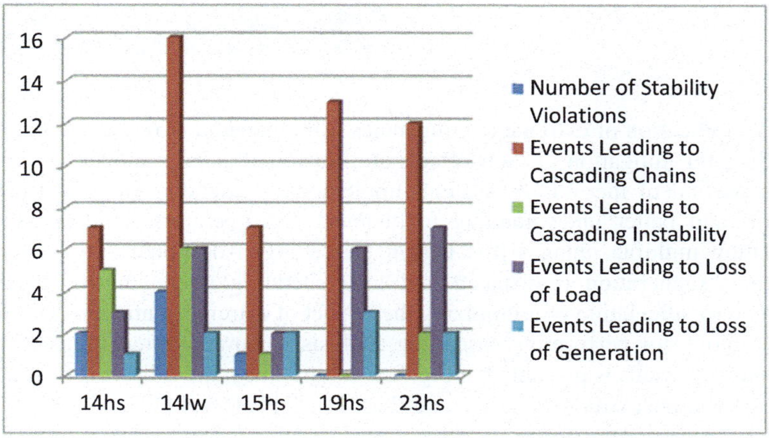

Fig. 8.8 Type of identified violations as a result of cascading simulation in five studied cases [28]

Number of studied cases	5
Total number of initiating events	107,287
Number of stability violations	7
Events leading to cascading chains	54
Events leading to cascading instability	14
Events leading to loss of load	27

8.6.4.2 Assessing the Cascading Effects of Extreme Contingencies with Respect to Standards TPL-001-4 and CIP 014-1

Paper [56] presents a comprehensive, practical approach to identify and analyze the effects of extreme contingencies (physical or cyber threats) that might lead to widespread power disruptions and cascading in IPC system. The proposed approach is primarily used in performing North American Electric Reliability Corporation (NERC) compliance studies. Understanding the effects of extreme contingencies on vulnerabilities of IPC system is needed to determine when a disruption of service is likely to occur and how to take appropriate steps to reduce the associated

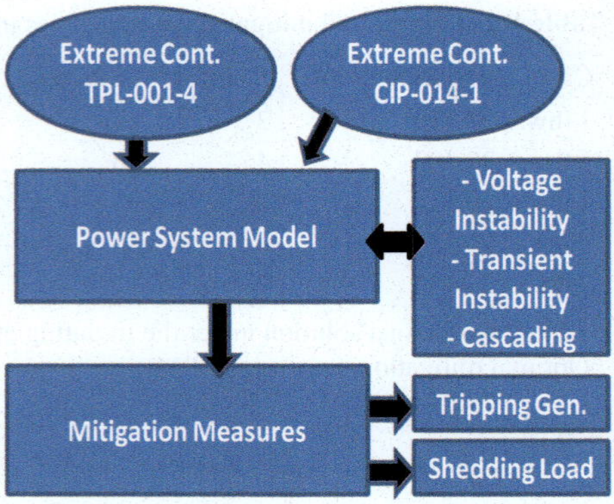

Fig. 8.9 General methodology for performing cascading analysis in power systems

risk. Identification of extreme contingencies using generation reallocation and load shedding for mitigation of their effects are illustrated using a model of the actual Idaho portion of the WECC system. Implemented cascading approach provides support and better understanding to planners and operators in identifying and evaluating multiple outages that lead to cascading. Additionally, two corrective strategies (generation re-dispatch and load shedding) were shown to be very effective for alleviating or minimizing the impact of extreme contingencies. General methodology for performing cascading analysis in power systems under extreme contingency events is presented in Fig. 8.9. In this study, the following criteria were used to generate extreme contingencies based on TPL-001-4 [26]:

1. All *n*-1 contingencies are defined as breaker-to-breaker contingencies.
2. *N*-2 extreme contingencies (Category 1) are automatically created from bulk transmission elements (lines and transformers) at 100 kilovolts (kV) and above and generating units equal to or greater than 20 megavolt amperes (MVA).
3. Loss of all transmission lines on a common right-of-way (ROW) (Category 2b).
4. Loss of a switching station or substation (Category 2c).
5. Loss of all generating units at a generating station with output 200 megawatts (MW) or greater (Category 2d).
6. Loss of a large load or major load center of 100 MW or greater (Category 2e).

Top-ranked initiating events/contingencies that lead to voltage stability violations and cascading instability for cases 15HS, 15LW, and 20HW were identified, evaluated, and presented in Table 8.3.

The results show the performance of three representative cases under extreme contingency events with respect to NERC TPL-001-4 and CIP-014-1. Special attention was paid to contingencies that lead to voltage stability violations and to cascading. The generation re-dispatch and load shedding were shown to be very effective for alleviating the majority of the voltage stability violations and cascading. Table 8.3 shows voltage stability violations (VS) and cascading (CS) for three Idaho cases, 15HS, 15LW, and 20HW. It indicates the 15LW case experienced the largest percentage of violations (voltage stability and cascading) in terms of TPL initiating

Table 8.3 The extreme events by categories in three WECC cases

Case name	Standard	Contingency category	Number of initiating events	Number of occurrences of voltage instability	Number of occurrences of cascading
15HS	TPL-001-4	1	23,871	2	10
		2b	6	1	3
		2c	270	2	2
		2d	4	0	0
		2e	1	0	0
	CIP-014-1		11	1	1
15LW	TPL-001-4	1	23,871	6	17
		2b	6	0	0
		2c	282	3	1
		2d	4	0	0
		2e	0	0	0
	CIP-014-1		11	1	1
20HW	TPL-001-4	1	23,871	6	1
		2b	6	0	0
		2c	280	3	0
		2d	4	0	0
		2e	1	0	0
	CIP-014-1		10	0	0

events. It also shows that 15HS case experienced the largest percentage of violations (voltage stability and cascading) in terms of CIP initiating events. Less than 0.1% of all initiating events cause voltage instability and cascading in Idaho Power cases, which means the system is robust in terms of steady-state voltage stability and cascading propagation.

8.6.4.3 IPC Experience of Implementing Cascade Analysis Study Using the Node/Breaker Model

Paper [57] presents a practical approach for performing the steady-state cascading analysis of a power system represented as a node/breaker (NB) model using real-time state estimator (SE) cases. This paper discusses advantages of using the NB model in cascading analyses compared to the bus/branch (BB) model. The paper presents how the breaker configuration can impact outages and contingencies. In the BB model, only branch limits can be modeled. But in the NB model breaker, disconnect and other switch limits can be modeled and can impact the outcome based on the configurations and equipment limits. The North American Electric Reliability Corporation (NERC) reliability standard CIP-014-2 [26] specifically requires proper methodology to be developed and used to perform cascading analyses. The methodology presented in this paper is tested using the 90,000 node West Wide System Model (WSM). The extensive simulations performed to validate the methodology demonstrate its practicality for performing cascading analyses in the steady-state domain. It is applied and successfully used by IPC to identify critical substations and weak links in the system.

The risk assessment includes a contingency, cascade analysis, and transient stability analysis. The objectives of a contingency (loss of substation), cascade analysis, and transient stability analysis are as follows:

- Identify transmission substations and their associated primary control centers that, if rendered inoperable or damaged due to physical attack, could result in widespread instability, uncontrolled separation, or cascading within an interconnection.
- Protect stations deemed critical after risk assessment.

The principal schematic diagram for a cascading analysis on a NB model is shown in Fig. 8.10. The following steps are taken when performing the cascading analysis:

1. Define and apply the potential outage to the NB model.
2. Model the operation of the existing remedial action schemes (RAS) associated with the initiating event.
3. Define the monitoring limits for thermal overloads and voltage violations to simulate the operation of protection systems.
4. Find generators/loads operating with a terminal voltage below 0.9 per unit (PU), as well as transformers, transmission lines, series capacitors, breakers, disconnects, and fuses functioning above a specified percentage of their short-term seasonal emergency thermal rating.
5. Trip all devices found to be in violation.
6. Solve power flow.
7. Reiterate with the updated system until cascade analysis criteria fail or no additional cascading occurs.
8. Report data, such as the buses in voltage violation, tripped load/generation or lines, breakers, transformers, etc., that were above their short-term seasonal emergency thermal rating during the cascade analysis.
9. Repeat the procedure for other potential outages, and reporting could prove helpful in identifying equipment prone to tripping, which may initiate further cascading.

As a result of steps 4 and 5, severe overloads and voltage deviations may occur that trigger actions of protection devices, and cascading could propagate across the interconnection, incurring a significant loss of load that might lead to system collapse. A number of factors, such as loss of generating units, circuit outages, breaker failures, relay actions, RAS actions, etc., may result in thermal overloads, voltage deviations, and other actions that could lead to cascading. Cascading usually implies the coexistence of overloads, voltage violations, and different stability phenomena during its development.

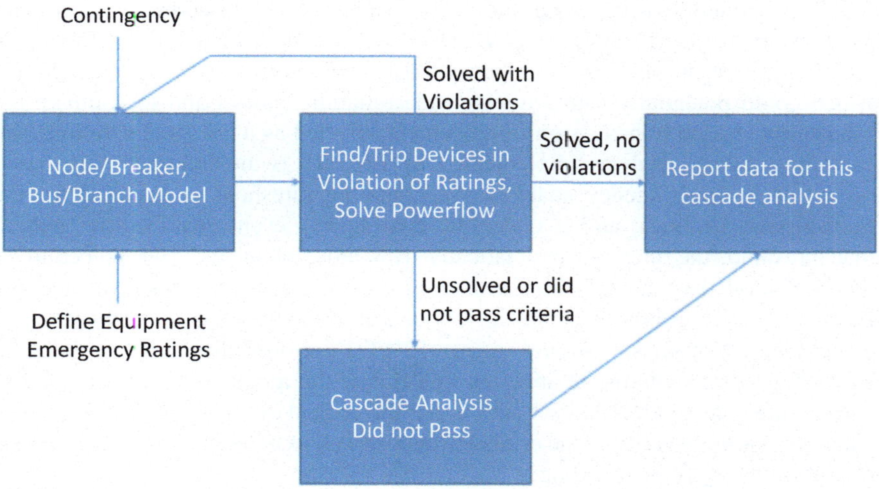

Fig. 8.10 Flowchart of a cascade analysis for a single potential outage

Table 8.4 A summary of cascading analysis

Total events	8
Total passed	7
Total failed	1
Number of diverging solution (immediately after station loss)	1
Number of cascading and lead to stable point	2
Number of cascading and lead to not stable point	0

In paper [57], identification of critical substations in IPC's system was performed using the NB model. In this example, eight stations were each simulated to be suddenly taken out of service from the peak summer 2016 WSM NB model, and the cascading analysis of the steady-state solution followed. For lines, breakers, disconnects, etc., over 125% of their short-term emergency seasonal rating were tripped. Loads and generators with terminal voltage dipping below 0.9 pu were also tripped. A summary of cascading results is presented in Table 8.4. A total of eight events were tested against cascading, one immediately reached a divergent solution and two propagate over two tiers before reaching a stable point.

8.6.5 ERCOT Experience in Analysis of Cascading Outages

Paper [59] presents the approach used by ERCOT in transmission planning process by taking into account cascading analysis. As an example of the planning process, paper [59] evaluates the Houston import project with addition of a large transmission line. This example introduces a more comprehensive and modernized assessment for transmission system planning. In planning assessment ERCOT performs the NERC category C and D contingency [26] analysis to identify critical events which could potentially cause system-wide reliability issues and to compare the robustness of each transmission improvement selected as a potential solution. The cascading outage analysis implemented by ERCOT is an iterative study based on a full AC contingency analysis. An operating threshold is assumed for the transmission facilities, and if under the contingency event any facilities exceed that threshold, the analysis automatically trips those elements. This procedure is repeated until there are no more violations exceeding the threshold, or until the number of cascades is above three tiers, or until the load flow does not converge. For any potential transmission improvements, ERCOT performs the cascading analysis. ERCOT performs additional analyses to estimate the amount of the load shed to address the potential reliability issues under the critical events. This can be used as a measure of the severity of the critical events as well as potential mitigation actions to prevent the system cascading.

In response to the new NERC Reliability Standard TPL-001-5 [26], ERCOT is updating its planning criteria and study methodology. The biggest challenges in the new standards are likely to be the number of sensitivity analyses required, including different stressed system conditions, and the increased performance requirements in several contingency categories. In addition to these new standards, other challenges and opportunities lie ahead for the ERCOT grid. These include increasing penetration of intermittent renewable resources and new technologies such as demand response programs and energy storage. As these challenges continue to evolve, more comprehensive approaches considering these complexities will be required to appropriately assess transmission improvements.

For cascading analysis, ERCOT tests the following NERC contingencies for cascading analysis:

- P2.2 (HV – bus section fault), P2.3 (HV – internal failure of non-bus tie breaker), P2.4 (internal failure of bus tie breaker), P4 (HV – fault plus stuck breaker), P4.6 (EHV – fault plus a stuck bus-tie breaker), P5 (HV – fault plus relay failure to operate), P6 (N-1-1 with system adjustment)
- EE1 (N-1-1 without system adjustment), EE2 (local-area events affecting the transmission system), EE3 (wide-area events affecting the transmission system based on system topology such as loss of two generating stations)

8.6.6 ISONE Experience with Online Cascading Analysis

Paper [60] describes the implementation of the Region of Stability Existence (ROSE) approach at ISO-NE for online estimation of the power system transfer capability based on voltage and thermal limitations and for security monitoring. The ROSE approach uses state estimator model and PMU measurements to determine steady-state transfer capability of the system per N-1 and N-2 security criteria, develops corrective action improving the limits, and alarms the operator when system conditions approach the limit. ROSE implementation is based on "hybrid" approach where SE solution (model) is used to compute voltage stability limits, and PMU data (measurements) is used to determine the position of the current operating point. This approach allows improving security of transmission system by continuously monitoring operational margin expressed in MW flow or in bus voltage angles and alarming the operator if the margin violates a pre-defined threshold.

Reference [61] presents the cascading analysis performed at ISO-NE. The analysis of IROL violations (thermal and voltage) is performed with the following steps:

- Use state estimator (SE) power flow case for cascading analysis of the critical contingency, solving power flow, checking transmission loading, and tripping the transmission element if the MVA flow is over-rating by a certain amount.
- Solve power flow again and repeat the process until:
  - Voltage collapse
  - There is no more overloading
- Confirm IROL violation if voltage collapses or uncontrolled cascading failures resulting in a loss of more than X MW of load or generation.

NERC Standard TOP-007 [26] requires the reliability coordinator (RC) to report IROL violations. The Standard TOP-007 states: "Following a Contingency or other event that results in an IROL violation, the Transmission Operator shall return its transmission system to within IROL as soon as possible, but not longer than 30 minutes." The RC shall report any IROL violation exceeding 30 minutes to the regional reliability organization and NERC within 72 hours. For identified IROL violations such as system instability, unacceptable system response or equipment tripping, voltage levels in violation of applicable emergency limits, loadings on transmission facilities in violation of applicable emergency limits, and unacceptable loss of load based on regional and/or NERC criteria, the reporting entity needs to propose mitigation measures to alleviate those violations. Cascading analysis identifies the critical cascade which may become the IROL violation based on measurable consequences. Existence of a critical cascade longer than 30 minutes means IROL violation reportable event. The key attributes of a critical cascade [61] are the following:

- System-wide voltage collapse occurs upon applying initiating contingency or as the result of cascading tripping.
- Islanding of the system and total MW of load in separated islands is greater than pre-defined threshold.
- Actual interface MW flow during cascade exceeds "stability" interface limit by pre-defined % level.
- Total MW loss of load exceeds pre-defined threshold.
- Total MW loss of generation exceeds pre-defined threshold.
- Cascade propagates beyond balancing area footprint.

8.6.7 Cascading Event Reported to NERC in 2018

Reference [62] presents a variety of disturbance events across North America reported to NERC. The initiating event reported to NERC [63] was a 138 kV line tree contact followed by the misoperation of a 345/138 kV transformer. The event resulted in two contingency overloads. Upon performing cascading analysis, the entity realized that one of the contingencies could cascade if left unmitigated. The operator took action to shed load pre-contingent to prevent the possible cascade.

Prior to the event, there were three planned outages in the area; see Fig. 8.11. Outage #1 was in-place for over a month as part of a capital project to reinforce the area. Outages #2 and #3 occurred the morning of the event. Earlier, the RC issued a hot weather alert due to increased temperatures that day. Under those circumstances, TOPs consider deferring noncritical transmission work on the system. In this case, the RC and TOP ran the next day studies to assess the risk and determined that they were N-1 secure. There were no single contingencies that could cause a thermal, voltage, or stability issue on the system. Furthermore, outages #2 and #3 were both considered recoverable within a few hours (if needed) as the work was to pour foundations under the lines. Shortly after noon, a 138 kV line tripped due to a tree contact (outage #4). Simultaneously, there was a misoperation (outage #5) that tripped the 345/138 kV transformer at Substation B. Following the tree contact and the misoperation, there were no base case violations; however, there were two contingencies identified:

- 140% of emergency rating on the 138 kV line from Substation D to Substation E for the loss of the 345/138 kV transformer at Substation A
- 132% of emergency rating on the 138 kV line from Substation D to Substation E for the loss of the 138 kV line from Substation A to Substation C

No controlling actions were available, so a cascading event began.

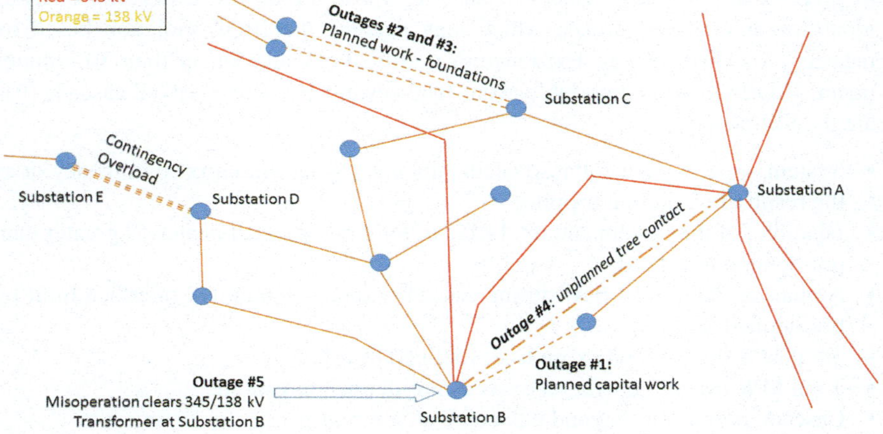

Fig. 8.11 An example of event reported to NERC

8.6.8 Practice of Cascading Analysis in Other US Companies

Paper [64] describes American Transmission Company (ATC) experience in performing cascading studies using an automated process run by POM/OPM software. The implemented process helps to conduct a comprehensive steady-state multiple element contingency screening and cascading analysis, finding mitigations to restore the system to be within emergency and/or normal performance limits, and automatically tabulating significant results and summaries in spreadsheets. This process is utilized as a part of ATC's NERC TPL-003 compliance studies starting from 2013.

Paper [65] presents Midcontinent Independent System Operator (MISO) experience in the development, testing, and implementation of a fast automated process for assessing power system performance following loss of two bulk transmission elements consecutively (*N*-1-1 contingency analysis) and simultaneously (*N*-2 contingency analysis). The implemented approach at MISO offers a flexibility to utilize various sets of system adjustments depending on types and values of post-contingency limit violations. It also incorporates sequential contingency simulation in order to identify potential cascading modes due to thermal overloads. The proposed process in [65] is utilized as a part of MISO's NERC compliance studies to assess and improve the reliability of a transmission grid and reduce its vulnerability to cascading outages.

Reference [66] presents PJM's experience in area of cascading analysis. It shows how cascading tree methodology is used to evaluate some of the transmission projects. Also, it was used to perform compliance studies related to the standard CIP-014 and to identify the critical substations based on any of the criteria (instability, uncontrolled separation, or cascading).

Paper [67] presents the Southern Company Services (SCS) methodology and experience in performing the analysis of cascading failures. Developed and implemented methodology by SCS allows screening of multiple initiating contingencies, simulating the cascading process, evaluating the system impacts, ranking the cascading scenarios based on their severity and likelihood, and identifying the tap contingencies that require primarily attention. It has already influenced real-life decision-making in transmission system enhancement projects' prioritization and selection.

Paper [68] presents a detailed analysis of a severe weather-related event in the Western Region of the Entergy System. As a result of the event, three critical lines in the same right-of-way are forced out which led to a fast voltage collapse in the area. Approximately, 1100 MW of load in the region was lost within minutes. This paper presents detailed analysis to replicate the events and determine the cause of the fast voltage collapse. The voltage collapse was verified using the steady-state and dynamic analysis. The cause of the voltage collapse was attributed to rapid tripping of critical transmission lines, limited amount of real power generation in the area, and high reactive power demand.

Paper [69] presents how phasor measurement units (PMUs) provide the capability for maintaining reliable operation of the bulk power system and enhancing the real-time monitoring and control room solutions. As the risks to the North American electrical grid from both natural disasters and potential malicious attacks increase, the need for improved reliability and resiliency of the power system network becomes higher. This paper indicates that over 1700 PMUs are deployed in the North American grid, and numerous PMU-based applications have been developed. This paper summarizes synchrophasor-based applications and presents synchrophasor experience by utilities such as BPA, ERCOT, IPC, and ISO-NE.

8.6.9 Practice of Cascading Analysis in Countries Outside of North America

Paper [70] presents the methodology used for risk and vulnerability analysis of a power system under extreme outage events in Norway. The developed methodology is described as a six-step process and can be used to identify, analyze, and monitor vulnerabilities in power transmission and distribution systems. The applicability of this methodology for the risk and vulnerability analysis is tested on two real systems in Norway: 420 kV transmission system and 132 kV regional distribution system. More details about methodology and vulnerability analysis of real systems in Norway can be found in [70].

Paper [71] presents the risk-based methodology to assess the risk of load loss caused by any single or multiple dependent contingencies that may lead to a cascading process. The proposed methodology is capable of quantifying the impact of each contingency via the amount of load loss at the end of the cascading process. The risk is evaluated as the product of the impact and the probability of each contingency. The proposed approach is applied to perform risk-based security assessment in operation of the Italian EHV transmission grid. The performed risk analyses highlight that the most probable cascading paths are often different from the paths which mainly contribute to the risk. It was emphasized that widely used deterministic methods for security assessment are not capable of taking into account the probability of occurrence and the quantitative characterization of the impact of the contingencies.

Paper [72] presents experience in China with design and implementation of an online intelligent alarming system for cascading failures of wind farms. It indicates that large amounts of alarms coming into control center when faults occur and create problems to operators especially when alarms are false or missing or multiple faults are happening. Authors of paper [72] have developed and implemented the online intelligent alarm-processing system to deal with the alarms automatically. The system is designed to apply rules reflecting cause-effect relationship between faults and alarms. The system is also using models and real-time data from SCADA/EMS and fault information data to prevent and minimize the impact of cascading events. The system is implemented into actual power system in Jiaxing, China.

8.7 Conclusions

Maintaining an adequate level of reliability of the modern power grid is a challenging problem that utilities face today due to many causes such as frequent extreme events (e.g., failure of multiple physical components, extreme weather events, and other natural disasters), high penetration of intermittent renewables sources, and the increasing complexity of energy system infrastructure.

Understanding the effects of cascading phenomena and determining vulnerabilities of power systems are tasks aimed to determine when a disruption of service is likely to occur and what appropriate mitigation measures to apply to minimize the cascading vulnerability impact. The ability of a grid to survive extreme outages and major disturbances that may lead to cascading is not comprehensively studied by planners and operators today. Therefore, there is a need that evaluating and predicting the risks and consequences of cascading outages become an integral part of planning and operation studies by utilities.

This chapter discussed key issues related to practical analysis of cascading outages such as development of cases, system operating states, potential chain of cascades starting with an initiating event and ending with final operating conditions, standards related to the cascading, a brief review of cascading methodologies, and tools and industry practices in the analysis of cascading outages. Through various case studies and results, power industry practitioners have demonstrated the benefits of performing cascading analysis that leads to economic planning decisions and ultimately results in a more reliable power system. The presented material in this chapter is primarily focused on steady-state cascading analysis.

Significant research has been conducted on the various aspects of cascading, but practical applications of developed methodologies by utilities have been lagging those efforts. The existing cascading tools exhibit various deficiencies, as outlined in previous chapters of this book. A recent IEEE CFWG survey found continuing dissatisfaction with the cascading tools available today. Utilities and research organizations need to determine the framework and requirements for developing the adequate tools to study cascading outage events.

The suggested approaches for cascading analysis of outage events in previous chapters are in fact making further enhancements to the current methodologies used by the utility industry today. Most large blackouts that happened throughout the world involve a sequence of cascading outages that are complex and require adequate methodologies and tools to be properly studied. Cascading outages expose transmission planners and operation engineers to new challenges in areas of data preparation, modeling issues, vulnerability metrics, methodologies, and computation tools. As a result of those challenges, further work is needed in the following areas:

- Analyzing and the identification of measures for preventing and mitigating cascading outages.
- Ranking and mitigating contingencies (initiating events) based on a probabilistic approach.

- Defining metrics to properly measure the impact of cascading outages.
- Use of PMU technologies for enhancing the real-time monitoring and control solutions by identifying root cause of cascading outages, reducing the frequency of power outages and restoration/recovery time.
- Developing and implementing maintenance and inspection proactive approaches to catch problems before they become emergencies that could lead to blackouts.
- Performing both steady-state and transient stability performances in cascading analysis.
- In order to properly capture the probabilistic nature of a bulk power system and uncertainty of the rapidly changing power grid, use of probabilistic cascading assessment is needed.

References

1. North American Electric Reliability Corporation, Glossary of terms used in NERC reliability standards, 2021. [Online]. Available: https://www.nerc.com/files/glossary of terms.pdf
2. S. Lee et al., *Mitigating Cascading Outages on Power Systems: Recent Research Approaches and Emerging Methods*. Electric Power Research Institute, Report, 2005. https://www.epri.com/research/products/1010701
3. R. Baldick, B. Chowdhury, I. Dobson, Z. Dong, B. Gou, D. Hawkins, H. Huang, M. Joung, D. Kirschen, F. Li, J. Li, Z. Li, C.-C. Liu, L. Mili, S. Miller, R. Podmore, K. Schneider, K. Sun, D. Wang, Z. Wu, P. Zhang, W. Zhang, and X. Zhang, Initial review of methods for cascading failure analysis in electric power transmission systems. in *Proceedings of the IEEE PES General Meeting 2008*, https://ieeexplore.ieee.org/document/4596430
4. N. Bhatt, S. Sarawgi, R. O'Keefe, P. Duggan, M. Koenig, M. Leschuk, S. Lee, K. Sun, V. Kolluri, S. Mandal, M. Peterson, D. Brotzman, S. Hedden, E. Litvinov, S. Maslennikov, X. Luo, E. Uzunovic, B. Fardanesh, L. Hopkins, A. Mander, K. Carman, M.Y. Vaiman, M.M. Vaiman, M. Povolotskiy, Assessing vulnerability to cascading outages. in *IEEE PSCE Conference*, 2009, https://ieeexplore.ieee.org/document/4840032
5. M. Vaiman, K. Bell, Y. Chen, B. Chowdhury, I. Dobson, P. Hines, M. Papic (chair), S. Miller, P. Zhang, Risk assessment of cascading outages: Part I - overview of methodologies. in *Proceedings on IEEE PES General Meeting 2011*, https://ieeexplore.ieee.org/document/6039405
6. M. Papic, K. Bell, Y. Chen, I. Dobson, L. Fonte, E. Haq, P. Hines, D. Kirschen, X. Luo, S.S. Miller, N. Samaan, M. Vaiman, M. Varghese, P. Zhang, Survey of tools for risk assessment of cascading outages. in *2011 IEEE Power and Energy Society General Meeting*. https://ieeexplore.ieee.org/document/6039371
7. M. Vaiman, K. Bell, Y. Chen, B. Chowdhury, I. Dobson, P. Hines, M. Papic, S. Miller, P. Zhang, Risk assessment of cascading outages: Methodologies and challenges. IEEE Trans. Power Syst. **2**, 631–641 (2012). https://ieeexplore.ieee.org/document/6112807
8. M. Vaiman, P. Hines, J. Jiang, S. Norris, M. Papic (Chair), A. Pitto, Y. Wang, G. Zweigle, Mitigation and prevention of cascading outages: Methodologies and practical applications. in *Proceedings of the IEEE PES General Meeting*, Vancouver, July 2013, pp. 1–6. https://ieeexplore.ieee.org/document/6672795
9. J. Bialek, E. Ciapessoni, D. Cirio, E. Cotilla-Sanchez, C. Dent, I. Dobson, P. Hines, J. Jardim, S. Miller, M. Papic, A. Pitto, J. Quiros-Tortos, D. Wu, Benchmarking and validation of cascading failure analysis tools. IEEE Trans. Power Syst. **31**(6), 4887–4900 (2016). https://ieeexplore.ieee.org/document/7404289

10. E. Ciapessoni, D. Cirio, E. Cotilla-Sanchez, R. Diao, I. Dobson, A. Gaikwad, P. Henneaux, S. Miller, M. Papic, A. Pitto, J. Qi, N. Samaan, G. Sansavini, S. Uppalapati, R. Yao, *Benchmarking Quasi-Steady-State Cascading Outage Analysis Methodologies* (PMAPS Boise, Idaho, 2018). https://ieeexplore.ieee.org/document/8440212
11. S.T. Lee, Estimating the probability of cascading outages in a power grid. Power System Computation Conference (PSCC), Glasgow, Scotland, 2008. https://www.semanticscholar.org/paper/
12. S.T. Lee, *Probabilistic Reliability Assessment for Transmission Planning and Operation Including Cascading Outages* (PSCE, 2009) https://ieeexplore.ieee.org/document/4840086
13. S.T. Lee, Probabilistic online risk assessment of non-cascading and cascading transmission outage contingencies. Eur. Trans. Electr. Power (2008). https://onlinelibrary.wiley.com/doi/abs/10.1002/etep.277
14. B. Otomega, T. Van Cutsem, Identifying plausible cascading events in system stability assessment. 3rd International Conference on Energy and Environment CIEM2007, Bucharest, 22–23 November. https://orbi.uliege.be/bitstream/2268/5650/1/CIEM2007_2.pdf
15. K. Sundar, M. Vallem, R. Bent, N. Samaan, B. Vyakaranam, Y. Makarov, N-k failure analysis algorithm for identification of extreme events for cascading outage pre-screening process. in *Proceedings of IEEE PES General Meeting 2019*. https://ieeexplore.ieee.org/document/8973425
16. Y. Liu, A. Zhang, P. Dehghanian, J.K. Jung, U. Habiba, T.J. Overbye, Modeling and analysis of cascading failures in large-scale power grids. IEEE Kansas Power and Energy Conference (KPEC), 2022, Manhattan, KS. https://ieeexplore.ieee.org/document/9814815
17. S.R. Khazeiynasab, J. Qi, Resilience analysis and cascading failure modeling of power systems under extreme temperatures. J. Mod. Power Syst. Clean Energy **9**(6) (2021). https://ieeexplore.ieee.org/document/9627860
18. M.J. Eppstein, P.D.H. Hines, A "random chemistry" algorithm for identifying collections of multiple contingencies that initiate cascading failure. IEEE Trans. Power Syst. **27**(3), 1698–1705 (2012). https://ieeexplore.ieee.org/document/6152191
19. P. Rezaei, P. Hines, M. Eppstein, Estimating cascading failure risk: Comparing Monte Carlo sampling and random chemistry. in *Proceedings of IEEE PES General Meeting 2014*. https://ieeexplore.ieee.org/document/6939392
20. Dobson, Estimating the extent of cascading transmission line outages using line outages from standard utility data and a branching process. in *Proceedings of IEEE PES General Meeting 2011*. https://ieeexplore.ieee.org/document/6039461
21. H. Ren, I. Dobson, Using transmission line outage data to estimate cascading failure propagation in an electric power system. IEEE Trans. Circuits Syst. II Express Briefs **55**(9) (2008). https://ieeexplore.ieee.org/document/4539786
22. I. Dobson, B.A. Carreras, D.E. Newman, J.M. Reynolds-Barredo, Obtaining statistics of cascading line outages spreading in an electric transmission network from standard utility data. IEEE Trans. Power Syst. **31**(6), 4831 (2016). https://ieeexplore.ieee.org/document/7407675
23. R.C. Hardiman, M. Kumbale, Y.V. Makarov, An advanced tool for analyzing multiple cascading failures. 8th' International Conference on Probabilistic Methods Applied to Power Systems, Iowa State University, Ames, Iowa, September 12–16, 2004. https://ieeexplore.ieee.org/document/1378760
24. TransCARE (Transmission Contingency and Reliability Evaluation) Tool, EPRI, 2012. https://www.epri.com/research/products/1025568
25. *Potential Cascading Modes (PCM) Program Manual* (V & R Energy Systems Research, Inc., Los Angeles, 2012). https://vrenergy.com/software-solutions/pom-suite/pcm/
26. NERC Reliability Standards for the Bulk Electric Systems of North America, May 13, 2021 [Online]. Available: https://www.nerc.com/pa/Stand/Pages/ReliabilityStandards.aspx
27. W. Li, *Risk Assessment of Power Systems -Models, Methods, and Applications (Book)* (IEEE Press, 2005)
28. M. Papic, O. Ciniglio, Prediction and prevention of cascading outages in Idaho power network. in *Proceedings of PES General Meeting 2014*. https://ieeexplore.ieee.org/document/6939101

29. Adequate Level of Reliability (ALR) Metrics, NERC, 2020 Online: https://www.nerc.com/pa/Stand/Resources/Documents/Adequate_Level_of_Reliability_Definition_(Informational_Filing).pdf
30. M. Papic et al., Research on Common-Mode and Dependent (CMD) outage events in power systems– a review. IEEE Trans. Power Syst. **32**(2), 1528–1536 (2017). https://ieeexplore.ieee.org/document/7508488
31. A.G. Phadke, J.S. Thorp, Expose hidden failures to prevent cascading outages in power systems. IEEE Comput. App. Power **9**(3), 20–23 (1996). https://ieeexplore.ieee.org/document/526849
32. S. Ekisheva, M.G. Lauby, M. Papic, *Analysis of CDM Events in NERC and WECC Power Systems Using TADS Outage Statistics* (PMAPS, 2022). https://ieeexplore.ieee.org/document/9810591
33. M. Papic, I. Dobson, Comparing a transmission planning study of cascading with historical line outage data. 2016 International Conference on Probabilistic Methods Applied to Power Systems (PMAPS), Beijing, China, 2016. https://ieeexplore.ieee.org/document/7764070
34. M. Papic, S. Ekisheva, E. Cotilla-Sanchez, A risk-based approach to assess the operational resilience of transmission grids. *Appl. Sci.* MDPI 2020. https://www.mdpi.com/2076-3417/10/14/4761
35. M. Papic, S. Ekisheva, J. Robinson, B. Cummings, Multiple outage challenges to transmission grid resilience. in *Proceedings of IEEE PES General Meeting 2019*. https://ieeexplore.ieee.org/document/8973606
36. M. Vaiman, M. Papic, M. Povolotskiy, Industry practices in the analysis of cascading outages: IEEE PES CFWG survey results. Panel Session: "Tackling cascading outages under emerging internal and external factors: methodologies, tools and industrial practices", Panel Session. in *Proceedings of IEEE PES General Meeting 2021*, July 26
37. H.H. Alhelou, M.E. Hamedani-Golshan, T.C. Njenda, P. Siano, A survey on power system blackout and cascading events: Research motivations and challenges. MDPI Energies **12**, 682 (2019). https://www.mdpi.com/1996-1073/12/4/682
38. M. Koenig, P. Duggan, J. Wong, M.Y. Vaiman, M.M. Vaiman, M. Povolotskiy, Prevention of cascading outages in Con Edison's Network. PES T&D Conference, 2010. https://ieeexplore.ieee.org/document/5484278
39. CIGRE WG C2.02.24, Defense Plan against extreme contingencies. *Electra*, pp. 46–61, April 2007, https://e-cigre.org/publication/316-defense-plan%2D%2Dagainst-extreme-contingencies
40. D. Karlsson, X. Waymel et al., System protection schemes in power networks. Task Force 38.02.19 Report, CIGRE, June 2001. https://e-cigre.org/publication/187-system-protection-schemes-in-power-networks
41. V. Madani, D. Novosel, S. Horowitz, M. Adamiak, J. Amantegui, D. Karlsson, S. Imai, A. Apostolov, IEEE PSCRC report on global industry experiences with System Integrity Protection Schemes (SIPS). IEEE Trans. Power Deliv. **25**(4), 2143–2155 (2010). https://ieeexplore.ieee.org/document/5565539
42. CIGRE WG39.05, P.M. Anderson, B. K. Lereverend et al., *Industry Experience with Special Protection Schemes*, Electra, No. 155, August, 1994. https://e-cigre.org/publication/ELT_155_5-industry-experience-with-special-protection-schemes
43. Western Systems Coordinating Council, Remedial Action Scheme Design Guide, July 2022. www.wecc.biz
44. C.W. Taylor et al., WACS-wide-area stability and voltage control system: R&D and online demonstration. in *Proceedings of the IEEE*, June 2005. https://ieeexplore.ieee.org/document/1428005
45. R. Ramanathan, B. Tuck, J. O'Brien, BPA's experience of implementing remedial action schemes in power flow for operation studies. in *Proceedings of IEEE PES General Meeting 2013*. https://ieeexplore.ieee.org/document/6672441

46. M. Varghese, J. Licheng, S. Ghosh, G. Lin, B. Pek, *The CAISO Experience of Implementing Automated Remedial Action Schemes in Energy Management Systems* (IEEE PES GM, 2009). https://ieeexplore.ieee.org/document/5275849
47. California Operating Studies Subcommittee (OSS) Handbook Rev. 9.0 Jan., 2007. www.caiso.com
48. ERCOT Operating Guides, August 2010. www.ercot.com
49. G. Trudel, S. Bernard, G. Scott, Hydro-Quebec defense plan against extreme contingencies. IEEE Trans. Power Syst. **14**(3), (1999). https://ieeexplore.ieee.org/document/780908
50. S.C. Pai, J. Sun, BCTC's experience towards a smarter grid – increasing limits and reliability with centralized intelligence Remedial Action Schemes. Electric Power Conference, IEEE Canada, Oct. 2008, pp. 6–7. https://ieeexplore.ieee.org/document/4763366
51. V. Arcidiacono, S. Corsi, A. Natale, C. Raffaelli, New developments in the applications of Enel transmission system automatic voltage and reactive power control. CIGRE Meeting - Paris, 1990. https://www.osti.gov/etdeweb/biblio/10107188
52. ENTSO-E, Technical background and recommendations for defense plans in the continental Europe synchronous area, Jan 31, 2011. www.entsoe.eu
53. P. Gomes, G. Cardospo Jr., Reducing Blackout Risk by System Protection Schemes-Detection and Mitigation of Critical System Conditions, CIGRE, C2–201, 2006. https://e-cigre.org/publication/C2-201_2006-reducing-blackout-risk-by-system-protection-schemes%2D%2Ddetection-and-mitigation-of-critical-system-conditions
54. P. Gomes, H.J. Chipp, Brazilian Defense Plan against Extreme Contingencies. CIGRE/IEEE OPES International Symposium 2003. https://ieeexplore.ieee.org/document/970160
55. M.D. Zhong, Maintaining System Integrity to Prevent Cascading Blackout, CIGRE B5–207, 2006. https://e-cigre.org/publication/B5-207_2006-maintaining-system-integrity-to-prevent-cascading-blackout
56. M. Papic, O. Ciniglio, M. Vaiman, Practical Experience in Assessing the Effects of Extreme Contingencies with Respect to Standards TPL-001-4 and CIP 014-1. in *Proceedings of IEEE PES General Meeting 2015*. https://ieeexplore.ieee.org/document/7285836
57. R. Ramanathan, A. Popat, M. Papic, O. Ciniglio, Idaho power experience of implementing cascade analysis study using the Node/Breaker Model. in *Proceedings of IEEE PES General Meeting 2017*. https://ieeexplore.ieee.org/document/8273978
58. M. Papic, O. Ciniglio, Prevention of NERC C3 category outages in Idaho Power's Network: Risk based methodology and practical application. in *Proceedings of IEEE PES General Meeting*, Vancouver, July 2013, pp. 1–6. https://ieeexplore.ieee.org/document/6672893
59. S.W. Kang, J. Boyd, X. Yu, G. Gnanam, J. Billo, Comprehensive regional transmission planning –ERCOT experience. in *Proceedings of IEEE PES General Meeting 2015*. https://ieeexplore.ieee.org/document/7286136
60. S. Maslennikov, E. Litvinov, M. Vaiman, M.M. Vaiman, Implementation of ROSE for on-line voltage stability analysis at ISO New England. in *Proceedings of IEEE PES General Meeting 2014*. https://ieeexplore.ieee.org/document/6939549
61. E. Litvinov, Cascading analysis for automated IROL assessment at ISO New England, CAMS CFWG Panel Session. in *Proceedings of IEEE PES General Meeting 2015*
62. NERC ERO Event Analysis Process (EAP), January 2020. https://www.nerc.com/pa/rrm/ea/Pages/EA-Program.aspx
63. NERC Lesson Learned "Cascading Analysis Identifies Need for Pre-Contingent Load Shed", December 2018, https://www.nerc.com/pa/rrm/ea/Pages/Lessons-Learned.aspx
64. C. Guo, C. Lawrence, M. Povolotskiy, M. Vaiman, Enhanced NERC TPL-003 steady state compliance studies at American transmission company. in *Proceedings of IEEE PES General Meeting 2015*. https://ieeexplore.ieee.org/document/7285690
65. D. Chatterjee, J. Webb, Q. Gao, M.Y. Vaiman, M. M. Vaiman, M. Povolotskiy, N-1-1 AC contingency analysis as a part of NERC compliance studies at Midwest ISO. in *Proceedings of IEEE PES General Meeting 2009*. https://ieeexplore.ieee.org/document/5484209

66. E. Bernabeu, Cascading trees & power system resilience, presentation. in *Proceedings of IEEE PES General Meeting 2018*
67. Y.V. Makarov, R.C. Hardiman, D.L. Hawkins, Risk, reliability, cascading, and restructuring. in *Proceedings of IEEE PES General Meeting 2004*. https://ieeexplore.ieee.org/document/1372816
68. S.M. Raymon, D. Powell, K. Vongkhamchanh, Simulation and analysis of a major disturbance in energy system that resulted in voltage collapse. in *Proceedings of IEEE PES General Meeting 2006*. https://ieeexplore.ieee.org/document/1709530
69. M. Vaiman, R. Quint, A. Silverstein, M. Papic, D. Kosterev, N. Leitschuh, A. Faris, S. Yang, B. Blevins, S. Rajagopalan, P. Gravois, O. Ciniglio, S. Maslenikov, E. Litvinov, X. Luo, P. Etingov, Using synchrophasors to improve bulk power system reliability in North America. in *Proceedings of IEEE PES General Meeting 2018*. https://ieeexplore.ieee.org/document/8586560
70. G.L. Doorman, K. Uhlen, G.H. Kjolle, E.S. Huse, Vulnerability analysis of the Nordic power system. IEEE Trans. Power Syst. **21**(1) (February 2006). https://ieeexplore.ieee.org/document/1708873
71. E. Ciapessoni, D. Cirio, S. Grillo, S. Massucco, A. Pitto, M.F. Silvestro, *Operational Risk Assessment and Control: A Probabilistic Approach* (ISGT Europe, 2010) https://ieeexplore.ieee.org/document/5638975
72. J. Mu, H. Sub, Q. Guo, W. Wu, F. Xu, B. Zhang, *Design of an Online Intelligent Alarming System for Cascading Failures of Group of Wind Farms* (IEEE PES GM, 2013). https://ieeexplore.ieee.org/document/6673043

Index

K. Sun (ed.), *Cascading Failures in Power Grids*, Power Electronics and Power Systems, https://doi.org/10.1007/978-3-031-48000-3

The manufacturer's authorised representative in the EU is Springer Nature Customer Service Centre GmbH, Europaplatz 3, 69115 Heidelberg, Germany. If you have any concerns regarding our products, please contact ProductSafety@springernature.com

Printed and bound by CPI Group (UK) Ltd, Croydon, CR0 4YY
01/10/2026
02225872-0009